Environmental Impact of the Offshore Oil and Gas Industry

Environmental Impact of the Offshore Oil and Gas Industry

Stanislav Patin

Translated from Russian by Elena Cascio

EcoMonitor Publishing
East Northport, New York

Stanislav Patin
Environmental impact of the offshore oil and gas industry

With 69 tables and 87 illustrations

Translated from Russian by Elena Cascio

Published by:
EcoMonitor Publishing
Post Office Box 866
East Northport, NY 11731-0523 U.S.A.

Publisher's Cataloging-in-Publication
(Provided by Quality Books, Inc.)

Patin, Stanislav Alexandrovich
    Environmental impact of the offshore oil and gas industry / Stanislav Patin ; translated from Russian by Elena Cascio. — 1st ed.
    p.   cm.
    Includes bibliographical references and index.
    ISBN 0-9671836-0-X
    Library of Congress Catalog Card Number 99-73791

1. Offshore oil industry—Environmental aspects. 2. Offshore gas industry—Environmental aspects. 3. Offshore oil industry—Environmental aspects—Russia. 4. Offshore gas industry—Environmental aspects—Russia. I. Cascio, Elena. II. Title.
    QH545.O38P38 1999   333.8`2314
                                                      QBI99-602

Contents

About the Author

Stanislav Patin, is a Doctor of Science (Biology, Chemistry); Professor (Ecology); Corresponding Member of the Contenant International Academy, Head Scientist at the Russian Federal Research Institute of Fisheries and Oceanography (VNIRO) in Moscow. During his 40-year scientific career, he has participated in a wide scope of research, including studies on marine pollution, environmental impact assessments, and sea protection, particularly in relation to the offshore oil and gas industry. As Chairman of the Aquatic Toxicology Committee under the Russian Academy of Sciences and head of a number of Research Councils and Working Groups under Russian Federal Ministries and Agencies, Dr. Patin has been responsible for realization of numerous national and international projects in the area of marine environmental protection. For more than 10 years, he has been involved in the work of the International Council for the Exploration of the Sea (ICES) as a member (Russia Representative) of the Advisory Committee on the Marine Environment. Dr. Patin is the author of over 200 publications (including six books) in the field of marine ecology, oceanography, ecotoxicology, biogeochemistry, ecological monitoring, environmental management, and sea protection.

Environmental Impact of the Offshore Oil and Gas Industry

Introduction

The silhouettes of the oil platforms and drilling rigs located dozens of miles offshore in waters hundreds of meters deep have become a typical feature of many shelf areas in the World Ocean. They can be found at all latitudes, from the coral reefs of Australia to the Arctic ice caps of Alaskan bays and seas. Many such platforms are large industrial plants. Thousands of them extract up to 30% of the world's oil and gas production. By taking into account the recent actions of the Russian government and oil industry circles, we can predict that soon, a large number of fixed offshore platforms are going to appear on the shelves of the Russian northern and far-eastern seas. The potential recoverable hydrocarbon resources of the Russian ocean margins are grandiose. They reach 90–100 billion tons of oil equivalent and about 40% of the world's reserves of natural gas [Mazur, 1993; Malovitski et al., 1994].

Thus, the shelf zone of the World Ocean, which some 20–30 years ago was used mostly for shipping and fishing, has become an arena for a relatively new, dynamic, and large-scale economic activity—offshore oil and gas production. The expansion of the oil and gas industry on the continental shelf has provoked many essential questions:

- What are the main factors of the offshore oil and gas industry's impact on the shelf ecosystems and marine biological resources?
- What are the ways this impact reveals itself and what are its consequences at different levels of the life hierarchy in the ocean?
- What is the offshore oil and gas industry's impact on other kinds of shelf activities, especially on fishing?
- Is it possible to ensure a balance of interests for the offshore oil and gas production and fishing industries and, if possible, under what conditions?
- Are the biological consequences of accidental oil spills into the marine environment reversible?

1

• What are the environmental requirements, national and international standards, and rules regulating the activity of the offshore oil and gas industry? How do they ensure the stability of the shelf ecosystems and reproduction of biological resources?

This list of questions could be extended. However, the previous is sufficient to conceive the complexity, variety, and urgency of emerging problems. The environmental significance of these problems, likewise of all problems of anthropogenic impact on the World Ocean [Patin, 1982; 1995a], is amplified by the phenomenon of uneven distribution of marine life and its concentration in the shelf and coastal zone. This zone, which comprises only about 10% of the total area of the World Ocean, is characterized by the most intensive bioproduction processes and major concentration of marine flora and fauna. This zone is the place where the main living resources of the ocean reproduce. It provides up to 80–90% of the worldwide landing of marine commercial organisms. At the same time, most of the known oil and gas fields are also located on the shelf. In many regions, their development has already reached a wide scope. This takes place in the background of the numerous other human activities that occur in the shelf waters and onshore areas. At present, over half of the world's population lives along the coasts and up to 50% of the gross national product of many countries is being produced in these areas.

These circumstances define the urgent, large-scale, and long-term nature of emerging environmental problems, some of which were previously mentioned. They can be solved only by creating conditions that would allow the maximum recovery of the offshore oil and gas without disturbing the vulnerable and vitally important shelf ecosystems. These conditions would ensure preserving the self-reproducing (and thus priceless) biological resources of the World Ocean. This task should certainly be interpreted and solved only within the context of the broader goal of environmental protection of the ocean from the impact of all other numerous types of human activity.

Different aspects of anthropogenic impact on the hydrosphere, including the ones associated with oil and gas industry's activities, have been the focus of the world's science for a long time. The studies of oil pollution, its sources, levels, biological effects, environmental consequences, and other related issues, have been especially active and comprehensive. The list of publications on this topic is very impressive and includes thousands of articles and books. At the same time, a review of this literature indicates that monographic studies of a summarizing and generalizing nature (especially those that directly deal with the subject of this book) are rather rare.

Well-known works devoted to these issues include Proceedings of the International Meeting *Ecological impacts of the oil industry* [Dicks, 1989]; collective monographs *North Sea oil and the environment* [Cairns, 1992] and *Long-term environmental effects of offshore oil and gas developments* [Boesch, Rabalais, 1987]. They also include a report of the Joint Group of Experts on Scientific Aspects of Marine Pollution (GESAMP) entitled *Impact of oil and related chemicals and wastes on the marine environment* [GESAMP, 1993] and an independent scientific review *Environmental implications of offshore oil and gas development in Australia* [Swan et al., 1994].

In Russia for the last 30 years, such comprehensive studies have hardly been published. At the same time, some works have been devoted to specific aspects of the problem, including marine oil pollution and engineer-technical issues of environmental safety during oil and gas developments. Such kinds of publications include the monograph *Biological resources of the sea and oil pollution* by O. G. Mironov [1972]; a translation of the book by A. Nelson-Smith [1977] *Oil pollution and marine ecology;* a collection of articles edited by A. I. Simonov and O. G. Mironov entitled *Impact of oil and oil products on marine organisms and communities* [1984]; the monograph by G. I. Guseinov and R. E. Alekperov *Environmental protection during oil and gas field development* [1989], and the book by I. I. Mazur *Ecology of oil and gas development* [1993]. Recently, some studies on the regional environmental and fisheries aspects of developing the offshore hydrocarbon resources of the Russian shelf have been published. In particular, these address the activities in the Barents Sea [Matishov, 1993; Borisov et al., 1995], Sakhalin [Eranov et al., 1993], and the Russian southern seas [Patin, 1995b].

Unlike many publications that focus mostly on specific aspects of the problem, this book attempts to give the environmental analysis of all main factors of the oil and gas industry's impact on marine life, fish stock, and fisheries in shelf waters. It also summarizes world and Russian experience in this field and substantiates the environmental and fisheries requirements for the oil and gas developments on the Russian shelf.

One of the motives for writing this book is the recently increased interest in the ecological problems of the offshore oil and gas industry's activity. We should note that Russia, possessing the richest offshore oil and gas resources in the world, has just begun their industrial exploitation. Russian environmental protection and fisheries circles and the general public are concerned by the growing amount of new tenders, projects, and licenses granted for developing Russian offshore hydrocarbon resources, particularly in the areas that have traditionally supplied the country with valuable fish and other marine products (the Caspian Sea, Barents Sea, Sakhalin, and others). From time to time, this concern is reinforced by alarming media reports about tanker accidents, tremendous oil spills, the death of marine animals, and ecological catastrophes on the shelves of other countries. In Russia, these either have been avoided thus far or have been kept in secret from the public for many years.

The diversity of positions on developing the offshore oil and gas resources (both in Russia and abroad) varies widely. Some have strong beliefs about the projects' environmental safety, while others totally reject them and offer the darkest prognoses. Significantly, such polarity of opinions is also seen in the scientific community and is reflected in many publications, reports, and discussions (see, for example, Blackman, 1988; Warren, 1989; Gray, 1988; 1991; Neff, 1993). Thus, the attempts to analyze and summarize the available data and materials on this urgent, complex, and controversial issue seem to be especially important. This issue is, in turn, only a part of a broader problem of anthropogenic impact on the World Ocean (especially on its coastal zone) associated with the rapid development of the world's economy [Danilov-Danilyan, 1995]. It explains why many issues discussed in the following chapters often lie beyond the limits defined by the title of this book.

This book is only an attempt to reflect the main environmental aspects of the extremely complex and multidisciplinary problem of developing offshore oil and gas resources. The conclusions and recommendations given in this book are by no means final or indisputable. The ultimate goal is to attract attention to some major environmental problems emerging on the world's shelves and to discuss their content. We hope to emphasize the need for a rapid and adequate response to this new situation. Finally, we intend to define some approaches and ways for such response by taking into account the results of the latest and the most comprehensive studies in this area.

References

Blackman, R. A. 1988. Offshore oil impact. *Mar.Pollut.Bull.*, 19(4):185–186.
Boesch, D. F., Rabalais, N. N., eds. 1987. *Long-term environmental effects of offshore oil and gas developments.* New York: Elsevier Applied Science, 708 pp.

Borisov, V. M., Ponamarenko, V. P., Osetrova, N. V., Semenov, V. N. 1995. Bioresources of the Barents Sea and the project of developing Shtokmanovskoe gas condensate field. *Rybnoe khozyaistvo* 1:12–28. (Russian)

Cairns, W. J., ed. 1992. *North Sea oil and the environment. Developing oil and gas resources, environmental impacts and responses.* London and New York: Elsevier Applied Science, 722 pp.

Danilov-Danilyan, V. I. 1995. Ecological orientation of economics—the foundation of sustainable development. In *Ecology. Economics. Business.*— Moscow: IRIS-Press, pp. 5–10. (Russian)

Dicks, B., ed. 1989. *Ecological impacts of the oil industry. Proceedings of the International Meeting organized by the Institute of Petroleum and held in London in November 1987.* New York, 316 pp.

Eranov, V. N., Velikanov, A. J., Mikheev, A. A. 1993. Ecological aspects of industrial development of the North-East Sakhalin shelf. *Vestnik Dalnevost.otd. RAN* 3:94–99. (Russian)

GESAMP. 1993. *Impact of oil and related chemicals and wastes on the marine environment. GESAMP Reports and Studies No.50.* London: IMO, 180 pp.

Gray, J. S. 1988. Environmental politics and monitoring around oil platforms. *Mar.Pollut.Bull.* 19(11):549–550.

Gray, J. S. 1991. Anthropogenic or ecocentric? *Mar.Pollut.Bull.* 22(11):529.

Gusseinov, G. I., Alekperov, R. E. 1989. *Environmental protection during oil and gas field development.* Moscow: Nedra, 230 pp. (Russian)

Malovitski, J. P., Martirosyan, V. N., Golovchak, V. V., Gumenyuk, J. N., Fedorovski, J. F., Zakalski, V. M. 1994. Oil and gas potential of the Russian ocean margins. *Neftyanoe khosyaistvo* 4:27–32. (Russian)

Matishov, G. G., ed. 1993. *Arctic seas: bioindication of the state of environment, biotesting and technology of pollution destruction.* Apatiti: KNTS PAN. (Russian)

Mazur, I. I. 1993. *Ecology of oil and gas development.* Moscow: Nedra, 494 pp. (Russian)

Mironov, O. G. 1972. *Biological resources of the sea and oil pollution.* Moscow: Pishchepromizdat, 105 pp. (Russian)

Neff, J. M. 1993. *Petroleum in the marine environment: regulatory strategies and fisheries impact.* Battelle Ocean Sciences Laboratory. Duxbery, 13 pp.

Nelson-Smith, A., trans. 1977. *Oil pollution and marine ecology.* Moscow: Progress, 302 pp. (Russian)

Patin, S. A. 1982. *Pollution and the biological resources of the oceans.* London: Betterworth Scientific, 320 pp.

Patin, S. A. 1995*a*. Global pollution and biological resources of the World Ocean. In *World Fisheries Congress Proceedings, Athens, 1994.*—New Delhy: Oxford IBH Publishing Co, pp. 69–75.

Patin, S. A. 1995*b*. Ecological and fisheries aspects of oil and gas developments on the shelf of the Russian southern seas. In *Theses of International Symposium on the Problems of Mariculture.*—Moscow: VNIRO, pp. 8–10. (Russian)

Simonov, A. I., Mironov, O. G., eds. 1984. *Impact of oil and oil products on marine organisms and communities.* Leningrad: Gidrometeoizdat, 205 pp. (Russian)

Swan, J. M., Neff, J. M., Young, P. C., eds. 1994. *Environmental implications of offshore oil and gas development in Australia—the findings of an independent scientific review.* Sydney: Australian Petroleum Exploration Association Limited, 696 pp.

Warren, R. P. 1989. Offshore oil and gas exploration: what are the environmental effects do to justify limitation of access to coastal waters. *APEA Journ.:* 84–94.

Chapter 1

Present State and Trends of Developing Offshore Oil and Gas Resources

1.1. General Characteristics and Prospects

Crude oil and natural gas have been playing the leading role in the world's fuel-energy balance for a long time. Their contribution to the total energy consumption has been increasing continuously. The start of the oil industry on a large scale is traced to the end of the nineteenth century. The level of gas consumption has been growing since the 1920s. The use of oil and gas resources has approximately doubled every decade. At present, they supply about 63% of the world-wide energy needs [Beck, 1993]. For some individual countries, this contribution reaches 80–90%.

Annual worldwide production of these energy resources totals about 3 billion tons of oil and 2,000 billion cubic meters of gas. Russia accounts for more than 10% and 20%, respectively. The estimates of geological and discovered oil and gas resources given in different publications vary considerably. However, most of them indicate that these resources are sufficient to supply humankind's energy needs for at least the next 100 years.

The historical development of the oil and gas industry has been remarkable for its high dynamics, rapid technical progress, and wide geography of exploration and production activities. One of the noticeable tendencies of the last 10–20 years is the growing contribution of natural gas and relative stabilization or even decreasing of oil production in a number of large regions. Table 1 suggests that this tendency is going to remain for the next several decades. By 2030, natural gas will take second place among the world's sources of energy [Krasilov, 1992].

This trend gives a basis for some environmental optimism, because natural combustible gas can be considered the cleanest of all fossil fuels. For example, in comparison with coal, burning gas

Table 1
World's Energy Resources (%) [Krasilov, 1992]

Sources of Energy	In 1989	Optimal in 2030
Oil	33	14
Coal	24	8
Gas	18	18
Renewable sources	20	60
Nuclear power	50	0

produces twice as less carbon dioxide. It is also virtually free from sulfur and particular emissions. This advantage of natural gas has been highly appreciated by the residents of many European countries. The offshore gas developments in the North Sea, which started about 30 years ago, made replacing coal and peat by natural gas possible. It became easier to breathe in many large cities. Many noteworthy historical buildings, such as Westminster Abbey in England, were cleaned from the age-long effects of air pollution. In fact, the famous smog, long recognized as a grave threat to humans and the environment, disappeared from London's atmosphere. Thus, from the environmental perspective, wider use of natural gas has obvious advantages, especially in urban and metropolitan areas.

Another significant trend in the world's oil and gas industry's development is connected with the geography of its expansion. As easily available inland hydrocarbon fields were being depleted, the attention of the gas and oil industry switched toward the shelf resources of the World Ocean. Experts long ago predicted the existence of prolific offshore oil and gas basins and fields related to adjoining onshore fields.

These prognoses have proved to be absolutely correct (see Figure 1). At present, about 1,000 known oil and gas fields exist in the World Ocean. Potentially prolific oil and gas areas cover 60–80 million square kilometers, and 13 million square kilometers are located on the continental shelf [Zalogin, Kuzminskaya, 1993].

The continental shelf is the underwater extension of the coastal part of the land, the gradually sloping submerged portion of continental margins that has the same geological structure as the adjoining land. The shelf zone is bordered by the ocean coastline on one side and by the steep bottom drop-off (continental slope) on the other (see Figure 2). The water depths at which the continental shelf begins

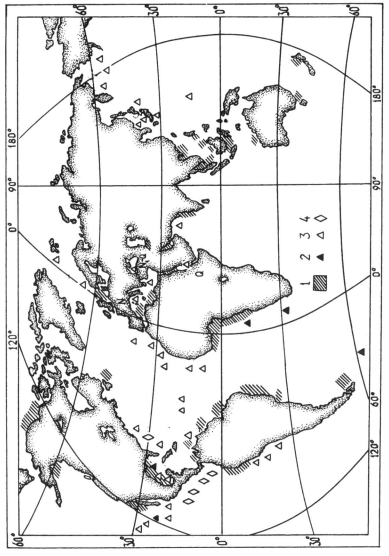

Figure 1 Worldwide distribution of major oil and gas hydrocarbon resources on the continental shelf: 1—zones of oil and gas accumulations; 2, 3, 4—regions of oil, gas, and gas hydrate manifestation in wells [Levin, 1994].

9

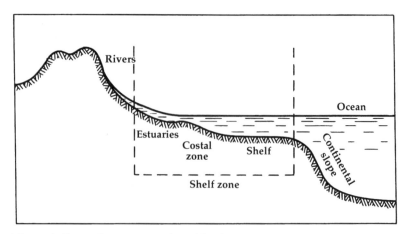

Figure 2 General structure and profile of the shelf zone.

to slope to the ocean floor are usually 100–200 m. In some regions, though, for example in the southern part of the Sea of Okhotsk, they reach more than 500 m. The total area of the world's continental shelves exceeds 30 million square kilometers, and the Russian shelves account for more than 5–6 million square kilometers [Granberg et al., 1993; Malovitski et al., 1994].

The largest shelf zones of the World Ocean include the northern margins of Eurasia, where they stretch for up to 1,500 km off shore, the Bering Sea, Hudson Bay, the South China Sea, and some other regions [Slevich, 1977]. The international regulation and management of shelf resources, including offshore oil and gas developments, are generally defined by the United Nations Convention on the Law of the Sea (1982) and by a number of other international agreements [Matthews, 1992].

Although available estimates of worldwide offshore hydrocarbon potential vary widely (from 320 to 2,000 billion tons of oil equivalent), they leave no doubt that these resources and the prospects of their industrial developments are grandiose [Zick et al., 1990]. Some authors [Baybulatova, Vostokov, 1989] note the gravitation of the largest oil and gas basins to the zones of tectonic belts in the Earth's crust known as plate margins. Typical features of these zones include increased seismic activity, intensive thermal flows, active lithodynamic processes, and highly developed bottom relief. Significantly, the regions of increased bioproductivity and active fishing are also located in the impact zones of these belts. Most likely, these depend not only on the geological structure of the seafloor but also on the general patterns of water circulation, distribution of biogenic substances, terrestrial flow, and other complex biogeochemical processes in the marginal areas of the ocean [Lisitsin, 1994].

Numerous studies show apparent attraction of the oil- and gas-bearing structures to the delta and estuary sedimentary deposits of paleorivers. In 40 of the largest oil and gas basins, more than 1,500 zones of hydrocarbon accumulation have been found in the deep structures of delta deposits [Artemiev, 1993].

Recently, the possibilities of the offshore oil and gas developments in the polar areas, primarily on the Arctic shelf, have become the focus of increased attention (Figure 3). Previous experience of exploration and production in these regions, as well as some preliminary estimates [Engelhardt, 1985], suggest that potential resources of

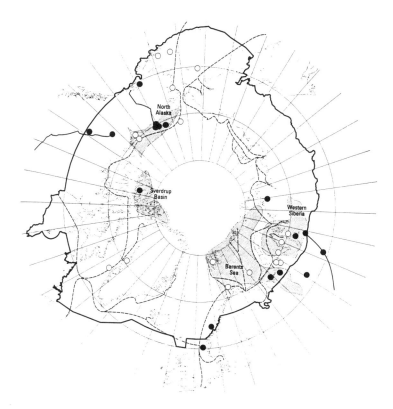

Figure 3 Major areas of oil and gas activities in the Arctic [AMAP, 1997].

oil hydrocarbons in the Arctic are comparable with the rest of the world's oil reserves.

Besides the exploration and extraction of the oil and gas resources located on the shelf at depths up to 200–400 m, the prospects of developing the oil and gas hydrocarbons from the deep ocean basins have attracted a growing interest over the last decade [Grigorenko, Suprunenko, 1990; Levin, 1994]. The results of geophysical studies and deep drilling suggest the presence of oil- and gas-bearing structures at the bottom of the Atlantic, Indian, and Pacific Oceans as well as in the Caribbean, Aegean, Black, and Bering Seas, the Sea of Japan, and other areas (see Figure 1) at depths of up to 2–5 km. The total area of deep ocean basins characterized by the presence of promising oil and gas accumulations adds up to about 10–40 million square kilometers [Levin, 1994]. The amount of hydrocarbons concentrated there can reach up to 35–40% of the total oil and gas potential of the World Ocean.

The offshore reserves of gas hydrates, that is, solid compounds of natural gas (mainly methane) and water formed under high pressure and in relatively low temperatures, are of special interest. In 1965, Russian scientists discovered the accumulations of gas hydrates in reservoirs under permafrost strata. Since then, it has been proved that gas hydrate accumulation in bottom sediments of the marine shelf is an ubiquitous phenomenon. Thermodynamic conditions favorable for stable gas hydrate formation are present in 25% of the land (mainly subarctic regions) and in about 90% of the World Ocean. This includes the entire area of the continental slope and about 10% of the shelf, especially the zones of shelf and slope conjunction [Zubuva et al., 1990]. Gas hydrates are found in the bottom deposits (up to 400 m from the seafloor surface) at depths from 480 to 5,500 m. They accumulate there due to the hydrocarbon flow from the zones of deep petroleum formation and as a result of biogenic methane production.

About 20 large regions of known gas hydrate accumulations are located off the coast of South and Central America, in particular in the Gulf of Mexico, and in other regions, including the Bering, Beaufort, and Caspian Seas, Sea of Okhotsk, and some other areas. Evidence indicates the existence of enormous gas hydrate reserves in the Black Sea [Kruglyakova et al., 1993]. The global estimates of these hydrocarbon resources vary widely—from 10^{15} to 10^{18} m^3. However, even the minimum estimates are an order higher than the worldwide potential gas resources of conventional fields [Zubuva et al., 1990]. For example, some of the recent studies suggest that the global gas hydrate reserve ranges from 14×10^{12} to 34×10^{15} m^3 in onshore, continental accumulations and from $3,100 \times 10^{12}$ to $7,600,000 \times 10^{12}$ m^3 in offshore, oceanic accumulations. At the same time, the unexploited conventional gas reserve is estimated at 260×10^{12} m^3 [Kellard, 1994].

The technology and equipment for developing the offshore gas hydrate resources have not been developed yet. It might be assumed that both mechanical methods and different thermal and chemical techniques of gas hydrate decomposition (e.g., hot water pumping, introducing inhibitors such as methanol, and so on) will be used. Undoubtedly, this can cause environmental problems. Most likely, these problems will be more serious than the ones generated by the extraction of other offshore resources (for example, ferrous-manganese nodules) from the sea bottom. Nevertheless, the huge potential of gas hydrate accumulations makes them a unique hydrocarbon reserve available for future development on a worldwide basis.

1.2. Historical and Regional Aspects

The offshore oil production activity started in the nineteenth century. For example, in 1824 in the Baku area (the Caspian Sea) at a distance of 20–30 m offshore, the first waterproof wells were built and oil was extracted from the shallow layers. In 1870 in Japan, a special island for oil derricks was created. In the 1890s, the industrial drilling of wells angling out up to 200 m seaward was started along the coast of California and Virginia. However, a really remarkable breakthrough toward the continental shelf was accomplished in the 1920s when numerous rigs, trestles, pipelines, and other structures appeared in the sea. First, these installations were located in the shallow waters close to the shore. Then they moved farther seaward on the continental shelf. This occurred along the coast of North America, in the Caribbean basin, and in some other regions.

In the former USSR, similar activities started in the 1920s in the Caspian Sea. Many technical methods and equipment for the offshore oil production were developed and tested there. In 1934 on Artem Island, the first multiple drilling (i.e., drilling several wells from the same platform) was performed. In 1935, the first offshore metal drilling structures were installed in the Caspian Sea. By 1972, 1,880 steel islands as well as trestles totaling in length over 300 km existed in the Caspian Sea [Orudzhev, 1974]. The hazardous consequences of these developments, mainly in the form of oil pollution of the marine environment and biota of the Caspian Sea, were obvious to everybody. However, they were not officially admitted in accordance with the usual practice of those days.

The geography of oil and gas production on the continental shelf is incredibly wide. The largest offshore oil industry centers appeared in the 1950s in the Persian Gulf, on the shelf of Venezuela (Lake Maracaibo), in the Gulf of Mexico, and off the California coast. Especially rapid and large-scale offshore oil and gas developments started in the 1970s. By that time, the large fields on the Atlantic shelf (the Gulf of Mexico, Caribbean Sea, and Gulf of Guinea), in the eastern part of the Pacific Ocean (Cook Inlet and California coast), near the coast of Alaska, Canada, Australia, and New Zealand, in the Persian and Suez Gulfs, North and Mediterranean Seas, on the shelf of South Asia, and in a number of other regions had been found and thoroughly explored. As a result, Great Britain, Norway, Italy, Malaysia, Indonesia, and Australia numbered among the leading oil-producing countries. Successful exploration activities were conducted on the continental shelves of the former USSR. About 70% of the shelves were found to be promising oil-bearing areas [Salnikov, Slevich, 1979].

By the early 1980s, more than 800 hydrocarbon fields were known in the World Ocean. In fact, many of them had unique concentrations of natural gas as well. By that time, an impressive success in developing the methods and equipment needed for oil and gas production at depths up to 400 m had been achieved. Technical difficulties of designing, assembling, transporting, and installing huge drilling structures in regions with extremely severe natural conditions (such as the areas affected by hurricanes, cyclones, increased seismic activity, low temperatures, or ice fields) had been overcome. Oil production complexes began to include drilling platforms, pipelines, barges, survey vessels, machinery for pipelaying operations, storage tanks, coastal terminals, and other units of different type and function [Samarski, Suvorov, 1979].

Technical progress made it possible in a relatively short time to increase offshore oil and gas production to 20–25% of their total world production by the beginning of the 1990s. This process continues at present with an increased speed. In 1987–1991 on the federal continental shelf of the USA, over 2,200 exploratory wells and over 2,100 development wells were drilled [MMS, 1995]. In 1992, over 1,000 exploratory and testing drillings were conducted, including the ones on the shelves of the Far East, Southeast Asia, and Australia (227 wells), North and Central America (214 wells), Western Europe (211 wells), and other regions. In 1993 on the shelf of the North Sea alone, 25 new oil and gas fields were put into operation [Knott, 1993; Neff,

1993]. In the Italian economic zone (mainly on the shelf of the Adriatic Sea), over 100 gas-bearing geological structures have been found
and about 900 exploratory and production wells have been drilled
[Dossena et al., 1995]. Modern, sophisticated equipment makes it
possible to drill wells up to 6,000 m deep (for example, in the North
Sea) and to angle them out stretching horizontally up to 1,600 m at
depths of about 4,500 m [Cardy, 1993].

The investments into the offshore oil and gas industry amount
to hundreds of billions of dollars. In Great Britain, for example, they
have totaled 60 billion pounds sterling since the mid-1960s. The
monetary worth of oil and gas hydrocarbons annually produced in
the British sector of the North Sea exceeds 20 billion pounds sterling
with an annual production of 50–110 million tons of oil and 50 billion
cubic meters of gas [Taylor, Turnbull, 1992].

The total number of offshore drilling platforms in the World
Ocean numbers in our day into the thousands. In the Gulf of Mexico
alone, more than 3,800 of them are present [MMS, 1995]. A production platform consists of a deck (or decks) secured to the ocean floor
by huge piles (Figures 4 and 5). It is built in segments onshore and is

Figure 4 Offshore drilling rig used by Exxon for drilling off of Sakhalin
Island.

Figure 5 Offshore drilling rig used by Exxon for drilling off of Sakhalin Island.

barged to the production site. Some of the production platforms are truly gigantic. The length of the underwater legs can measure up to 300 m, and the height can total up to 400 m. About 600 platforms presently installed offshore have a substructure weight in excess of 4,000 tons. One platform can have up to 100 production and injection wells. It can remain in place for over 30 years while withstanding extreme dynamic impacts. These include hurricane winds of 200 km/hour or more, cyclonic storms and 30-meter waves, strong earthquakes, and powerful ice fields. In the severe Arctic conditions, a practice of dispersing out the seawater is used to create a protective ice cover around the drilling platforms. The drilling is completed in the winter, and these artificial ice islands totally disappear when the Arctic spring and summer come.

All evidence suggests that the worldwide trend of oil and gas complex expansion onto the continental shelf is going to get stronger over the next decades. Simultaneously, one might expect that enforcing stricter national and international environmental regulations will lead to equipment and technology improvement,

including achieving zero discharges during exploration and production operations.

1.3. Oil and Gas on the Russian Shelf

The general trend to expand oil and gas production onto the continental shelf can be seen in Russia as well. Depletion of the largest oil
and gas fields in the Volga region and Western Siberia forces the industry to pay more attention to the shelves of the Russian northern
and far-eastern seas. Profound exploration in these regions has been
in progress for a long time (in some areas, since the 1930s). It has resulted in the discovering of promising oil- and gas-bearing fields.
70% of these fields are located at depths less than 100 m.

Recent estimates [Granberg et al., 1993; Malovitski et al., 1994]
suggest that promising oil- and gas-bearing areas are found on about
90% of all Russian shelves. They cover 5.2–6.2 million square kilometers. The West Arctic region (the Barents and Kara Seas) accounts
for about 2 million square kilometers and the East Arctic region (the
Laptev, East-Siberian, and Chukchi Seas)—for 1 million square kilometers of oil- and gas-bearing areas. Around 0.8 million square kilometers of these areas are located in the Russian Far East (the Bering
Sea, Sea of Okhotsk, and Sea of Japan), 0.1 million square kilometers
belong to the Russian southern seas (some areas of the Black and
Caspian Seas and the Sea of Asov), and a small area is found in the
Baltic Sea (Kaliningrad region). Potential recoverable hydrocarbon
resources of the Russian continental shelves are estimated within 90
to 100 billion tons of oil equivalent. Natural gas resources account for
80% of them.

Practically everywhere on the Russian shelf, the affinity between the offshore petroleum-bearing provinces and corresponding
geological structures of the adjoining inland areas is found. Global
experience indicates that in such cases, the oil and gas potential of the
shelf fields is higher than that of the onshore accumulations.

The high potential of oil and gas fields of the Russian shelves
and thus the security of the hydrocarbon supply in Russia in the foreseeable future are beyond any doubt. At the same time, it is important to stress that the main part of these resources are located
in remote areas characterized by severe climatic conditions (the Arctic and Far East). The geological and geophysical explorations of
these resources have not been extensive enough. They cover, on average, only 0.17 km/km². This is several times less than the scope of

exploration on the shelves of the North Sea, Gulf of Mexico, and a number of other regions. So far, only 100 exploratory deep wells have been drilled on the Russian shelves, including 65 wells near Sakhalin and 35 in the West Arctic [Malovitski et al., 1994].

Two giant oil- and gas-bearing basins found in the West Arctic (on the shelves of the Barents and Kara Seas) cover a total area of 2 million square kilometers. They contain potential resources of at least 50–60 billion tons of conventional fuel (in oil equivalent). The region exploration, including drilling, revealed the existence of ten prolific oil and gas fields [Dubinin, 1994; Malovitski et al., 1994]. The resources of only two of them in the Kara Sea (Rusanovskoe and Leningradskoe) are estimated at 5 trillion cubic meters of natural gas. This amount is very impressive when taking into consideration that worldwide gas production at present totals 2 trillion cubic meters a year.

In the Barents Sea, the Shtokmanovskoe gas condensate field and the Prirazlomnoe oil field in the area of Pechora Bay are of special interest. The Shtokmanovskoe gas condensate field is probably the world's largest known offshore gas field. Its reserves total about 3 trillion cubic meters of gas and more than 20 million tons of gas condensate. Geological oil reserves of the Prirazlomnoe field amount to more than 200 million tons. The oil and gas developments in the region have been in progress since 1992. The large-scale industrial exploitation of these fields is planned to begin between 1998–2000. In the future, these activities might lead to significant changes in the world's system of oil and gas transportation [Parfenov, 1994]. Today in the Barents Sea, Norway and Russia have already installed over 10 drilling platforms.

Similar large-scale activities are planned in some other areas of the Russian northern shelves. For instance, in the Jamal area, a gas condensate field is estimated to produce up to 80–100 billion cubic meters of natural gas a year [Mazur, 1993]. The laying of a main gas pipe across Baidaratskaya Bay is planned for transportation of this gas.

The shelves of the Far East and Eastern Siberia have especially good prospects for large-scale and long-term developments of the offshore oil and gas fields. The promising areas in these regions (excluding Sakhalin and its shelf) are estimated at about 1.5 million square kilometers. Potential recoverable resources are estimated at billions of tons of conventional fuel. These reserves are concentrated mostly in the Sea of Okhotsk and the Bering, Chukchi, and East-Siberian Seas. Here, more than 20 oil- and gas-bearing and potentially oil- and gas-bearing basins of different geotectonic nature have been discovered.

The development of these unique natural resources of the Far East and Eastern Siberia is regulated within the framework of the federal Far-eastern program on licensing of the use of potentially oil- and gas-bearing areas until the year 2000 [Far-eastern program, 1994]. This program is targeted to attract national and foreign investments in order to accelerate the socioeconomic development of the region. Figure 6 and Table 2 reflect the scale and timetable of the exploration and production activities within the frame of this program. A system of international auctions for the rights to explore and develop the offshore resources in the Sea of Okhotsk and Bering Sea were put into effect by a governmental directive.

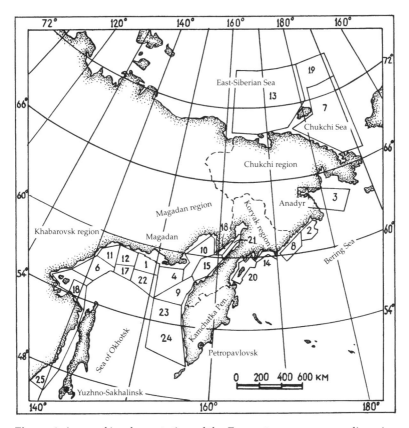

Figure 6 Areas of implementation of the Far-eastern program on licensing of the use of potentially oil- and gas-bearing areas until the year 2000 [Far-eastern program, 1994]. (Figures—areas of future developments, see Table 2.)

Table 2
Timetable of Tenders Stipulated by the Far-eastern Program on Licensing of the Use of Potentially Oil- and Gas-bearing Areas until the Year 2000 [Far-eastern program, 1994]

	Tender		
	Geophysical Surveys	Surveys, Exploration, and Development	
# and Area on Figure 6	Announcement	Announcement	Realization
1. Primagadanski	1994	1993	1994
2. Vostochno-Khatyrski	1994	1993	1994
3. Anadyrski	1994	1994	1995
4. Pyaginski	1994	1995	1996
5. Yuzhno-Shantarski	1994	1994	1995
6. Severo-Shantarski	1994	1995	1996
7. Yuzhno-Chukotski	1994	1996	1997
8. Zapadno-Khatyrski	1994	1997–2000	
9. Utkholokski	1994	1995	1996
10. Gizhiginski	1994	1997–2000	
11. Zapadno-Kukhtuiski	1994	1996–2000	
12. Vostochno-Kukhtuiski	1994	1996–2000	
13. Vostochno-Sibirski	1994	1996–2000	
14. Olyutorski	1994	1997–2000	
15. Shelikhovski	1994	1997–2000	
16. Severo-Shelikhovski	1994	1997–2000	
17. Koshevarovski	1994	1996–2000	
18. Yuzhni	-	1994	1994
19. Severo-Chukotski	1994	1996	1997
20. Ilpinski	1994	1997–2000	
21. Mametchinski	1994	1997–2000	
22. Tsentralni	1994	1996	2000
23. Zapadno-Kamchatski	1994	1995	1996
24. Yuzhno-Kamchatski	1994	1995	1996
25. Tatarski	1994	1995	1996

Similar activities are in progress on the shelf of Sakhalin. Its rich oil and gas resources discovered a long time ago have been intensively explored for the last 20 years. A number of large oil companies (Exxon, Shell, Marathon, and others) have already accomplished the first preliminary steps of oil and gas developments in the eastern and northeastern areas of the Sakhalin shelf within the frame of the Sakhalin-1 and Sakhalin-2 projects.

Geological reserves of only one of five giant fields located here (Odopinskoe) amount to over 200 million tons of oil and 0.1 trillion cubic meters of gas. The projects of oil and gas development of the Sakhalin shelf stipulate to install large offshore drilling platforms, construct undersea pipelines going to the shore, build onshore complexes for the storage and processing of oil and gas hydrocarbons (including a plant for producing liquefied natural gas), and to lay an inland gas main on the island. Realization of these projects may result in Sakhalin becoming a large oil and gas production center and an important oil and, especially, liquefied natural gas supplier.

Exploratory and production developments of the offshore oil and gas fields on the other Russian shelves (the Caspian, Black, Baltic Seas, and the Sea of Asov) have been in progress for a relatively long time although in a smaller scale. After the disintegration of the former USSR, the possibilities of oil exploration for Russia have become considerably narrower. Nevertheless, it does not exclude their activation and expansion in the future. In particular, this may occur as a result of the partnership between Russia and other countries.

The shallow shelf of the Caspian Sea, particularly its northern part, deserves a special mention. The recoverable hydrocarbon resources of this area are estimated to reach up to 2–3 billion tons, with oil accounting for 70%. At the same time, the exploration and drilling activities here are complicated by the very deep location of petroleum deposits (over 4 km), high reservoir drive, and large amounts of aggressive sulfurous gases [Granberg et al., 1993]. Besides, the Northern Caspian Sea has a unique fisheries value as a region for reproduction of the world's largest population of sturgeon.

For thousands of years, the continental shelf has been used mainly for landing various valuable sea products, including fish. At present, it provides annually over 80 million tons of sea products (in Russia—about 5 million tons). At the end of the twentieth century, the continental shelf has become a place for another large-scale and vitally important activity—oil and gas production. How can these two industries coexist together? Is it possible to extract oil and gas from the sea bottom to the fullest extent without disturbing the ecology of the shelf zone and its priceless biological resources? These and other related issues discussed in the following chapters are extremely important for the stable and environmentally safe development of many countries. The answers to these questions are vital for Russia, which is in the very beginning of developing the world's richest offshore hydrocarbon reserves.

Conclusions

1. One of the main trends in the world's oil and gas industry's development is the expansion of oil and gas production on the continental shelf of the World Ocean. For a long time, the continental shelf has been an arena mainly for shipping and fishing. At present, it has become a place for another large-scale activity— offshore oil and gas production. Offshore oil and gas production, being a very dynamic branch of one of the world's most important industries, is already supplying 25–30% of people's needs for hydrocarbon fuel.

2. The potential offshore oil and gas resources are estimated by different authors from 300 to 2,000 billion tons of oil equivalent. Available data indicate that they are sufficient to supply the energy needs of humankind in the twenty-first century. Accumulations of offshore oil and gas resources are usually found in the shelf zones as well as in deep ocean basins and zones of tectonic junctions. The locations of the offshore oil and gas fields often overlap with the zones of high biological productivity of the World Ocean.

3. Gas hydrates are extremely interesting and promising but practically undeveloped sources of hydrocarbon gases (mainly methane). They are found in many marine regions. The resource estimates are enormous, at least an order higher than gas resources recoverable from conventional fields. At the same time, industrial development of offshore gas hydrate fields may lead to serious environmental consequences.

4. The potential recoverable hydrocarbon resources are found on 90% of the Russian continental shelves. They total up to 90–100 billion tons of conventional fuel and consist mostly of natural gas (about 80%). The main part of these resources are located in remote regions of the West Arctic and Far East in the areas with severe climatic conditions.

5. In Russia, the oil and gas industry's activities are gradually expanding northward and eastward. They are reaching toward the shelves of the northern and far-eastern seas. The success of current and future large projects of offshore hydrocarbon development in Russia (as in many other countries) depends upon the degree of mastering modern technologies. Essential condition would be the ability of Russian specialists to use the available international experience in order to prevent environmental hazards on the shelf.

References

AMAP. 1997. *Arctic Monitoring and Assessment Programme. Arctic pollution issues: a state of the Arctic environment report.* Oslo: AMAP, 186 pp.

Artemiev, V. E. 1993. *Geochemistry of the organic substance in the system riversea.* Moscow: Nauka, 205 pp. (Russian)

Baybulatova, K. B., Vostokov, E. N. 1989. Geological aspects of the problem of pollution of the World Ocean. In *Complex studies of pollution of the World Ocean in connection with developing its mineral resources.* —Leningrad, pp.28–41. (Russian)

Beck, R. J. 1993. Natural gas to start long period of growth during next 3 years. *Oil and Gas Journ.* 91(44):58–63.

Cardy, S. 1993. Angling for oil at Wytch Farm. *Petroleum Review* 47(561):446–447.

Dossena, G., Ratti, S., Cannavacciuolo, A., Sebastiani, G. 1995. Research on the impact of hydrocarbons. Exploration and production activities in the Adriatic Sea. *Les mers tributaires de Mediterrannee. Bulletin de 15, Institut oceanographique, Monaco.* Numero special 15. CIESM Science Series No.1: 203–214.

Dubinin, I. B. 1994. Prospects of developing Shtokmanovskoe and Prirazlomnoe fields in the Barents Sea. *Gidrotekhnicheskoe stroitelstvo* 28(3):15–17. (Russian)

Engelhardt, F. R., ed. 1985. *Petroleum effects in the Arctic environment.* London and New York: Elsevier Applied Science, 281 pp.

Far-eastern program. 1994. *Far-eastern program on licensing of the use of potentially oil- and gas-bearing areas until the year 2000.* Moscow–Anadir–Magadan–Khabarovsk, 31 pp. (Russian)

Granberg, I. S., Sorokov, D. S., Suprunenko, O. I. 1993. Oil and gas resources of the Russian shelf. *Razvedka i okhrana nedr* 8:8–11. (Russian)

Grigorenko, J, Suprunenko, O. 1990. Modern understanding of oil and gas formation and accumulation in the geological structures of the World Ocean. In *Geology and mineral resources of the World Ocean.*—Warsaw: INTERMORGEO, pp.357–361. (Russian)

Jin, Di, Grigalunas, Th.A. 1993. Environmental compliance and energy exploration and prediction: application to offshore oil and gas. *Land Econ.* 69(1):82–97.

Kellard, M. 1994. Natural gas hydrates: energy for the future. *Mar.Pollut.Bull.* 29(6–12):307–311.

Knott, D. 1993. More North Sea oil flowing in despite of stormy disruption. *Oil and Gas Journ.* 91(44):23–28.

Krasilov, V. A. 1992. *Environmental protection principles, problems, priorities.* Moscow: Institut okhrany prirody i zapovednogo dela, 174 pp. (Russian)

Kruglyakova, R. P., Prokoptsev, G. N., Berlizeva, N. N. 1993. Gas hydrates of the Black Sea—potential source of hydrocarbons. *Razvedka i okhrana nedr* 12:7–10. (Russian)

Levin, L. E. 1994. Oil and gas potential of deep basins of the World Ocean. *Priroda* 6:24–28. (Russian)

Lisitsin, A. P. 1994. Marginal filter of the oceans. *Oceanologia* 34(5):735–747. (Russian)

Malovitski, J. P., Martirosjan, V. N., Golovchak, V. V., Gumenyuk, J. N., Fedorovski, J. F., Zakalski, V. M. 1994. Oil and gas potential of the Russian ocean margins. *Neftyanoe khosyaistvo* 4:27–32. (Russian)

Matthews, G. J. 1992. International law and policy on marine environmental protection and management: trends and prospects. *Mar.Pollut.Bull.* 25(1–4):70–74.

Mazur, I. I. 1993. *Ecology of oil and gas development.* Moscow: Nedra, 494 pp. (Russian)

MMS. 1995. *Minerals Management Service. Outer continental shelf natural gas and oil resource management program: cumulative effects 1987–1991. OCS Report MMS 95-0007.* Herdon, Va.: U.S.Department of the Interior (USDOI).

Neff, J. M. 1993. *Petroleum in the marine environment: regulatory strategies and fisheries impact. Report to Exxon Company.* Houston, TX, 13 pp.

Orudzhev, S. A. 1974. *Deep-water large-block foundations for the offshore drilling rigs.* Moscow: Nedra, 205 pp. (Russian)

Parfenov, A. F. 1994. Transportation system in the West Arctic. *Gidrotekhnicheskoe stroitelstvo* 28(3):17–22. (Russian)

Salnikov, S. S., Slevich, S. B. 1979. Industrial development of the ocean resources. In *Problems of exploration and exploitation of the World Ocean resources.*—Leningrad: Sudostroenie, pp.235–247. (Russian)

Samarski, V. N., Suvorov, K. G. 1979. Developing technical means for the offshore oil and gas production. In *Problems of exploration and exploitation of the World Ocean resources.*—Leningrad: Sudostroenie, pp.265–268. (Russian)

Slevich, S. B. 1977. *Shelf. Exploration and development.* Leningrad: Gidrometeoizdat, 250 pp. (Russian)

Taylor, B. G. S., Turnbull, R. G. H. 1992. Development. In *North Sea oil and environment: developing oil and gas resources, environmental impacts and responses.*—London and New York: Elsevier Applied Science, pp.57–88.

Zalogin, B. S. 1984. *Economic geography of the World Ocean.* Moscow: MGU, 321 pp. (Russian)

Zalogin, B. S., Kuzminskaya, K. S. 1993. Change trends in the nature of the World Ocean under anthropogenic impact. In *Life of the Earth: nature and society.*—Moscow, pp.66–72. (Russian)

Zick, V., Dimov, G., Malinovski, J., Berchia, I., Gramberg, I., eds. 1990. *Geology and mineral resources of the World Ocean.*—Warsaw: INTERMORGEO, 756 pp. (Russian)

Zubuva, M., Ginsburg, G., Solovyev, V., Shadrina, T. 1990. Gas hydrates. In *Geology and mineral resources of the World Ocean.* Warsaw: INTERMORGEO, pp.461–469. (Russian)

Chapter 2

Anthropogenic Impact on the Hydrosphere and the Present State of the Marine Environment

An objective assessment of the environmental consequences of any kind of human activity should take into consideration both the general ecological situation and specific effects of other types of anthropogenic impact. Hence, it would be relevant to start analyzing the consequences of the offshore oil and gas activities from a short review of the present state of the marine environment.

The ecology of the shelf zone deserves special attention within the framework of this problem. This zone and a narrow strip of adjoining land are characterized by the concentration of grandiose natural resources (including oil and gas). These areas also have conditions favorable for the exploitation of these resources as well as for development of many other activities. These activities include those that, to a large extent, have ensured the emergence and progress of the world's economy and, in a broader context, the origin and development of civilization itself. At present, about 80% of the Earth's population and about 50% of all large cities (those with a population of over 1 million) are located in a coastal area within 50 km onshore [Zalogin, Kuzminskaya, 1993].

Experts predict that urbanization and different forms of economic activities in the coastal zone will continue to grow and expand in upcoming decades [Charlier, 1989]. So it is not by accident that many countries and international organizations have started paying increased attention to the environmental problems of this relatively small but extremely important zone [Pearce, 1991; GESAMP, 1992; ECOPS, 1993; Aybulatov, 1994; ICES, 1995]. In a sense, the well-being of all humankind and the world's progress in the twenty-first century will depend to a great extent on the success of solving these problems.

2.1. Structure and Scale of Anthropogenic Impact

The term *anthropogenic impact* is used rather often in environmental literature when assessing the state of the hydrosphere and water ecosystems. At the same time, studies especially focused on the analysis of anthropogenic impact in all diversity of its manifestations occur very rarely [Kuderski, 1991; Pearce, 1991; Parsons, 1993]. Many more publications provide a description and classification of the sources, composition, and consequences of pollution of the hydrosphere [Patin, 1979; GESAMP, 1987; Izrael, Tziban, 1988; Laane, 1992; Bruegmann, 1993]. Pollution is undoubtedly a leading factor of anthropogenic impact on the water environment (including the impact of offshore oil and gas production). However, it certainly is not the only one. Moreover, pollution itself can be fully assessed only within the framework of describing all other impacts of human activity on the hydrosphere.

Table 3 gives some idea about the structure and factors of anthropogenic impact on water ecosystems. It also shows the nature, scale, and degree of environmental hazards under different kinds of anthropogenic impact. Here, sanitary-hygienic, ecological, and fisheries consequences are those changes in the water environment that threaten human health, lead to the imbalance in the structure and function of the water ecosystems, or undermine the stock of commercial species and interfere with fishing, respectively. As to the scale of these consequences, the local level is usually limited to the zone of obvious and easily registered effects (usually not more than several kilometers from the impact source). The regional scale includes the changes within large bodies of water (e.g., enclosed seas, bays, estuaries, and basins of big rivers). The global level applies to the effects and consequences (almost always hypothetical and difficult to prove) in areas distant from the source of direct anthropogenic impact (e.g., ocean pelagic zones and polar zones).

Obviously, all these gradations are rather relative and overlap to some extent. Practically any impact simultaneously causes changes and consequences of a sanitary-hygienic, ecological, and fisheries nature. It is difficult and not always possible to distinguish them clearly. This is also true when speaking about the scales of these phenomena. For example, regional consequences usually result from combined local effects. At the same time, sometimes one local event (for example, a discharge of a highly toxic effluent or an accidental oil spill) can change the ecological situation of an entire region. Regional ecological anomalies, in turn, gradually form the global background. Clear boundaries between them are impossible to reveal.

Similarly, the ranks of environmental hazard (high, considerable, weak, uncertain) used in Table 3 are not absolute. Specific aspects of using such ranks are the subject of debate about the methodology used to assess the states of natural systems. For our purposes, it is important to stress the absence of any commonly accepted methods of such assessments as well as conventionalism of any attempts to describe the state of ecosystems or impact factors in terms of good/bad or strong/weak. Good or bad ecosystem states do not exist at all. Our notion about the hazard of one or another impact is relative by its own nature. It reflects only the demands of different kinds of water use (e.g., water supply, recreation, and fisheries).

Nevertheless, such a qualitative approach has several important advantages. It allows us to differentiate impact factors in the hydrosphere. It also gives a relative assessment of the possible consequences of different impacts. One such application of this approach is given in Table 3. Of course, the given assessments are subjective. They reflect situations that, from our point of view, are the most widespread and typical in the marine and freshwater systems.

The first thing in Table 3 that attracts attention is the extreme variety of factors involved, diversity, and mosaic nature of anthropogenic impact on the hydrosphere. It includes such multifactorial phenomena as changes in temperature regime and radioactive background, discharges of toxic effluents and inflow of nutrients, irretrievable water consumption and damage of water organisms during seismic surveys, landing of commercial species and their cultivation, destruction of the shoreline and construction of drilling rigs.

Underestimation of the striking complexity of anthropogenic impact on the water ecosystems and the use of a single-factorial approach to analyze their state, focusing on some single aspect of human activity, generally lead to a distorted picture of the consequences of such activity. It is important to take into consideration that simultaneous impacts of several factors can cause synergetic effects. That is, the consequences can exceed the mere sum of the effects caused by each factor separately. Such situations are quite possible, for example, when radioactive, chemical, and thermal impacts are combined.

Another important circumstance is that many kinds of economic activities are rather difficult to differentiate based on their effects in the marine and freshwater systems. Many pure inland activities can lead to the ecological changes in the marine environment.

Table 3
Structure, Factors, and Scale of Anthropogenic Impact on Marine, Coastal, and Freshwater Ecosystems

Types of Activity and the Impact Factors	Nature, Scale, and Degree of Environmental Hazard								
	Sanitary-hygienic			Ecological			Fisheries		
	L	R	G	L	R	G	L	R	G
Industrial Activities									
Effluent input*	+++	+	−	+++	+	−	+++	+	?
Solid waste disposal*	+	−	−	+	−	−	+	−	−
Atmospheric emission and fallout*	++	−	−	+++	++	+?	++	+	?
Water consumption	−	−	−	+	−	−	+	−	−
Accidents*	+++	−	−	+++	+	−	+++	+	−
Agriculture									
Nutrient runoff and eutrophication*	+++	+	−	++	++	−	+	+	−
Pesticide runoff*	+++	++	−	++	++	+?	++	+	−
Animal farm effluent input*	++	−	−	++	−	−	+	−	−
Shoreline transformation	−	−	−	++	−	−	+	−	−
Irrigation	−	−	−	+++	++	−	++	+	−
Steam-electrical Power Plants									
Atmospheric emission and fallout*	++	−	−	+++	++	+?	++	−	?
Water consumption	−	−	−	+	−	−	+	−	−
Thermal pollution	−	−	−	+	−	−	+	−	−
Hydroelectric Power Plants									
Transformation of water bodies; water regime and ecosystem disturbances	+++	++	−	+++	+++	?	+++	+++	−
Dam Construction									
Transformation of water bodies; water regime and ecosystem disturbances	++	++	−	+++	+	−	+++	+	−

Table 3 (*continued*)

Types of Activity and the Impact Factors	Nature, Scale, and Degree of Environmental Hazard								
	Sanitary-hygienic			Ecological			Fisheries		
	L	R	G	L	R	G	L	R	G
Nuclear Power Plants									
Radioactive contamination*	++	−	−	+	+?	−	+	−	−
Thermal pollution	−	−	−	+	−	−	+	−	−
Accidents*	+++	+++	−	++	+	−	+++	++	−
Mining									
Tailings and chemical wastes	++	+	−	+++	+	−	++	+	−
Onland Transportation									
Atmospheric emission*	+++	++	−	++	+	?	+	?	−
Municipal Activity and Human Settlement									
Sewage input*	+++	++	−	+++	++	−	++	+	−
Water consumption	−	−	−	+	−	−	+	−	−
Seashore Construction and Development									
Shoreline and coastal zone transformation	+	+	−	++	+	−	++	+	−
Offshore Oil and Gas Production									
Liquid and solid waste discharge*	−	−	−	+	+	−	+	+	−
Geophysical surveys	−	−	−	+	−	−	+	?	−
Subsea pipeline emplacement*	−	−	−	++	?	−	++	+	−
Offshore structure abandonment	−	−	−	+	+	−	++	++	−
Accidents*	++	+	−	+++	+	−	+++	+	−
Onland Oil and Gas Production									
Oil pollution*	++	−	−	++	+	−	++	+	−
Pipeline emplacement	−	−	−	+	−	−	++	−	−
Accidents*	+++	−	−	+++	−	−	++	+	−
Mineral, Sand, and Gravel Extraction; Bottom Deepening Activities									
Shoreline and bottom relief disturbances, increased turbidity*	++	−	−	+++	+	−	+++	+	−

Table 3 (*continued*)

Types of Activity and the Impact Factors	Nature, Scale, and Degree of Environmental Hazard								
	Sanitary-hygienic			Ecological			Fisheries		
	L	R	G	L	R	G	L	R	G
Sea Dumping									
Bottom habitats disturbance, increased turbidity	—	—	—	+++	+	—	+++	+	—
Oil input*	++	+	—	++	+	?	+	—	—
Shipping									
Ship waste discharge*	++	—	—	++	+	—	++	+	—
Physical disturbances of water habitats	—	—	—	+	—	—	+	—	—
Nonindigeneous organism invasion	+++	++	—	+++	+++	—	+++	+++	—
Accidents*	+++	—	—	+++	+	?	++	+	—
Fisheries									
Ecosystem transformations	—	—	—	+++	+++	+?	+++	++	+?
Bottom biotope disturbances	—	—	—	+++	+++	—	+++	++	—
Aquaculture									
Ecological disturbances in the water column and on the bottom*	+	—	—	+++	—	—	+	—	—
Invasion of exotic species	—	—	—	+	+	—	?	?	—
Recreation									
Sanitary-hygienic and ecological disturbances in the coastal zone*	+++	++	—	+++	+	—	+++	+	—
Forestry									
Complex impact on water drainage system*	+	—	—	+++	++	—	+++	++	—
Technological Use of Seawater									
Water consumption	—	—	—	+	—	—	+	—	—

Notes: 1. Scale of impact and its consequences: L—local, R—regional, G—global.
2. Degree of environmental hazard: +++—high, ++—considerable, +—weak, —absent, ?—uncertain. 3. *—activities and impacts accompanied by chemical pollution.

Examples include dam construction, removal of river water for irrigation, cutting of forests, use of chemicals in agriculture, atmospheric emissions from factories and automobiles, sewage discharges into lakes and rivers, and many other impacts that take place hundreds and thousands of kilometers away from the seashore. Sooner or later, these activities affect the ecology of estuaries, bays, coastal waters, and sometimes of entire seas. The situations that developed, for example, in the Aral and Caspian Seas, the Sea of Asov, the Baltic and Black Seas clearly show that dividing the ecological problems and environmental protection programs into marine and freshwater ones is artificial and inept. From a broader perspective, we may state that effective protection of the water environment is impossible without protection of the inland ecosystems and vice versa. Such approaches get practiced more and more in a number of countries and at the international level [Pearce, 1991; GESAMP, 1991].

We should note that sanitary-hygienic consequences are clearly associated with the local scale. Ecological and fisheries disturbances also spread to the regional level. Sometimes, they can be found even on the global scale (see Table 3). Such tendency is quite understandable if we take into consideration both the specificity of sanitary-hygienic requirements and the localization of direct water usage (drinking water supply, recreation, and others) in limited areas (such as, water reservoirs, beaches, and coastal zones). At the same time, environmental and fisheries requirements for water quality are broader and more rigid than sanitary-hygienic standards. Therefore, deviations of many parameters from these requirements with high frequency and probability are registered in practically all water bodies [Patin, 1979; Rules for Water Protection, 1991].

Anthropogenic impact on marine and freshwater ecosystems should thus be defined as a combined manifestation of any form of industrial or other human activity that causes obvious or hidden disturbances of the natural structure and function of water communities, changes in composition and characteristics of biotopes, alternations in hydrological regime and geomorphology of water bodies, diminishing fisheries and recreational value, and any other negative ecological, economic, or socioeconomic consequences. This definition is based on the concept of a multifactorial nature of anthropogenic impact on the hydrosphere. This impact cumulatively results in structural and functional responses of the water ecosystems and biota.

The concept of anthropogenic impact is extremely important for analyzing the ecology of coastal and shelf zones which, as mentioned before, have been the center of various human activities for centuries. These include urbanization, construction of seaports and harbors, development of natural resources (including oil production and fishing), marine aquaculture, shipping, recreation, and many others. Anthropogenic activities that are in progress in the narrow area on both sides of the shoreline provide 50% and more of the gross national product of many countries. All of the activities and impact factors listed in Table 3 affect (usually directly and hazardously) the shelf ecology. At present, the anthropogenic disturbances of the shelf zone are found on a global scale. In many areas, they have reached critical limits. This is the prize for the unjustifiably rapid economic growth and shortsighted environmental policy (or rather for its absence).

The first obvious symptoms of anthropogenic press on the coastal zone and continental shelf appeared about 50–60 years ago. By now, anthropogenic impact has become so intense, diverse, and dynamic that the decision makers at last seem to realize its danger. Offshore oil and gas production is only one fragment of the heterogeneous mosaic of human activities occurring in the shelf zone (as can be seen in Table 3). The nature, structure, and relative contribution of this fragment in the general picture of anthropogenic impact on marine ecosystems and biological resources are going to be discussed in the following chapters.

2.2. Marine Pollution

At least two reasons allow us to consider pollution as the main, most widespread, and most dangerous factor of anthropogenic impact on the hydrosphere. First, pollution accompanies most (15 out of 20, as listed in Table 3) kinds of human activities, including offshore oil and gas production and marine oil transportation. Second, in contrast with land ecosystems where pollutants get fixed in the soil and plants, in the water environment, they quickly spread over large distances from the sources of pollution. In the freshwater and inland ecosystems, the effects of pollution are obvious. They literally appear right in front of our eyes. In contrast, the World Ocean has a large inertia of response to all forms of external impact. It requires a long hidden (latent) period to manifest the evidence of nonobvious consequences of this impact. The danger of the situation is complicated by the fact that when it happens, it will be too late to do anything.

2.2.1. Pollutant Input into the Marine Environment

Among all the diversity of human activities and sources of pollution (Table 3), we can distinguish three main ways that pollutants enter the marine environment:

- direct discharge of effluents and solid wastes into the seas and oceans (industrial discharge, municipal waste discharge, coastal sewage, and others);
- land runoff into the coastal zone, mainly with rivers;
- atmospheric fallout of pollutants transferred by the air mass onto the seas' surface.

Certainly, the relative contribution of each of these channels into the combined pollution input into the seas and oceans will be different for different substances and in different situations. Quantitative estimates of these processes are difficult because of the lack of reliable data and the extreme complexity of the natural processes, especially at the sea–land and sea–atmosphere boundaries.

For a number of pollutants (metals, nitrates, phosphates, oil and some other hydrocarbons), this task is even more complicated. They are distributed in the marine environment in the background of natural biogeochemical cycles of the same substances. Numerous evidence exists of extremely high concentrations of oil and gas hydrocarbons, heavy metals, radionuclides, nutrients, and suspended substances that are not connected with human activity at all. Examples of natural processes causing such situations include volcanic activity on the bottom; oil and gas seepage; splits and breaks of the earth's crust; algae blooms; mud flows; river floodings; and many others. Such phenomena should be taken into consideration in order to get the objective assessment of anthropogenic impact and its consequences in the hydrosphere.

Recognizing these complications explains why many earlier conclusions about the levels, flows, and balance of many substances in the hydrosphere are currently under revision. Developing new approaches and more precise analytical methods to determine trace amounts of contaminants allowed the Joint Group of Experts on Scientific Aspects of Marine Pollution (GESAMP) to get more reliable estimates of the contribution of different channels into the total contamination of the marine environment [GESAMP, 1990]. These data given in Figure 7 show that land-based and atmospheric sources account for about two-thirds of the total input of contaminants into the marine environment, constituting 44% and 33%, respectively. The

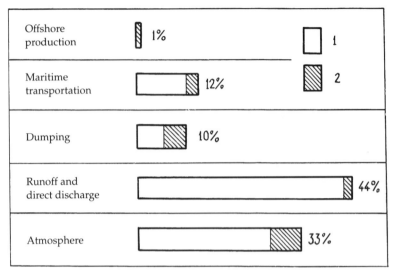

Figure 7 Relative contribution of contaminants to the marine environment in relation to transport pathways: 1—portion restricted to near shore; 2—portion delivered to open sea [Windom, 1992].

main pollution press undoubtedly falls on the shelf zones and especially on the coastal areas.

Certainly, given ratios can radically differ in different regions and for different pollutants. Nevertheless, atmospheric emissions always lead to dilution and spreading of pollutants over wide areas, in particular, over open (pelagic) parts of the oceans. It often results in regional and global distribution of pollutants. Sources other than atmospheric emissions usually form zones of local (or, more seldom, regional) pollution. Shelf activities (mainly developments of the offshore oil and other resources) generally contribute insignificantly (about 1–2%) to the total input of all pollutants in the marine environment. At the same time, as Chapter 4 will show, in the areas of intensive oil and gas production, the contribution of the offshore activities into total input of oil hydrocarbons can be considerably higher—up to 30%.

Some estimates [Panov et al., 1986; Mazur, 1993] indicate that the world's oil and gas industry annually discharges over 3 billion tons of solid wastes, about 500 km^3 of liquid wastes, and about 1 billion tons of aerosol. These discharges contain over 800 substances, and oil and oil products certainly dominate among them. Total worldwide annual loss of oil at all stages of its extraction and transportation exceeds 45 million tons. About 22 million tons are lost on land, about 7 million tons are lost in the sea, and about 16 million tons enter the atmosphere due to incomplete combustion of the liquid fuel [Panov et al., 1986]. Chapter 4 will give other available estimates of the oil input into the World Ocean.

2.2.2. Sources, Composition, and Degree of Hazards of Pollution Components

Table 4 shows the extreme diversity of marine pollution components, variety of their sources, scales of distribution, and degree of hazards. These pollutants can be classified in different ways, depending on their composition, toxicity, persistence, sources, volumes, and so on.

In order to analyze large-scale pollution and its global effects, it is common to distinguish a group of the most widespread pollutants. These include chlorinated hydrocarbons, heavy metals, nutrients, oil hydrocarbons, surface-active substances, and artificial radionuclides. These substances form the so-called background contamination that exists at present in any place in the hydrosphere. They are the subject of extensive research within the framework of national, regional, and international programs.

Depending on the type of impact on the water organisms, communities, and ecosystems, the pollutants can be grouped in the following order of increasing hazard:

- substances causing mechanical impacts (suspensions, films, solid wastes) that damage the respiratory organs, digestive system, and receptive ability;
- substances provoking eutrophic effects (e.g., mineral compounds of nitrogen and phosphorus, and organic substances) that cause mass rapid growth of phytoplankton and disturbances of the balance, structure, and functions of the water ecosystems;
- substances with saprogenic properties (sewage with a high content of easily decomposing organic matter) that cause oxygen deficiency followed by mass mortality of water organisms, and appearance of specific microphlora;
- substances causing toxic effects (e.g., heavy metals, chlorinated hydrocarbons, dioxins, and furans) that damage the physiological processes and functions of reproduction, feeding, and respiration;
- substances with mutagenic properties (e.g., benzo(a)pyrene and other polycyclic aromatic compounds, biphenyls, radionuclides) that cause carcinogenic, mutagenic, and teratogenic effects.

Table 4

Composition, Sources, and Degree of Hazard and Scale of
Distribution of Marine Pollution Components

Types of Impact, Groups of Substances	Scale of Distribution	Hazard		Sources
		Sanitary-hygienical	Eco-fisheries	
Physical				
Persistent debris (plastics, glass, metals)	L, R, G	–	+	Coastal disposal, shipping, fishing, offshore activities
Dredged sediments, drilling cuttings	L	+	+	Sea dumping
Oil slicks, tar balls	L, R	++	++	Oil production, transportation, river runoff, sewage
Suspended solids	L, R	++	++	Bottom dredging, offshore structure emplacement/ removal, drilling, river runoff, sewage
Thermal impacts	L	+	+	Heated water discharges from power plants and other industries
Chemical				
Synthetic Organic Chemicals				
DDT, aldrin, toxa-phene, lindane, endrin, etc.	L, R	+++	++	Biocides (pesticides, herbicides, fungicides) in agriculture and forestry
Polychlorinated biphenils (PCBs)	L, R	++	++	Discharges of electronic industry, insulators and plasticizers production
Dioxins and furanes	L, R	+++	++	Combustion process, chemical dumps, pulp and paper effluents

Table 4 (*continued*)

Types of Impact, Groups of Substances	Scale of Distribution	Hazard		Sources
		Sanitary-hygienical	Eco-fisheries	
Chlorophenols	L	++	+	Wood preservatives
Alkylbenzol-sulfonates, sulfanols, etc.	L	+++	+	Detergents, emulsifiers, and other components of industrial and municipal discharges
Freons	L, R, G	?	–	Propellants, refrigerants, aerosol products
Organotin	L, R	+	+	Stabilizers, catalysts in the plastics industry, fungicides, disinfectants
Organophosphates (e.g., malathion, parathion)	L, R	+++	+++	Pesticides in agriculture and forestry
Organoarsenicals	L, R	+	+	Pesticides, industrial wastes
Phthalic acid esters	L, R	++	+	Discharges of textile, plastics, and other industries
Aromatic amines	L	+	+	Dyes in textile production
Carbamates	L, R	++	++	Pesticides in agriculture and forestry
Dinitrophenols	L, R	+	+	Pesticides in agriculture
Oil Hydrocarbons Crude oil and oil products	L, R, G	++	++	Oil production, storage, marine transportation, river runoff, industrial discharges, atmospheric fallout, natural sources

Table 4 (*continued*)

Types of Impact, Groups of Substances	Scale of Distribution	Hazard		Sources
		Sanitary-hygienical	Eco-fisheries	
Benzo(a)pyrene, anthracene, and other polyaromatic hydrocarbons	L, R, G	++	+	Incomplete combustion of fuels, natural biosynthesis
Natural Gas				
Hydrocarbons of methane series (e.g., methane, ethane), sulfur dioxide, and others	L, R	+?	++	Natural gas production, technological losses, natural sources
Inorganic Chemicals				
Metals (e.g., mercury, lead, cadmium, copper)	L, R, G	++	++	Industrial and municipal discharges, atmospheric fallout, river runoff, ocean dumping
Nutrients (e.g., nitrates, phosphates)	L, R	++	+++	Agricultural and river runoff, sewage
Acids and bases	L	+	+?	Burning of fossil fuels, atmospheric emissions, industrial effluents
Radionuclides				
Strontium-90, caesium-137, plutonium, carbon-14, tritium, etc.	L, R, G	+	–	Nuclear testing, nuclear energy plants, nuclear waste disposal
Antibiotics	L	+	?	Aquaculture
Biological				
Pathogenic microorganisms (bacteria, viruses, fungi)	L, R	+++	+	Sewage, beach runoff, coastal recreation

Table 4 (*continued*)

Types of Impact, Groups of Substances	Scale of Distribution	Hazard		Sources
		Sanitary-hygienical	Eco-fisheries	
Biochemical Oxygen Demand (BOD)	L	+++	+++	Sewage, coastal water eutrophication
Nonindigeneous organisms	L, R	+++	+++	Accidental and intentional introducing of organisms from other regions

Notes: see Table 3.

Some of these pollutants (especially chlorinated hydrocarbons) cause toxic and mutagenic effects. Others (decomposing organic substances) lead to eutrophic and saprogenic effects. Oil and oil products are a group of pollutants that have complex and diverse composition and various impacts on living organisms—from physical and physicochemical damage to carcinogenic effects. Chapter 4 gives a more detailed ecotoxicological analysis of such processes for oil and gas hydrocarbons.

To estimate the hazard of different pollutants, we should take into account not only their hazardous properties but other factors, too. These include the volumes of their input into the environment, the ways and scale of their distribution, the patterns of their behavior in the water ecosystems, their ability to accumulate in living organisms, the stability of their composition, and other properties. In some cases, for example, while establishing the standards for water and seafood quality, while screening pesticides and other compounds by degree of their ecological hazard, or while calculating the assimilation capacity of water bodies, rather rigid procedures of quantitative estimates are used [GESAMP, 1986; Patin, Lesnikov, 1988; Izrael, Tziban, 1988; ICES, 1994].

However, the analysis of the global ecological situation and substantiation of marine environmental protection at the international or large-scale regional levels are impossible without using comprehensive expert assessments. These assessments take into consideration the combination of all available heterogeneous information and experience accumulated in different countries. One of the last assessments, conducted by the Joint Group of United Nations' experts on marine pollution [GESAMP, 1992], gives the following ranking of the main sources and components of marine pollution. They are listed in order of priority:

- input into the coastal water of large amounts of nutrients (mainly nitrates and, to a lesser extent, phosphates) due to sewage disposal and fertilizer runoff from the land;
- microbial contamination from sewage, which pollutes water and sea products and causes many diseases among people, including cholera and hepatitis;
- littering of beaches by plastic waste disposed on land and from ships, which causes serious damage to some marine organisms, especially mammals and birds;
- input of chlorinated hydrocarbons from rivers and land runoff, currently rising in tropical and subtropical regions;
- oil contamination, which exists everywhere, often fouls beaches and negatively affects the coastal recreational areas;
- heavy metals (cadmium, lead, mercury, and others) which, at present, are hazardous mainly in the vicinity of contamination sources;
- artificial radionuclides, which represent an insignificant portion of natural radioactivity and do not pose any special danger to people and marine organisms.

Such ranking is generally similar to the priority assessments given in Table 4. However, some differences reflect the specifics of the ecological situation in the Russian seas. For example, it is difficult to agree with the classification of chlorinated hydrocarbons, oil, and artificial radionuclides as less-important pollutants. Without going further into a detailed discussion about this issue, we will just mention the problems caused in Russia by the Chernobyl accident as well as the state of the Russian southern and northern seas. One of the main factors of anthropogenic impact on these seas is the pollution by chlorinated and oil hydrocarbons, respectively. In some other countries, for example Poland, oil and oil aromatic hydrocarbons are also considered priority toxicants in the national programs of marine monitoring [Andrulewich, 1992]. The following chapters will discuss this topic further. Let us just stress once more that neither our ratings (Table 4) nor the previously given priority list (based on the assessments of the group of UN experts) claim to be universal. The materials given in Table 4 are a summary of our views and judgment about the most typical, current situations of global contamination of the marine environment. It takes into account the specifics of the anthropogenic ecology of Russian seas.

2.2.3. Global and Regional Aspects

At present, the signs and consequences of human activity can be found everywhere on the Earth. Since the beginning of the century, the concentration of carbon dioxide in the atmosphere has increased 1.3 times, the concentration of methane—2.4 times, the concentration of nitrogen oxides—1.4 times, and so on. Current annual atmospheric emission of carbon dioxide from the burning of fossil fuel equals about 20 billion tons. As a result of global warming (greenhouse effect), transforming and destroying the ozone layer in the stratosphere (ozone holes), and a number of other processes, global anomalies in the biosphere have emerged. Although the analysis of these processes is beyond the framework of this book, they deserve a special mention. In particular, global warming can cause very rapid and radical climatic changes on the shelves of the arctic and far-eastern seas [Budyko et al., 1992; Patin, 1997]. It can affect the large projects of oil and gas developments in these regions discussed in Chapter 1. Therefore, the possibility of global climatic changes should become a focus of special research within the context of the strategy of the offshore oil and gas industry's development.

Numerous publications, including the ones mentioned in this book, undoubtedly indicate the existence of a large-scale (global) field of background contamination of the hydrosphere. The annual amounts of wastewaters discharged into the marine environment account for thousands of cubic kilometers. These represent up to 20% of the annual flow of all rivers [Morosov, 1990]. In Russia, these volumes in 1991 reached over 70 km^3.

In spite of the abundance and variety of the quantitative data on the concentrations and distribution of contaminants in the marine environment, it is rather difficult to use any average figures. There are at least two reasons for that. First, these data are obtained at different times and in different regions. Second, many chemical analyses of trace components in the marine environment give unreliable results. The last circumstance, noted in many recent publications (see, for example, Topping, 1992; ICES, 1994), is so serious that it makes the data of many past research questionable. In any case, the data on the concentrations of anthropogenic traces in seawater can be used mainly for comparing the contamination levels between individual areas. In very rare situations, they can help to reveal temporal trends in these levels. The last task can be accomplished more reliably using data regarding contaminant concentrations in bottom sediments and benthic biota. The results of their analyses provide a more accurate cumulative reflection of temporal changes of the contaminant levels in the marine environment.

These circumstances were taken into consideration while preparing Table 5 and Figures 8–12. They summarize the most reliable results of recent studies and give the ranges of typical levels of the main components of global pollution in the surface waters. The data from the latest reviews [Izrael, Tziban, 1988; Fowler, 1990; Monina, 1991; Laane, 1992; Bruegmann, 1993; Davis, 1993; Izrael et al., 1993; Patin, 1995] and publications of international organizations [GESAMP, 1986, 1987, 1992, 1993; UNEP, 1990; ICES, 1994, 1995] were used for this purpose.

As can be seen from the data in Figures 8–12 and Table 5, the concentration ranges for most substances are rather wide—up to 4–5 orders of magnitude. This phenomenon does not allow us to make reliable estimations of average concentrations for large regions and ecological zones of the World Ocean. Probably, it reflects not only analytical errors but the diversity of pollutant inputs and sources—atmospheric fallout, river input, and others. The ranges of the pollution levels are wider in the enclosed seas, coastal waters, and zones of local impact as compared with the waters of the open ocean.

In spite of all these variations, limitations, and difficulties in interpreting the seawater analysis data, evidence (which was first noted in the 1970s [Patin, 1976; 1979]) allows us to make a conclusion about the existence of increased pollution levels in the enclosed seas and coastal waters as compared with the open ocean. This conclusion, substantiated by a team of UN experts [GESAMP, 1992], was criticized by some researchers [Taylor, 1993; Davis, 1993]. In particular, they argue that available data reveal contaminants that sometimes are present in the same concentrations in the coastal and open waters or have even higher levels in the open waters as compared with the coastal zone. We do not want to disregard the importance of this discussion. Its results will define, to a certain extent, the general orientation of the international agreements and efforts to prevent marine pollution. However, we want to emphasize once more the unreliability of any averaging and quantitative comparing of concentrations of individual

Table 5
Typical Levels ($\mu g/l$) of the Worldwide Contaminants in the Surface Waters

Ecological Zone	Oil Hydrocarbons	Chlorinated Hydrocarbons	Metals		
			Mercury	Lead	Cadmium
South Ocean	$<10^{-1}$–1	$<10^{-4}$–10^{-3}	10^{-4}–10^{-2}	10^{-3}–10^{-2}	10^{-4}–10^{-2}
Ocean pelagic area					
Southern part	$<10^{-1}$–1	$<10^{-3}$–10^{-2}	10^{-4}–10^{-2}	10^{-3}–10^{-2}	10^{-4}–10^{-1}
Northern part	<1–10	10^{-3}–10^{-2}	10^{-5}–10^{-2}	10^{-3}–10^{-2}	10^{-4}–10^{-1}
Enclosed seas Open waters	<1–10^{2}	$<10^{-3}$–10^{-1}	$<10^{-3}$–10^{-2}	10^{-3}–10^{-1}	10^{-3}–10^{-1}
Coastal zones and estuaries	10–10^{2}	10^{-3}–1	10^{-3}–10^{-1}	10^{-2}–1	$<10^{-2}$–10^{-1}
Fresh waters	>10	$<10^{-2}$–1	10^{-2}–10^{-1}	1–10	10^{-2}–10^{-1}
Zones of local pollution	10^{2}–10^{4}	1–10^{3}	10^{-1}–10	10–10^{2}	10^{-1}–10

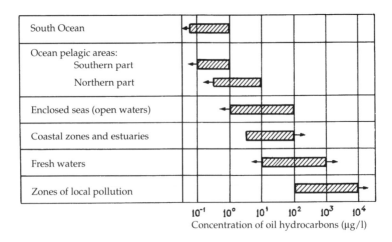

Figure 8 Ranges of typical levels of oil hydrocarbon concentration in the surface waters (arrows indicate possibility of widening the ranges).

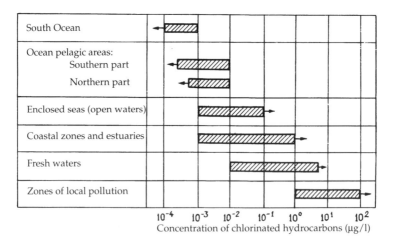

Figure 9 Ranges of typical levels of chlorinated hydrocarbon concentration (DDT, PCBs, aldrin, and others) in the surface waters.

contaminants for large areas. By considering the modern level of our knowledge, we believe that another way is more relevant—globally defining the general trends and characteristics of contamination in the World Ocean. One of these characteristics we have already noted, namely, the generally elevated concentrations of major contaminants

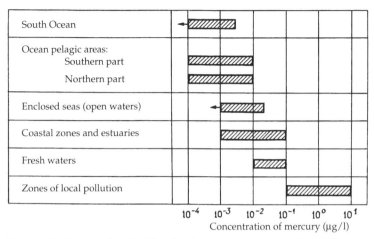

Figure 10 Ranges of typical levels of mercury concentration in the surface waters.

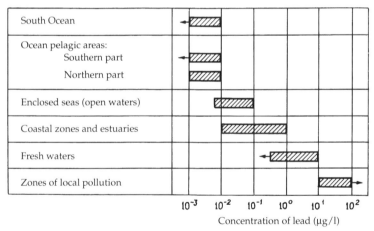

Figure 11 Ranges of typical levels of lead concentration in the surface waters.

in the enclosed seas and coastal zones as compared with the ocean pelagic areas.

Another trend is not so obvious but can still be seen in the combination of all presently available data. This trend manifests itself in the gradual increasing of contamination during the transition from the southern parts of all oceans to the north, where the

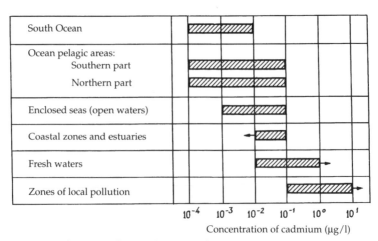

Figure 12 Ranges of typical levels of cadmium concentration in the surface waters.

main industrial centers and main pollution sources are concentrated. The relative stability of the differences in contamination levels in the surface waters of the northern and southern parts of the pelagic areas was revealed in the 1960s for artificial radionuclides [Shvedov, Patin, 1968] and later for chemical toxicants [Patin, 1976]. Besides the general pattern of distribution of pollution sources, there are two other factors explaining the relative stability of the global pollutant distribution in the World Ocean. Those are the relative confinement of large-scale water circulation within the limits of each hemisphere and the predominance of the zonal transport (that is, moving along the geographic parallels) of the traces in the atmosphere.

Another distinctive and repeatedly registered feature of the general picture of contaminant distribution in the marine environment is their localization at the water–atmosphere and water–bottom sediment boundaries. Practically everywhere and for all trace components (primarily for oil hydrocarbons), their concentrations are considerably (usually hundreds and thousands of times) higher in the surface microlayer of water and in the upper layer of bottom sediments. These boundaries provide the biotopes for the communities of hyponeuston and benthos, respectively.

The last feature of global pollution of the World Ocean is extremely ecologically alarming. It is the existence of elevated levels of contaminants in the zones of high bioproductivity where the main biomass of marine organisms is concentrated. Here, we mean the water layer up to 100 m from the water surface (photic layer) and boundaries of natural environments (water–atmosphere and water–bottom sediment, as previously mentioned) as well as enclosed seas, estuaries, coastal waters, and shelf zones. In these areas, which take only 10% of the World Ocean surface and less than 3% of its volume, the most intense processes of bioproduction, including the self-reproduction of the main living resources of the sea, take place [Moiseev, 1979]. The main press of anthropogenic impact, as can be seen from the data previously given and from many other published sources, is also concentrated here.

It is significant that at the regional and local levels, the intensity of anthropogenic press on the marine environment generally increases. The number and diversity of pollution components is growing as well. The contaminants with global distribution are combined here with hundreds and thousands of ingredients of local and regional distribution. Most of these substances are wastes and discharges from different local industries and activities. Often they are not included in the sphere of chemical-analytical control and monitoring. We usually get to know about their existence in the water environment from various signs of environmental trouble. These include the decline of abundance and various pathologies among fish and other organisms, poisoning or diseases among people, degradation of coastal ecosystems, fouled beaches, unusual algae blooms, and so forth.

Different marine regions are subjected to various and specific impact factors. The characteristics of these factors and their combination under specific conditions ultimately define the ecological situation in a given area, including the effects caused by offshore oil and gas development. The following chapters will discuss it in detail. Now we want to stress the alarming features of the ecological situation in the marine areas of Russia, especially the high pollution levels that very often exceed the maximum permissible limits. This fact, reflected in many scientific publications and official materials [Dubinina et al., 1993; Izrael et al., 1993; Laskorin, Lukyanenko, 1993], was one of the reasons why during the United Nations Conference on Environment and Development in 1992, Russia was rated as one of the most polluted countries of the world [Rybalski, Zaslavski, 1993].

Conclusions

1. Anthropogenic impact on the water environment is a cumulative manifestation of all kinds of human activity. It causes obvious and/or hidden disturbances in the natural structure and functions of water biotic communities, anomalies in composition and characteristics of their habitats, changes in the hydrology and geomorphology of water bodies, diminishing their fisheries and recreational value, and other negative effects of ecological, economic, or socioeconomic nature.

2. Sanitary-hygienic consequences of anthropogenic impact on the marine environment are revealed mostly at the local level. Ecological and fisheries disturbances also spread to the regional level and sometimes can be found even on the global scale.

3. Pollution is the leading factor of anthropogenic impact on the marine (mainly coastal) ecosystems. Land-based and atmospheric sources account for two-thirds of total pollution input into the marine environment, while offshore activities give only about 1–2%.

4. Typical features of marine pollution include:

- global spreading of a number of contaminants;
- their accumulation in the transit habitats (water–atmosphere and water–bottom boundaries);
- increase in contamination levels while moving from the open ocean (pelagic areas) to the coastal zones and enclosed seas and while moving from the southern marine regions to the north.

5. The main press of anthropogenic impact, in all its cumulative manifestations, is concentrated in the marine coastal areas, shelf zones, and the adjoining land. This is where 80% of the earth's population and a considerable part of the world's industry, including the oil and gas complex, are located. Ecological disturbances in these most bioproductive marine zones have become common phenomena reaching critical limits in some areas. The success in solving the problems of environmental protection of coastal and shelf areas while developing their natural resources is going to define to a considerable degree the well-being and progress of humankind in the twenty-first century.

References

Andrulewich, E. 1992. Harmful substances. *Stud.Mater.Oceanol.* 61:131–148.
Aybulatov, N. A. 1994. Anthropogenic expansion onto the coastal-shelf zone. *Vestnik RAN* 64(4):340–348. (Russian)
Bruegmann, L. 1993. *Meeresverunrelngung*. Berlinn: Academie Verlag, 294 pp.

Budyko, M. I., Borzenkova, I. I., Menzhulin, G. V., Selyakov, K. I. 1992. Forthcoming changes of the regional climate. *Izv.RAN. Ser.geogr.* 4:36–52. (Russian)

Charlier, R. H. 1989. Coastal zone: occupancy, management and economic competitiveness. *Ocean and Shoreline Management* 12(5):383–402.

Davis, W. J. 1993. Contamination of coastal versus surface waters. A brief meta-analysis. *Mar.Pollut.Bull.* 26(3):128–134.

Dubinina,V. G., Semenov, A. D., Nikonorov, I. V. 1993. Eco-fisheries problems and the concept of protection and restoration of water ecosystems. *Izvestia Russkogo Geograficheskogo Obshchestva* 125(3):23–29. (Russian)

ECOPS. 1993. *European Committee on Ocean and Polar Science. Prediction of change in the coastal seas. Grand challenges for European cooperation in coastal marine science.* 53 pp.

Fowler, S. W. 1990. Concentration of selected contaminants in water, sediments, and living organisms. In *UNEP Technical annexes to the report on the state of the marine environment. UNEP Regional Sea Reports and Studies No.114/2.* UNEP, pp.209–230.

GESAMP. 1986. *Environmental capacity. An approach to marine pollution prevention. GESAMP Reports and Studies No.30.* Rome: FAO, 51 pp.

GESAMP. 1987. *Land/sea boundary flux contaminants: contribution from rivers. GESAMP Reports and Studies No.32.* Paris: UNESCO, 174 pp.

GESAMP. 1990. *The state of the marine environment. GESAMP Reports and studies. No.39.* UNEP, 112 pp.

GESAMP. 1991. *Global strategies for marine environmental protection. GESAMP Reports and Studies No.45.* London: IMO, 36 pp.

GESAMP. 1992. *The state of the marine environment.* Oxford: Blackwell Scientific Publications Ltd., 128 pp.

GESAMP. 1993. *Impact of oil and related chemicals and wastes on the marine environment. GESAMP Reports and Studies No.50.* London: IMO, 180 pp.

ICES. 1994. *International Council for the Exploration of the Sea. Report of the ICES Advisory Committee on the Marine Environment. ICES Cooperative Research Report No.204.* Copenhagen: ICES, 122 pp.

ICES. 1995. *International Council for the Exploration of the Sea. Report of the ICES Advisory Committee on the Marine Environment. ICES Cooperative Research Report No.214.* Copenhagen: ICES, 135 pp.

Izrael, J. A., Tziban, A. V. 1988. *Anthropogenic ecology of the ocean.* Leningrad: Gidrometeoizdat, 528 pp. (Russian)

Izrael, J. A., Tziban, A. V., Panov, G. V. 1993. On the ecological situation in the Russian seas. *Metereologia i gidrologia* 8:15–21. (Russian)

Kuderski, L. A. 1991. Distinguishing different groups of anthropogenic factors according to their impact on the ecological processes in the water bodies. In *Theses of the Second Ull-Union Conference on Fisheries Toxicology* Vol.1.—Spb., pp.307–310. (Russian)

Laane, R. W. P. M., ed. 1992. *Background concentrations of natural compounds in rivers, sea water, atmosphere and mussels. Report DGW-92.033.* The Hague, 78 pp.

Laskorin, B., Lukyanenko, V. 1993. Strategy and tactic of the water bodies protection from pollution. *Mir nauki* 37(2):66–72. (Russian)

Mazur, I. I. 1993. *Ecology of oil and gas development.* Moscow: Nedra, 494 pp. (Russian)

Moiseev, P. A. 1979. Biological resources of the ocean and the prospects of their development. In *Problems of exploration and exploitation of the World Ocean resources.*—Leningrad: Sudostroenie, pp.287–298. (Russian)

Monina, J. I. 1991. Geography of oil pollution of the ocean. *Priroda* 8:70–74. (Russia)

Morosov, N. P. 1990. Pollution of the World Ocean by heavy metals (new views on the problem). *Journal Vsesoyusnogo Khimicheskigo Obshchestva im.Mendeleeva* 39(5):600–643. (Russian)

Panov, G. E., Petryashin, L. F., Lisyani, G. N. 1986. *Environmental protection in the oil and gas industry.* Moscow: Nedra. (Russian)

Parsons, L. S. 1993. *Management of marine fisheries in Canada.* Ottawa: Nat.Res.Council of Canada, 764 pp.

Patin, S. A. 1976. Ecological aspects of global pollution of the marine environment. *Oceanologia* 16(4):322–330. (Russian)

Patin, S. A. 1979. *Pollution impact on the biological resources and productivity of the World Ocean.* Moscow: Pishchepromizdat, 305 pp. (Russian)

Patin, S. A. 1988. Establishing fisheries standards for the water environment quality. In *Water toxicology and optimization of bioproductive processes in aquaculture.*—Moscow: VNIRO, pp.5–18. (Russian)

Patin, S. A. 1995. Global pollution and biological resources of the World Ocean. In *World Fisheries Congress Proceedings, Athens.*—New Delhy: Oxford IBH Publishing Co, pp.69–75.

Patin, S. A. 1997. Marine ecosystems, bioresources and global climate in the twenty-first century. *Rybnoe khosyaistvo* 2:45–55. (Russian)

Patin, S. A., Lesnikov, L. A. 1988. Main principles of establishing the fisheries standards for the water environment quality. In *Methodical guidelines on establishing the maximum permissible concentrations of pollutants in the fisheries.* Moscow: VNIRO, pp.3–10. (Russian)

Pearce, J. B. 1991. Collective effects of development on the marine environment. *Oceanologica Acta* SP(11):287–298.

Rules for water protection. 1991. *Rules for surface water protection (general principles).* Moscow: Goskompriroda SSSR, 34 pp. (Russian)

Rybalski, N. G., Zaslavski, E. M. 1993. Second UN Conference on Environment and Development. In *Proceedings of the 15th Mendeleev's Congress.*—Moscow. (Russian)

Shvedov, V. P., Patin, S. A. 1968. *Radioactivity of the seas and oceans.* Moscow: Atomizdat, 287 pp. (Russian)

Taylor, P. 1993. The state of the marine environment: a critique of the work and role of the Joint Group of Experts on Scientific Aspects of Marine Pollution (GESAMP). *Mar.Pollut.Bull.* 26(3):120–127.

Topping, G. 1992. The role and application of quality assurance in marine environmental protection. *Mar.Pollut.Bull.* 25(1–4):61–65.

UNEP. 1990. *The state of the environment. UNEP Regional Seas Reports and Studies No.15.* UNEP, 85 pp.

Windom, H. L. 1992. Contamination of the marine environment from land-based sources. *Mar.Pollut.Bull.* 25(1–4):32–36.

Zalogin, B. S., Kuzminskaya, K. S. 1993. Change trends in the nature of the World Ocean under anthropogenic impact. In *Life of the Earth: nature and society.*—Moscow, pp.66–72 (Russian)

Chapter 3

Factors of the Offshore Oil and Gas Industry's Impact on the Marine Environment and Fishing

Offshore oil and gas production includes multistage activities based, to a considerable extent, on the experience of analogous work on the land. At the same time, the specifics of marine conditions radically change the content and nature of most steps and operations of hydrocarbon resource development on the continental shelf.

Unlike many of the now available environmental reviews of the oil and gas industry's activities that discuss the engineering-technical aspects of ensuring ecological safety of the offshore developments [Vekilov, Geranin, 1989; Gusseinov, Alekperov, 1989; Gritzenko et al., 1993; Mazur, 1993], this chapter focuses mainly on the nature, scale, and possible environmental consequences of various impacts on different stages of developing the hydrocarbon resources on the continental shelf.

3.1. General Characteristics

We can distinguish four main stages of oil and gas development based on its organization and sequence of operations:

- geological and geophysical survey (seismosurveys, test drilling, and so on);
- exploration (rig emplacement, exploratory drilling, plugging the well, and others);
- development and production (platform emplacement, pipe laying, and drilling; hydrocarbon extraction, separation, and transportation; well and pipeline maintenance; and so forth); and
- decommissioning (disassembling, structure removal, well plugging, and others).

Each of these stages involves a certain set of activities that impose environmental impact. Tables 6 and 7 list the most important impact factors. It is easy to see that this impact has a complex nature and manifests itself in the form of physical, chemical, and biological disturbances in the water column, on the bottom, and partially in the atmosphere.

As the new resources are being developed and the old ones are being depleted (their exploitation can reach 20–40 years), local oil developments relocate and expand into new areas within large oil and gas basins. The sequence of the stages, given in Table 6 for development of an individual field, becomes unrecognizable in the background of the oil and gas industry's activity in the region. As a result, 5–10 years after beginning the development of a large oil- and gas-bearing basin, it is possible to see simultaneously newly built oil platforms, abandoned drilling structures, oil tankers, ships for seismic survey, and so on. Thus, local impacts overlap and interact with each other, forming wide areas of possible disturbances. The nature and intensity of these disturbances can certainly vary considerably depending on a combination of many natural and anthropogenic factors.

Figure 13 gives some idea about the scale of regional production activities. It shows the locations of drilling operations in the North Sea. The number of wells here has reached 4,000, with 160 producing fixed platforms and over 5,000 km of subsea pipelines [ICES, 1995].

Several large-scale oil and gas developments have started recently on the Russian shelves. The project of developing the Shtokmanovskoe gas condensate field in the Barents Sea stipulates drilling 120 wells up to 3 km deep, removing about 340 thousand tons of rock cuttings over 6–9 years, and installing deep-water ice-resistant platforms and underwater modules at depths up to 300–350 m. It also proposes laying 510–590 km of gas pipelines with diameters of 1,020–1,220 mm and building onshore facilities. Similar activities are planned in the zone of the Prirazlomnoe oil field in the Pechora Bay area. There, up to 60 wells about 2,500 m deep are going to be drilled, 2 production platforms (one of them will be ice resistant) at a depth of about 20 m are going to be installed, subsea oil pipelines are going to be laid, and underwater oil storage terminals are going to be built [Borisov et al., 1994].

The Sakhalin-1 project stipulates installing six platforms and drilling more than 600 wells up to 3,000 m deep. These platforms are going to produce over 60,000 tons of oil and about 80 million cubic meters of gas per day. The project proposes constructing a system of subsea pipelines with diameters up to 720 mm. The activities will

Table 6

Main Factors of Environmental Impact at Different Stages of Offshore Oil and Gas Production

Stage	Activities	Type and Nature of Impact
Geological and geophysical survey	Seismic surveys	Interference with fisheries and other users, impact on water organisms and fish populations
	Test drilling (core and shallow drilling; deep stratigraphic drilling)	Disturbances on the bottom, sediment resuspension, increase in turbidity, discharge of drilling muds and cuttings
Exploration	Rig emplacement and Exploratory drilling	Emissions and discharges of pollutants, interference with fisheries and other users, accidental blow-outs, and others (see Test drilling)
	Plugging the well and abandonment	Interference with fisheries and other users
Development and production	Platform emplacement, pipelaying operations, and support facilities construction	Physical disturbances, construction and commissioning discharges, interference with fisheries and other users
	Drilling of producing and injection wells	See Test drilling and Exploratory drilling
	Production operations and maintenance	Operational discharges, accidental spillage, interference with fisheries and other users, physical disturbance
	Support vessel traffic	Operational emissions and discharges, disrupting marine birds, mammals and other organisms, spilling oil
Decommissioning	Platform/structure removal, plugging, abandonment, use of bulk explosive charges	Operational discharges and emissions, interference with fisheries and other users, impact on water organisms when explosive charges are used

Table 7
General Characteristics of the Offshore Oil and Gas Industry's Impact on the Water Ecosystems and Bioresources

Nature of Impact and Its Consequences	Seismic Surveys and Explosions	Structure Emplacement	Drilling and Production Activities	Transportation by		Accidents
				Pipelines	Tankers	
Physical impact on:						
Pelagic species	+	+	–	–	+	+
Benthic communities	+	+	+	+	–	+
Benthic biotopes	+	+	+	+	–	+
Atmospheric emissions	–	–	+	–	+	+
Interference with fisheries	+	+	+	+	–	+
Disturbances of fish migrations	+	+	–	+	–	+
Chemical pollution	–	–	+	–	+	+
Increased turbidity	+	+	+	–	+	+
Tainting or contamination of commercial organisms	–	–	+	–	+	+

Note: + and – mean, respectively, presence or absence of impact and its consequences during different activities.

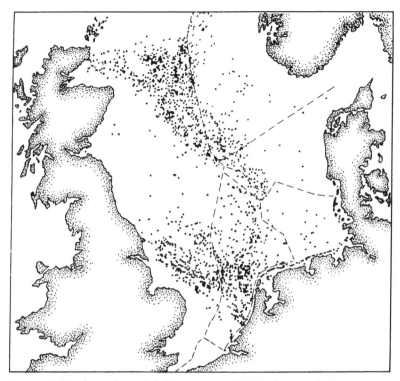

Figure 13 Total number of drillings up to 1990 in the North Sea between 50°N and 59°N [ICES, 1992].

involve conducting bottom dredging with removal of up to 70,000 m³ of ground. The project also stipulates building large onshore supporting facilities [Sakhalin-1, 1994].

Realization of such projects is inevitably going to cause environmental disturbances in the water column and on the bottom. The stipulated activities are accompanied by removal and resuspension of bottom sediments, discharges of drilling cuttings and produced waters, and many other impacts at all stages of field development.

The most intense and diverse environmental impacts are usually seen during the stage of development and production (see Table 6). At this stage, construction, assembling, drilling, and other activities in the sea and on the land are accompanied by intense support-vessel traffic. For example, during the development and exploitation of the Shtokmanovskoe and Prirazlomnoe fields on the shelf of the Barents Sea, over 150 vessels of 22 types are going to be used. These include different barges, pipe carriers, pipe layers, tug-boats, icebreakers, and collectors of drilling and other wastes. Gas carriers and oil shuttle tankers with a carrying capacity of up to 35,000 tons are going to be used for working in the field area and su-pertankers with a carrying capacity of up to 200,000 tons are going to be used for long-distance transportation, including oil export [Borisov et al., 1994].

Tankers transport a considerable part (possibly up to 50%) of the oil extracted from the shelf. The scale and ecological conse-quences of such transportation, as Chapter 4 will show, are rather striking. Besides the oil pollution, the tanker fleet is responsible for the appearance of nonindigenous organisms in some areas. At pres-ent, there are known more than 3,200 marine species of such organ-isms. Their rapid development under new conditions can lead to an ecological catastrophe [ICES, 1994]. One of such uninvited guests, the ctenophore *Mnemiopsis leidyi* (American comb jelly), brought in with tanker ballast waters in the Sea of Asov, the Black Sea, and the Mediterranean Sea, caused radical changes in the trophic structure of the large areas. In the Black Sea, it undermined the feeding base of commercial fish and thus the fishing potential of the area.

The most widespread and dangerous companion of the off-shore oil and gas complex (as well as any other industry) is pollution. Pollution is associated with practically all activities at any stage of oil and gas production. Liquid, solid, gaseous, and aerosol discharges and emissions during the drilling, technological, construction, and transportation operations include more than 800 substances, cer-tainly dominated by oil and oil products.

Table 8 gives information about the sources and volumes of oil pollution in the North Sea. These data indicate that the contribution of the offshore oil and gas industry in the regional balance of oil pol-lution can be very significant—over 30% of the total oil input into the basin. The production operations in the sea account for the over-whelming majority of this contribution. Figure 14 shows the vol-umes and sources of the oil input during the drilling activity in the North Sea.

At present, no standardized or commonly accepted methodol-ogy exists for the integral quantitative assessment of the environ-mental impact of the offshore oil and gas industry's activities. Different countries use different procedures and methods for this pur-pose. These include textual descriptions, experts' grading, and appli-cations of hazard and risk theory. Ultimately, solving this problem

Table 8
Total Oil Input to the North Sea (1985) (Institute of Offshore
Engineering, 1985, from [Somerville, Shirley, 1992])

Sources	Amount ($\times 10^3$ tons/year)
Natural seeps	1
Atmospheric	7–15
Rivers, land runoff	16–46
Coastal sewage	3–15
Coastal refineries	4
Oil terminals and reception facilities	1
Other coastal industrial effluent	5–15
Offshore oil and gas production	29
Sewage sludge	1–10
Dumped industrial waste	1–2
Dredged spoils	2–10
Operational ship discharges	1–2
Accidental or illegal discharges from shipping	No agreed estimate
Total	**71–150**

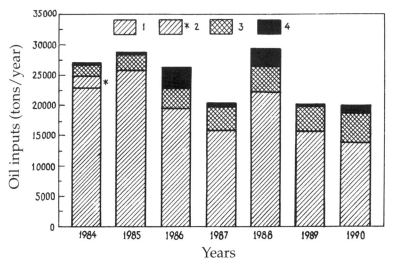

Figure 14 Total quantities of oil discharged by the offshore industry via cuttings, produced waters, and accidental spills into the North Sea: 1—drilling cuttings; 2—diesel-based drilling cuttings; 3—produced water; 4—accidental spills [ICES, 1995].

includes developing both the system of requirements for each activity, given in Table 6, and the system of environmental control and monitoring in the areas of these activities. These issues will be discussed in detail in Section 3.7 and in Chapter 8.

3.2. Geological and Geophysical Surveys on the Shelf

The impact on marine organisms and ecosystems starts during the first stage of offshore oil and gas production. This stage entails geological and geophysical surveys of the sea bottom aimed at revealing the geological oil- and gas-bearing structures. A geological survey includes bottom sampling, shallow coring activities, and deep stratigraphic testing. These cause sediment suspension and a local increase in turbidity. Deep stratigraphic drilling operations also involve the discharge of drilling muds and cuttings similar to what occurs in the drilling of exploratory and developmental wells.

Geophysical exploration includes methods of seismic, electrical, gravity, magnetic, and side-scan sonar surveys. The ecological consequences of large-scale geophysical activities are not as obvious as, for example, the effects of accidental oil spills. Besides, they are studied to a considerably lesser extent. At the same time, we are speaking about rather intense and specific impacts that appear in the water column and on the bottom. They result from generating powerful sound waves during seismic explorations and strong electrical fields during electrical surveys. Other geophysical methods, such as gravity and magnetic surveys, pose less ecological threat for the marine organisms than do seismic and electrical explorations.

3.2.1. Seismic Disturbances

A seismic survey is based on generating seismic waves and recording their reflection off the seafloor and subseafloor strata. The seismic profiles give an idea about the structure of the subsurface rock layers and the oil and gas potential of the explored area. In the past, geophysicists used explosives for generating sound waves. This practice was associated with high environmental risk. In particular, it led to a catastrophic ecological situation in the Caspian Sea in the 1960s. I was a member of the special Governmental Committee on this issue and witnessed firsthand the dramatic ecological consequences of the explosive use, including mass mortality of Caspian sturgeons (up to 200,000 large specimens).

However, even the modern technology of generating seismic waves poses a certain threat to marine biota, including commercial species. These impulses can create sound pressure of up to 150 atmospheres. They are sent every 20–60 seconds by sound wave generators (air and water guns) in numbers up to 30 and more. The sound waves are reflected off the seafloor and subseafloor strata to seismographic equipment that is towed behind a survey ship. Special receivers (hydrophones) detect these signals and, in turns, send them to the ship. There, they are processed by special systems.

The scale of such surveys and volumes of scanned water are very impressive. During 2–3 weeks of an average geophysical exploration, a survey ship, followed by a towed train (several kilometers in length) consisting of strings, cables, seismic generators, and hydrophones, usually covers a distance from 500 to 1,000 km [Grogan, Blanchard, 1992]. The number of seismic impulses executed during exploration of an area of 100 km^2 is not less than 5–8 million [Matishov, 1991]. When the route of a survey ship overlaps with the areas of trawling fishing, the tangling of trawls with towed strings of equipment might cause interference with commercial fishing.

Unfortunately, available information about the biological effects of high-energy waves of seismic signals on marine organisms is rather limited and contradictory. The incomplete data [Balashkand et al., 1980; Protasov et al., 1982; O'Keeffe, 1985; Davies, Kingston, 1992; Side, 1992; Pavlov et al., 1994] indicate the possibility of hazardous (up to lethal) effects of seismosignals on most water fauna species. However, they do not give the quantitative assessments of these effects on the total stock and reproduction of their populations.

In general, there seems to be no reason for the optimism that is sometimes expressed [Kriksunov et al., 1993] regarding the ecological safety of seismic surveys and their harmlessness to fish resources. The environmental and fisheries circles in a number of countries (for example, in Great Britain, Norway, and Canada) consider geophysical shelf exploration as a serious factor that can cause damaging effects on commercial organisms (especially during the periods of spawning and growth). In particular, the results of research conducted by Norwegian specialists indicate that school pelagic fish (especially herring) are able to respond to a seismic signal at a distance up to 100 km from the signal source [Dalen et al., 1996]. An example of marine fish responsiveness to sound pressure levels is given in Figure 15. Another study showed that damaging effects of intensive seismic exploration on the fish population caused a 70% decline of commercial catch near the Norwegian shore [Anonymous, 1994b].

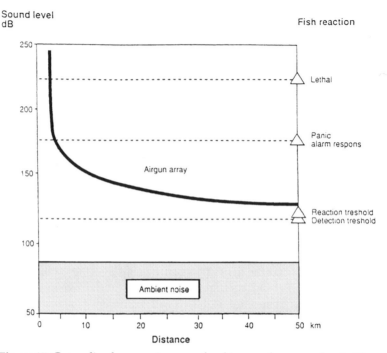

Figure 15 Generalized responsiveness of cod to sound pressure levels. The sound pressure levels versus distance of an array of airguns are given as well as the ambient noise level. The sound pressure level from the airgun is reduced to less than 200 dB within the first kilometer [NERI, 1996].

This is why some countries strictly regulate seismosurveys and require a fishing representative to be aboard the survey ships. In the North Sea, seismic surveys are not allowed from July through September in the areas where herring are spawning [Davies, Kingston, 1992]. In the Norwegian economic zone, in order to limit the areas and periods of seismic surveys, special studies of the region were conducted. As a result, the region was mapped with allocation of areas where seismic surveys would pose an increased environmental threat. These included the places of spawning, migration, and development of eggs, larvae, and fry of commercial fish [Bjoerke et al., 1991].

Mechanisms and manifestations of biological effects of high-energy waves of seismic signals on living organisms can differ. They range from damage of orientation and food search systems to physical damage of organs and tissues, disturbance of motor activity, and death. Early stages of fish development—larvae, fry, and probably developing eggs—are especially vulnerable.

One of the recent studies [Kriksunov et al., 1993] gives the radiuses of safe and lethal effects of seismic signals generated by different sources (pneumatic systems, gas detonation systems, electrospark generators). These radiuses are, respectively, 1–2.5 and 0.5–1 m. To compare, the radiuses of safe and lethal effects of small explosive charges (25 g and less), given in this study, are 2–18 m and 1–10 m, respectively. The authors of this publication note that in some cases, depending on the water depth and intensity of seismic surveys, seismic impacts causing fish fry mortality can be comparable to the impacts of natural hydroclimatic factors.

At the same time, other researchers (e.g., Dalen, Knutsen, 1987; Matishov, 1991) report mortality of 90% of larvae, fry, and even adult fish within a radius of up to 2 m from seismic sources. They also indicate that severe fish damages (hemorrhage, paralysis, loss of vision, and others) were found within the zone of up to 4 m from the sources of the seismic waves. V.M. Muraveyko with colleagues [Muraveyko et al., 1991] summarized the results of the original research and literature data on the impacts of seismic signals. They concluded that a safe radius within the impact zone in case of a pneumatic (air) gun with a pressure level of 140 atmospheres is about 5–7 m for zooplankton and ichthyoplankton and about 2 m for bottom organisms and phytoplankton [Muraveyko et al., 1991]. The researchers also noted disturbances of migration routes of salmon and other anadromous fish in the areas of seismic surveys.

The consequences of simultaneous combined impacts of large numbers (up to 30–40 and more) of seismic sources are poorly studied. These consequences as well as effects caused by new generating systems—water guns that can kill fish fry at a distance up to 7 m [Dalen, Knutsen, 1987]—are of special concern.

3.2.2. Electrosurveys

Electrosurveys are not as common in the offshore geophysical practice as seismic exploration. In some shallow shelf areas, however, they are used rather widely and effectively [Zick et al., 1990]. There are several varieties of electrosurveys. From the ecological perspective, the most dangerous are the methods that involve induction of artificial electrical fields in the marine environment (in the water column or on the bottom) with an intensity several orders higher than the corresponding parameters of natural fields.

Electroexploration techniques are rather complicated and diverse. Usually, they require the use of several ships. They include scanning the area with the help of recording and generating systems towed behind a ship or placed on the seafloor. The artificial electrical fields are induced with the help of direct current of 200 to 1,000 amp. The current polarity and working regime change during the survey.

Some reviews on this issue [Voronin et al., 1989; Muraveyko et al., 1991] indicate that direct current fields in the marine environment cause hazardous effects on fish. The potential gradient with 1–10 mV/cm causes primary responses, including behavioral changes, appearing of galvanotaxis, and others. As the current power increases, major fish damages are observed. Fish death occurs under the field intensity of 1–1.5 V/cm during 60 seconds.

At the same time, many experts believe that under actual conditions of marine electrical exploration, the damaging effects of electrical fields on adult fish are not significant. For developing eggs, however, the consequences can be hazardous due to increased electrosensitivity of some early stages of embryogenesis [Voronin et al., 1989; Muraveyko et al., 1991].

In conclusion, we want to emphasize once more both the lack of research about the mechanisms of biological effects of seismic impulses and induced electrical fields on marine organisms and the potential possibility of negative ecological and fisheries consequences of geophysical explorations on the continental shelf. The latter is especially alarming if we consider the increasing scale of oil and gas developments on the Russian shelves in areas of high bioproduction and fishing.

3.3. Drilling and Production Activities

Many aspects of the previously discussed impacts of geophysical surveys on marine ecosystems and biological resources have not been studied and still remain unknown. On the contrary, visible and obvious consequences of drilling and other production operations have been thoroughly researched from the very beginning of wide developments of the offshore oil and gas fields, especially in the Gulf of Mexico and the North Sea.

3.3.1. Steps and Operations

Drilling activities start during geological and geophysical survey. At this stage, core and shallow drilling operations are conducted to obtain information about geological parameters and drilling conditions. After locating the most promising pockets of hydrocarbons, the exploratory drilling activities start to determine whether the area contains commercial quantities of natural gas and oil. Depending on climatic conditions, the depth of the water, and the cost involved, different types of drilling units are used. These include mobile floating units (drill ships, drill barges, and semisubmersible rigs) and mobile bottom-founded units (jack-up rigs and submersible rigs). In the Arctic areas, wells are often drilled from artificial islands made from gravel or ice.

Regardless of the type of drilling unit used, all drilling methods are similar. Pipe sections and rotating bits that gradually decrease in diameter while the well gets deeper are used. The sides of the hole are supported by steel casing cemented in the formation for hermetic isolation of all systems. As drilling progresses, special drilling fluids (muds) are continuously pumped down the drill pipe and then pumped back up to the surface with the well cuttings. At the surface, the drilling cuttings are removed, and the drilling mud is returned to the circulating system (see Figure 16).

Drilling fluids (often called drilling muds when in use) are essential elements of modern drilling technology. They are designed to lubricate and cool the working drill bit and drill pipe, remove cuttings from the bottom of the well, control and regulate hydrostatic pressure, stabilize and seal the sides of the well, and prevent accidental blowouts in the well when encountering abnormally high pressures in the rock layers. The latter is also conducted with the help of special systems of blowout preventers that control the oil flow (circulation and safety valves, remote catches, gauges, chokes, and so forth).

After the well is drilled, it is tested for oil and gas by studying the rock cuttings and measuring the geophysical parameters of the formation. If no hydrocarbons are present, the hole is plugged and cut off below the seafloor. If exploratory drilling results confirm the existence of hydrocarbons, the well is plugged and bordered with the help of special armature. Developmental drilling may be planned. The technological procedure of drilling a producing well is the same as an exploratory one. The difference is that drilling is usually conducted from a rig sitting atop a fixed production platform.

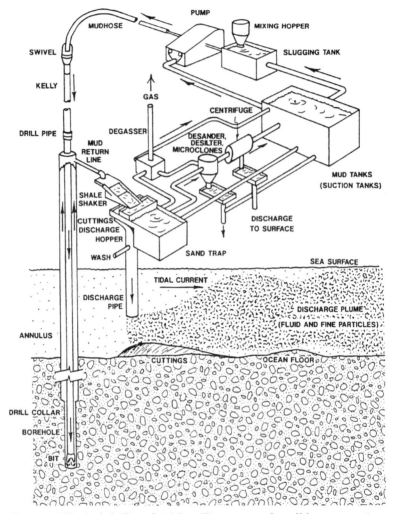

Figure 16 Typical drilling fluid handling system for offshore operations [Neff, 1981].

Offshore oil and gas developments are practically always based on drilling several wells from a single platform to develop the surrounding area. The number of wells per platform varies and can reach up to 70–100. Besides, for maximum product extraction, deviational drilling is widely used. In this case, the wells are angled out up to 5,000 m and more from the vertical (Figure 17).

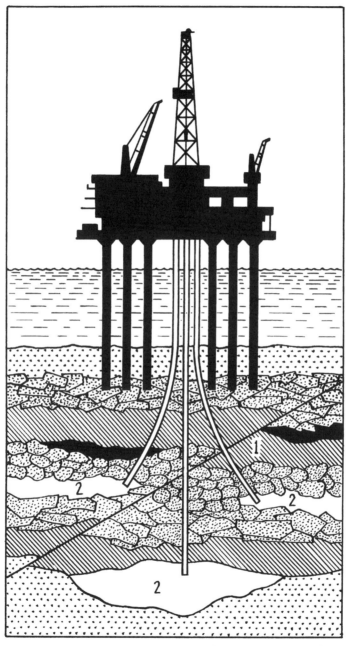

Figure 17 Deviational drilling from an offshore platform: 1—fault;
2—oil or gas zone [Chevron Corporation Publication, 1993].

Only about one-fourth of the oil in a reservoir is recovered by natural flow and pumping. To recover some of the remaining oil, gas injection, water, steam, fire, or chemical floods are used. These methods often require drilling new injection wells or converting old wells into injection wells.

Hydrocarbons are extracted from the well in a complex mixture of water, oil, and natural gas. They must be separated and treated. The gas either is removed from the site through a pipeline or is reinjected into the reservoir for maintaining pressure. At the same time, flaring the excessive amounts of gas produced with the oil still remains a dubious element of oil and gas production in many countries.

Each drilling platform has power units (generators that use gas or diesel fuel) and systems for cooling the power and technological equipment. The temperature of the seawater used in these systems (in amounts up to 30,000 m³/hour) is usually 10°C higher at the time of discharge than its original temperature.

3.3.2. Sources, Types, and Volumes of Waste Discharges

Practically all stages and operations of offshore hydrocarbon production are accompanied by undesirable discharges of liquid, solid, and gaseous wastes. Tables 9–11 give some idea about the most widespread discharges and their typical volumes.

While speaking about possible discharges of future oil and gas developments on the Russian shelves, we want to note that one platform alone on the eastern Sakhalin shelf is estimated to discharge about 60,000 m³ of drilling fluids, 15,000 m³ of drilling cuttings (during drilling of 50 wells), and 640 m³ of produced waters a day. On some platforms, the volumes of discharged produced waters can reach up to 20,000 m³ a day [Sakhalin-1, 1994]. The project of exploration and development of the Shtokmanovskoe field in the Barents Sea during the first 6–9 years stipulates the drilling of at least 120 wells with total removal of up to 340,000 tons of cuttings [Borisov et al., 1994].

The proportions and amounts of discharged wastes can change considerably during production. For example, the amount of solid drilling cuttings usually decreases as the well gets deeper and the hole diameter becomes correspondingly smaller. The volumes of produced waters increase as the hydrocarbon resources are being depleted and production moves from the first stages toward its completion. Drilling in the upper layers of bottom sediments (up to

Table 9
Typical Discharges During Oil and Gas Exploration and Production

Source, Activity	Discharge
Exploratory drilling	Drilling muds (mostly water-based), drilling cuttings
Developmental drilling	Drilling muds (oil- and water-based); drilling cuttings, well-treatment fluids
Well completion	Well-completion fluids
Well workover	Workover fluids
Production operations	Produced water (including formation water and injection water); ballast water; displacement water; deck drainage; drilling muds; drilling cuttings; produced sand; cement residues; blowout-preventer fluid; sanitary and domestic wastes; gas and oil processing wastes; slop oil; cooling water; desalination brine; test water from the fire-control system; atmospheric emissions
Accidental discharges	Oil spills; gas blowouts; chemical spills

Table 10
Typical Quantities of Wastes Discharged During Offshore Oil and Gas Exploration and Production Activities [GESAMP, 1993]

Discharges	Approx. Amounts (tons)
Exploration Sites (Ranges for a Single Well):	
Drilling mud—*periodically*	15 to 30
—*bulk at end*	150 to 400
Cuttings (dry mass)	200 to 1,000
Base oil on cuttings	30 to 120 (a)
Production Site (Multiple Wells):	
Drilling mud	45,000 (b)
Cuttings	50,000 (b)
Produced water	1,500/day (c)

Notes: (a) Actual loss to the environment may be higher; (b) Estimate based on 50 wells drilled from a single offshore production platform, drilled over 4 to 20 years; (c) From a single platform.

Table 11
Volumes of Treated Produced Water Discharged to the Ocean
in Different Parts of the World [Neff, 1998]

Location	Discharge Rate (m³/day)
U.S. Gulf of Mexico	549,000
Offshore California	14,650
Cook Inlet, Alaska	22,065
North Sea	512,000
Australia	100,000
West Java Sea (3 offshore facilities)	192,000

approximately 100 m) can be done without using complex drilling fluids. In such cases, seawater with additives of special clay suspensions can be used instead.

As can be seen from the data in Table 10, the discharges of produced waters considerably dominate over other wastes. Produced waters include formation water, brine, injection water, and other technological waters. Formation water and brine are extracted along with oil and gas. Injection water is pumped into the injection wells in hundreds of thousands of tons for maintaining the pressure in the system and pushing the hydrocarbons toward the producing wells. All of these waters are usually polluted by oil, natural low-molecular-weight hydrocarbons, inorganic salts, and technological chemicals. These waters need to be cleaned before they are discharged into the sea. Such cleaning under marine conditions is a complicated technical task. Special separation units on the platforms are used for oil separation. Depending on its quality, the produced water is either discharged into the sea or injected into the disposal well. Sometimes the oil-water mixtures are transported along the pipelines to onshore separation units [Kuznetsov, Samarskaya, 1993].

Produced waters, including injection waters and solutions of chemicals used to intensify hydrocarbon extraction and the separation of the oil-water mixtures, are one of the main sources of oil pollution in the areas of offshore oil and gas production. For example, in the North Sea, this source is responsible for 20% of all oil discharges from the oil and gas complex's activity in that region [PARCOM, 1991]. The worldwide annual input of these discharges in 1990 was estimated at 47,000 tons [GESAMP, 1993]. It is significant that, as a hydrocarbon reservoir is being depleted, the ratio between the water and oil fraction in the extracted product increases, and water be-comes the prevailing phase. At the same time, both the volumes of discharged waters and the difficulties of their treatment increase. An example of known and predicted quantities of water and oil dis-charged from North Sea production platforms is given in Figure 18.

Inevitably, all kinds of drilling are associated with drilling wastes, including drilling muds and cuttings. Drilling cuttings are re-moved from drilling muds and cleaned in special separators. The amount of oil left on cuttings after cleaning is much higher when us-ing oil-based fluids. Separated drilling muds and cleaning fluids

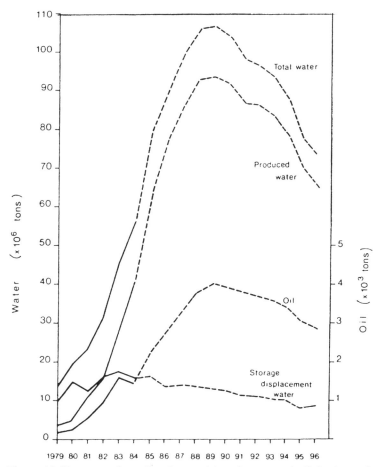

Figure 18 Known and predicted quantities of water and oil discharged from North Sea production platforms [Somerville et al., 1987].

used to treat cuttings are partially returned to the circulating system. Drilling cuttings and the rest of the drilling muds are either dumped overboard or transported to the shore for further treatment and disposal, depending on the situation and ecological requirements. The first variant is the most usual and is practiced almost everywhere, while the second one still remains an unrealized (or seldom realized) ecological requirement. Drilling cuttings covered by oil and often by toxic drilling fluids were the main source of oil pollution during drilling operations in the 1980s (see Figure 14). Their discharge into the North Sea in 1985 alone was estimated at 260,000 tons.

At the end of the 1980s, UNEP conducted an analysis of the methods to remove and dispose of drilling wastes (via boats/barges and pipelines, by burning, by burying, and so on) [UNEP, 1988]. The analysis showed that most of these methods were too expensive or posed an unacceptably high risk of accidents. Unfortunately, this rather pessimistic conclusion still correctly describes the general situation at present. Most often, drilling wastes go overboard at the offshore well sites. Disposing of drilling cuttings near the seafloor instead of dumping them on the water surface can limit the spread of polluted suspense and thus decrease the scale of its harmful impact.

Many countries and oil companies are looking for effective systems of cleaning and detoxification of oil-polluted drilling cuttings. One such system was developed and successfully tested in the USA [Minton et al., 1991]. Some methods to detoxify drilling cuttings use electrothermal treatment. The results of biotesting show that it considerably decreases the toxicity of discharged drilling waste [Alekperov, Kasimov, 1988].

Recently, a technology was developed to remove the drilling wastes, especially cuttings, by reinjecting their slurry into a geological formation. The method was successfully tested on the shelves of Alaska and the Gulf of Mexico [GESAMP, 1993]. It is planned to be used in the Barents Sea [Borisov et al., 1994]. This gives some hope to achieving zero discharge of oil-containing wastes during offshore oil and gas production. Some other measures (such as slim-hole drilling) to reduce discharges, particularly in environmentally sensitive locations, are being investigated by the industry [Swan et al., 1994].

The environmental hazard of drilling muds is connected, in particular, with the presence of lubricating materials in their composition. These lubricating substances usually have a hydrocarbon base. They are needed for effective drilling, especially in case of slant holes or drilling through solid rock. The lubricants are added into the drilling fluids either from the very beginning as a part of the original formulations or in the process of drilling when the operational need emerges. In both cases, the discharges of spent drilling muds and cuttings coated by these muds contain considerable amounts of relatively stable and toxic hydrocarbon compounds and a wide spectrum of many other substances. In the 1980s, from 60% to 78% of all chemicals that had been used by the offshore industry in the British zone of the North Sea entered the marine environment. It totaled from 117,000 tons to 138,000 tons annually. The wastes from drilling operations were the main part of these discharges [Somerville, Shirley, 1992].

One of the potential sources of oil pollution is produced sand extracted with oil. The amount of produced sand coated by oil can vary a lot in different areas and even during production in the same area. In some cases, it constitutes a considerable part of the extracted product. Most often, this sand is cleaned of the oil and dumped overboard at the well site. Sometimes, it is baked or calcified and transported to the shore.

The other discharges into the marine environment, shown in Table 9 (deck drainage, sanitary and domestic wastes, and so on), do not play essential roles in the environmental situation in the areas of oil and gas developments. They are treated and disposed in accordance with the norms regulating discharges from the ships.

3.3.3. Chemical Composition of Discharged Wastes

As noted earlier, the spectrum of chemicals entering the marine environment at different stages of oil and gas production is very wide. They include many hundreds of individual compounds and their combinations. Broadly speaking, all can be divided into two large groups. The first group consists of the extracted oil and gas hydrocarbons, which the following chapters will discuss in detail. The second group, which this section will review, unites the rest of the natural and technological components used at different technological stages.

Drilling fluids and cuttings. Drilling wastes deserve special attention. The volume of drilling wastes usually ranges from 1,000 to 5,000 m^3 for each well. Such wells can number into dozens for one production platform and many hundreds for a large field. For example, in the North Sea in 1988, the discharges of drilling cuttings alone led to the input of 22,000 tons of oil and about 100,000 tons of chemicals (mostly inorganic and non-toxic) into the marine environment. These discharges also included 4,900 tons of potentially toxic substances such as biocides, corrosion and scale inhibitors, detergents, emulsifiers, and oxygen adsorbents [Davies, Kingston, 1992].

Drilling cuttings separated from drilling muds have a complex and extremely changeable composition. This composition depends on the type of rock, drilling regime, formulation of the drilling fluid, technology to separate and clean cuttings, and other factors. However, in all cases, drilling fluids (muds) play the leading role in forming the composition of drilling cuttings.

No precise, standard formulation exists for drilling fluids. Their composition depends on the needs of the particular situations. These differ considerably in different regions and may even radically change during each drilling process while drilling rocks of very different structure (from solid granite formations to salt and slate strata). At present, two main types of drilling fluids are used in offshore drilling. They are based either on crude oil, oil products, and other mixtures of organic substances (diesel, paraffin oils, and so on) or on water (freshwater or seawater with bentonite, barite, and other components added). The main components of oil- and water-based drilling fluids are given in Tables 12 and 13, respectively. During the last 10 years, the preference is given to using the less-toxic water-based drilling fluids. However, in some cases, for example during drilling of deviated wells through hard rock, using oil-based fluids is still inevitable. The oil-based fluids, in contrast with the water-based ones, are usually not discharged overboard after a single application. Instead, they are regenerated and included in the technological circle again.

Originally, the oil-based drilling muds included diesel fuel as their base component due to its availability and low cost. However, starting in the 1980s, especially after many countries prohibited the use of diesel in drilling muds, the oil companies started to develop new formulations that replaced diesel oil with less hazardous substances. Alternative drilling fluids are composed mainly from low-molecular-weight, less toxic and more water-soluble, aromatic compounds and substances of paraffin structure. Research in this direction continues at present. Products of animal, vegetable, or synthetic origin are tested in order to find the optimal base for drilling fluids [GESAMP, 1993].

Recently, a new generation of drilling fluids based on the products of chemical synthesis with ethers, esters, olefins, and polyalphaolefins has been developed [Burke, Veil, 1995]. Such

Table 12
Main Components of Oil-based Drilling Fluids:
Amounts of Chemicals Used to Drill a Typical
North Sea Oil Well [Davies, Kingston, 1992]

	Amount Used	
Chemical	Tons	%
Barite	409	60.8
Base oil	210	31.3
CaCl	22	3.3
Emulsifier	15	2.2
Filtrate control/wetting agent	12	1.8
Lime	2	0.2
Viscosifier	2	0.2

Table 13
Representative Drilling Fluid Compositions [Smith et al., 1997]

Lignosulfonate Drilling Fluid		Polymer Drilling Fluid	
Component	Weight %	Component	Weight %
Seawater	76	Salt water	80
Barite	15	Barite	17
Bentonite	7	Bentonite	2
Lignosulfonate	1	Partially hydrolyzed polyacrylamide (PHPA)	0.2
Lignite	1	Xanthan gum biopolymer	0.2
Starch	0.2	Starch	0.6

drilling fluids allow highly deviational or horizontal drillings to be conducted. From the environmental perspective, the most important fact is that they have low toxicity as compared with other drilling formulations. In spite of the relatively high cost of the synthetic-based drilling fluids, their technological and environmental advantages open wide possibilities for their effective use in oil and gas production.

Each component of a drilling fluid has one or several chemical and technological functions (Table 14). For example, barite ($BaSO_4$) is

Table 14

Drilling Fluid Components and Their Technological Functions

Groups	Components	Functions
Lubricants	Paraffins, naphthalenes, and their derivatives, sulfanol, diesel and mineral oils, graphite, derivatives of fatty acids, lanolin, sprint, and others	To reduce friction and heat in the drilling zone
Weighting agents	Barite ($BaSO_4$), calcite, and others	To control and regulate hydrostatic pressure in the bore hole
Viscosifiers	Bentonite and other organophilic clays, carboxymethylcellulose (CMC), oxymethylcellulose (OMC), polyacrylates, lignite, high temperature polymers, starch, xanthan gum, guar gum, and others	To reduce the loss of fluids, control viscosity, stabilize the well bore
Thinners	Sodium tetraphosphates and other polyphosphates, methylated tannin, chromium lignosulfonates, calcium lignite, lignosulfonates, calcium sulfonate, low-molecular-weight acrylates, polyacrylamides, silicon-organic compounds, and others	To control viscosity and dispersion at different stages of drilling to prevent flocculation and excessive thickening of technological fluids
Stabilizers	Polyamides, sulfonated lignite, sulfonated phenolic resins, sodium chloride, granular and fibrous materials, and others	To ensure osmotic balance and stabilization of the well bore at different stages of drilling
Emulsifiers	Alkylated sufonates, derivatives of fatty acids, ethers, esters, and others	To form and maintain emulsions during drilling and other technological procedures
Electrolyte pH controllers	Potassium chloride, sodium chloride, gypsum ($CaSO_4$), caustic soda, lime, and others	To maintain the pH of drilling fluid, reduce corrosion, and stabilize emulsions
Corrosion inhibitors	Sodium sulfite, zinc carbonate, ammonium bisulfite, zinc chromate, diammonium phosphate (DAP), phosphoxit-7, and others	To prevent corrosion and scaling of pipes, drilling equipment, and other machinery
Solvents	Isopropanol, isobutanol, butanol, ethylene glycol, diesel oil, esters, ethers, and others	To prepare solutions of agents and technological fluids
Biocides	Sodium hypochlorite, biguanidine salt, quaternary ammonium salts, aliphatic dialdehydes, oxyalkylated phenols, fatty diamines, thiazolines, carbamates, paraformaldehyde, dichlorophenols, and others	To prevent microorganism development and microbial degrading of drilling fluid components and other agents

used to control and regulate hydrostatic pressure in the well. Emulsifiers (alkyl-acrylate sulfonate, alkylacryl sulfate, and others) form and maintain emulsions. Sodium and calcium chlorides create conditions for maintaining an isotonic osmotic balance between the water phase of the emulsion and surrounding formation water. Organophilic clays (such as amine treated bentonite clay) as well as organic polymers and polyacrylates ensure the optimal fluid viscosity necessary for drilling under different geological conditions. Sodium sulfite, ammonium bisulfite, zinc carbonate, and other oxygen scavengers are pumped into the well to prevent the corrosion of drilling equipment in the oxidizing environment. Lime is added to increase the pH of drilling fluids, which helps to reduce corrosion and stabilize the emulsions in the muds.

As a result of many technological operations and procedures, drilling muds and cuttings are saturated with hundreds of very different substances and compounds. It is their discharges into the sea that pose one of the main ecological threats during offshore oil production. In particular, many countries express concern regarding biocides, which are used to suppress microflora in the drilling and other circulating fluids. The list of such compounds includes over one hundred names. The most widespread biocides used in the oil and gas production practice include sodium salts of hypochlorite, formalin releasers, and glutaraldehyde as well as biguanidine and quaternary ammonium, and a number of other compounds. The composition of some compounds is not always known. Some biocides are highly toxic. Many countries either discourage (for example, in case of carbamates and thiocarbamates) or prohibit (for example, in case of dichlorophenols and pentachlorophenates) their use by the offshore oil and gas industry [GESAMP, 1993].

Drilling discharges also contain many heavy metals (mercury, lead, cadmium, zinc, chromium, copper, and others) that come from components of both drilling fluids and drilling cuttings (Table 15). Chapter 6 gives the ecotoxicological assessments and comparison of different drilling fluids and drilling cuttings.

Produced waters.   Produced waters usually include dissolved salts and organic compounds, oil hydrocarbons, trace metals, suspensions, and many other substances that are components of formation water from the reservoir or are used during drilling and other production operations. Besides, produced waters can mix with the extracted oil, gas, and injection waters from the wells. All of the above make the composition of the discharged produced waters very complex and

Table 15
Example Metal Content of Drilling Waste [NRC, 1983]

Metal	Effluent Concentration (mg/kg)	
	Shale Shaker Discharge (Drilling Cuttings) (77% wt solids)	Discharged Drilling Fluid (21% wt solids)
Barium	3,160	37,400
Chromium	44	191
Cadmium	<2	<1
Lead	10	3
Mercury	<1	<1
Nickel	15	4
Vanadium	11	5
Zinc	80	50

changeable (Figure 19). It is practically impossible to speak about some average parameters of this composition, especially because reliable and complete analytical studies of these wastes are very rare.

It is known, for example, that produced waters from Buccaneer field in the Gulf of Mexico contained high concentrations of benzenes, toluene, and xylenes (10–31 mg/kg total), considerable amounts of benzo(a)pyrene (1–2 µg/kg), biocides (a few mg/kg), corrosion and scale inhibitors (1–10 mg/kg), and many other components, including organic molecules and heavy metals [Middleditch, 1981]. Studies in the North Sea showed that produced water discharged from oil platforms had 20–40 mg/kg of dissolved hydrocarbons (mainly benzenes, toluene, and xylenes), large amounts of low-molecular-weight organic acids and esters, and variable concentrations of technological chemicals [Somerville et al., 1987].

Some data also indicate the presence of high levels of naphthalenes both in the oil fraction of produced waters and in the discharged waters (1.7 mg/l) after their treatment and oil separation on the drilling platforms near the Texas shore in the Gulf of Mexico [Anderson, 1985].

Chromatographic analyses of the discharge water from the production platforms on the Norwegian shelf [Grahl-Nielson, 1987] showed very high and relatively stable levels of phenol and its alkylated homologues in these discharges (5.5–5.9 mg/l) (Figure 20). It must be remembered that phenols are polar, very water-soluble,

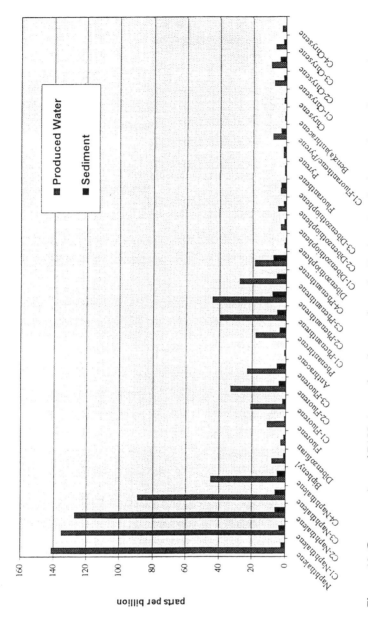

Figure 19 Concentrations of PAHs in produced water from the Trading Bay facility and in sediments from near the produced water discharge [Neff, 1998].

79

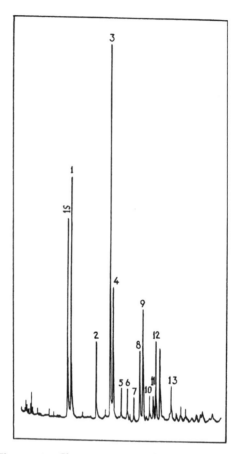

Figure 20 Chromatogram of derivatives of phenols in discharge water from a production platform. The various peaks present: IS—p-fluorophenol as international standard; 1—phenol; 2—o-cresol; 3—m-cresol; 4—p-cresol; 5—2, 6-xylenol; 6—ethylphenol; 7—methylnaphthalene; 8—2, 5-xylenol (+methylnaphthalene); 9—2, 4-xylenol; 10—ethylphenol; 11—2, 3-xylenol; 12—3, 5-xylenol; 13—3, 4-xylenol. The total amount of phenols in this particular sample was 5.4 mg/l [Grahl-Nielson, 1987].

and highly toxic compounds. Their maximum permissible concentration (MPC) in the natural waters according to Russian standard is 0.001 mg/l.

Petroleum hydrocarbons are always present in produced waters, especially when the latter are mixed with other technological waters and solutions. For example, the concentration of oil products in produced waters from the platforms on the Tengiskoe field in the northern part of the Caspian Sea in 1991 exceeded 25 g/l [Stradomskaya, Semenov, 1991]. However, the levels of oil in discharges vary extremely. They depend not only on the specific technological situation but on the fractional composition of the oil and the effectiveness of the oil/water separation methods as well. The oil separators mainly remove particulate and dispersed oil, while dissolved hydrocarbons in concentrations from 20 mg/l to over 50 mg/l go overboard as part of the discharged waters [Somerville et al., 1987; GESAMP, 1993]. The volumes of such discharges, as can be seen from Figure 14, reach thousands of tons of oil a year. This problem might be solved with the help of new effective technologies to treat oil-containing produced waters, in particular the ones based on using membranes from natural cellulose materials [Larson et al., 1991].

Another characteristic of the chemical composition of most produced waters is their very high mineralization. It is usually higher than the seawater's salinity reaching up to 300 g/l. Such mineralization is caused by the presence of dissolved ions of sodium, potassium, magnesium, chloride, and sulfate in produced waters. Besides, produced waters often have elevated levels of some heavy metals [Neff et al., 1987] as well as corrosion inhibitors, descalers, biocides, dispersants, emulsion breakers, and other chemicals.

Recent studies have revealed that produced waters frequently contain naturally occurring radioactive elements and their daughter products, such as radium-226 and radium-228. They are leached from the reservoir by formation waters and are carried to the surface with produced waters, oil, and gas. During contact with seawater, these radionuclides interact with sulfates, precipitate, and form a radioactive scale. In spite of a relatively low level of radioactivity, concern exists that this process can create centers of increased radioactive risk. This phenomenon has become a focus of attention in a number of countries. Applying the regulations defined by some international agreements, such as the London Dumping Convention (1972), that do not allow discharges of radioactive material into the marine environment are considered to be justified in this case [GESAMP, 1993].

Other wastes.   Large quantities of produced waters, drilling muds, and drilling cuttings, discussed above, as well as discharges of storage displacement and ballast waters are the source of regular and long-term impacts of the offshore industry on the marine environment. Besides these discharges, sometimes the need arises to conduct a one-time discharge of short duration. Such situations include, in particular, chemical discharges during construction, hydrostatic testing, commissioning, pigging, and maintenance of the pipeline systems. The pipeline discharges usually contain corrosion and scale inhibitors, biocides, oxygen scavengers, and other agents. The volumes of these wastes can be rather considerable. In the North Sea, they reach up to 300,000 m^3 of treated water discharged over a short period (hours to days) [GESAMP, 1993]. The discharge regime usually ensures that the dilution decreases the concentration and toxicity of the wastes to safe levels beyond a 500-meter radius from the place of discharge [Davies, Kingston, 1992]. The original acute toxicity of the wastes is supposed to decrease at least 100 times after the dilution.

Similar situations emerge during other technological and maintenance activities. Examples include cleaning and anticorrosion procedures, discharging the ballast waters from the hydrocarbon storage tanks, well repairing, well workover operations, replacing the equipment, and others. These discharges often contain surface-active substances, such as lignosulfonates, lignites, sulfo-methylated tannins, and many other chemicals with about a hundred names.

Table 16 gives some idea about the volumes of the most hazardous substances discharged during oil and gas production. Use of such compounds in different situations requires not only considering their technological effectiveness but taking into account other factors as well. These include the compounds' toxicity and rate of degradation, which define their ecological hazard in the marine environment. Chapter 6 will discuss it in detail.

3.3.4. Atmospheric Emissions

Although the atmospheric emissions accompany most of the oil and gas operations, this factor has not gained any special attention in the context of offshore developments. The available information is very limited and controversial. At the same time, in some areas of onland production, for example in Western Siberia and near Astrakhan in Russia, this source of pollution poses a serious threat to the water and

Table 16
Discharges of Individual Chemicals and Agents During the Oil
and Gas Production in the North Sea [NSTF, 1994]

Chemicals and Agents	Total Amount of Discharges (tons/year) in 1987–1988
Biocides	1,176 (439)*
Corrosion inhibitors	230
Oxygen scavengers	205
Scale inhibitors	3,996
Emulsion breakers	327
Oil removing agents	290
Defoamers	6
Flocculating agents	166
Dispersing agents	None reported
Diluents	0
Fluid loss agents	0 (3,266)
Viscosity control agents	12
Tencides (surfactants)	43 (2,276)
Cleaning agents	186
Gas treatment agents	789
Asphaltenes	(489)
Other chemicals	1
Total	7,442

Note: *—The figures in parentheses indicate the data for 1990.

onland ecosystems and to human health. For example, in the Nizh-nevartovki region (Tumen area), the atmospheric emission of hazardous substances from the Samotlorskoe oil field development in 1989–1992 varied from 0.38 million to 1.1 million tons a year [Krupinin, 1995]. The high content of hydrogen sulfide (6–30%) and other toxic substances in the natural gas and atmospheric emissions on the Orenburgskoe and Astrakhanskoe gas condensate fields created situations close to ecological catastrophes [Karamova, 1989].

Atmospheric emissions take place at all stages of oil and gas industry's activities. The main sources of these emissions include:

- constant or periodical burning of associated gas and excessive amounts of hydrocarbons during well testing and development as well as continuous flaring to eliminate gas from the storage tanks and pressure-controlling systems;

- combustion of gaseous and liquid fuel in the energetic units (diesel-powered generators and pumps, gas turbines, internal combustion engines) on the platforms, ships, and onshore facilities; and
- evaporation or venting of hydrocarbons during different operations of their production, treatment, transportation, and storage.

In spite of the fact that some countries now prohibit flaring of oil-associated gases, it remains one of the major sources of atmospheric emissions in the world. These gases are dissolved in the crude produced oil. As the pressure goes down, they bubble out in amounts up to 300 m^3 for each ton of extracted oil. The associated gases give about 30% of the gross world production of gaseous hydrocarbons. However, because of the undeveloped technology and lack of required capacities and equipment on many field developments, up to 25% of all associated gases are flared. In Russia alone, the volumes of annually burned (flared) oil-associated gases reach up to 10–17 billion cubic meters [VNIIP, 1994]. Astronauts have witnessed that the view of the gas-burning torches, for example above Western Siberia or the Persian Gulf, is an impressive proof of the large scale of human economic activity and, we would add, of its bad management as well.

Components of atmospheric pollution caused by oil and gas development include gaseous products of hydrocarbon evaporation and burning as well as aerosol particles of the unburned fuel. From the ecological perspective, the most hazardous components are nitrogen and sulfur oxides, carbon monoxide, and the products of the incomplete burning of hydrocarbons. These interact with atmospheric moisture, transform under the influence of solar radiation, and precipitate onto the land and sea surfaces to form fields of local and regional pollution.

Clear evidence of the impact of atmospheric emissions on the marine environment from the offshore flaring was found, in particular, during well testing in the Canadian zone of the Beaufort Sea. Here, the ice surface around the test site where intensive flaring of combustible wastes occurred was polluted by atmospheric fallout of heavy oily residue. The chemical composition of the residue was similar to one of the higher-molecular-weight fractions of produced oil [GESAMP, 1993].

According to some estimates [Kingston, 1991], up to 30% of the hydrocarbons emitted into the atmosphere during well testing precipitate onto the sea surface and create distinctive and relatively unstable slicks around the offshore installations. The results of the aircraft observations in the North Sea indicate that such slicks are found with an average frequency of 1–2 cases per every hour of flight [ICES, 1995].

The approximate calculations [Somerville, Shirley, 1992] show that in 1988, about 10% of the total offshore gas production in the Great Britain was flared and used for covering the energetic needs of field developments. Total atmospheric emissions (excluding methane) in the British section of the North Sea were estimated at about 20,000 tons a year. This takes into account its fugitive emissions, losses, and evaporation during extraction, storage, and transportation. The methane emission in this region was estimated at about 75,000 tons a year. These numbers, respectively, are less than 0.02% and 0.08% of total atmospheric emission of volatile organic compounds in Great Britain. Analogous estimates for the offshore oil and gas developments on the Norwegian shelf [Lindberg et al., 1990] showed the prevalence of carbon dioxide (88%) in atmospheric emissions. Available information about flaring emissions of other products in the Norwegian sector, including the most dangerous nitrogen oxides and volatile organic compounds, is extremely uncertain.

Technical means to rectify and prevent atmospheric pollution during offshore oil and gas production are practically identical to the analogous methods that are widely and often effectively used on land and in other industries. However, offshore atmospheric emissions thus far have not gotten the deserved attention, probably due to the remoteness of these developments from densely populated places.

3.4. Accidental Situations

3.4.1. Statistics and Causes

Accidents inevitably accompany hydrocarbon development. They are the sources of environmental pollution at all stages of oil and gas production. The estimates show that 1–2% of the oil produced in Russia is lost because of accidental spills and blowouts. For example, in the Tumen area and Komi Republic, the number of such accidents reaches hundreds and thousands of episodes every year. The corresponding losses reach into millions of tons of spilled oil [Mikhaylova, 1995; Yablokov, 1995].

Of course, this alarming situation on the Russian onland developments should not be extrapolated to the whole world's offshore oil industry. In many countries from the very beginning of its existence, this industry has paid increased attention to environmental safety and accident prevention under severe and sometimes extremely severe marine conditions. At the same time, the commercial optimism often expressed by the oil companies and their experts regarding the safety of offshore operations is not always justified.

In the Gulf of Mexico on the USA shelf, an estimated number of accidental oil spills (spills of over 1,000 barrels) for each billion barrels of extracted or transported oil equals, on average [Ray, 1985]:

- 0.79 accidental oil spills for each billion barrels of extracted oil during drilling operations on the platforms;
- 1.82 accidental oil spills for each billion barrels of transported oil during pipeline transportation; and
- 3.87 accidental oil spills for each billion barrels of transported oil during tanker transportation.

Of course, in other regions, such estimates can be different. The important thing to know is the proportion between the amount of hydrocarbons extracted during production and the amount lost during the accidents. For example, according to official data [Monk, Cormack, 1992], the volumes of oil spilled during drilling accidents on the British shelf from 1980 to 1988 represent from $1.1 \times 10^{-4}\%$ up to $2.9 \times 10^{-3}\%$ of the annual volumes of oil produced on the same platforms. Even if we consider that these data can be a bit underestimated and the figures previously given for Siberian oil developments can be exaggerated, the comparison is still not in favor of Russian oil practices.

The causes, scale, and severity of the accidents' consequences are extremely variable. They depend on a concrete combination of many natural, technical, and technological factors. To a certain extent, each accidental situation develops in accordance with its unique scenario. The most typical causes of accidents include equipment failure, personnel mistakes, and extreme natural impacts (seismic activity, ice fields, hurricanes, and so on). Their main hazard is connected with the spills and blowouts of oil, gas, and numerous other chemical substances and compounds previously discussed. The environmental consequences of accidental episodes are especially severe, sometimes dramatic, when they happen near the shore, in shallow waters, or in areas with slow water circulation. The same outcome is often inevitable when oil spilled in the open sea drifts toward the coastal shallow zone.

3.4.2. Drilling Accidents

Drilling accidents are usually associated with unexpected blowouts of liquid and gaseous hydrocarbons from the well as a result of encountering zones with abnormally high pressure. No other situations but tanker oil spills can compete with drilling accidents in frequency and severity. For example, even in the North Sea where, as some authors believe, the oil and gas industry has ensured meeting the strictest ecological and safety requirements from 1979 to 1987, 516 operational accidents were recorded. These led to the release of 21,530 tons of oil into the sea, 48% of it due to blowouts and spills on offshore drilling rigs [Shears, 1989]. In 1977, an accident on the oil field Ekofisk in the North Sea caused the spill of 22,500 tons of oil. A little bit earlier, a number of large accidents happened in the Gulf of Mexico and near Southern California. The volumes of spilled oil reached 0.4–1.4 million tons. One of the world's largest spills happened in 1979 near the shore of Mexico after an exploratory well blowout on the drilling rig Ixtoc-1. As a result, every day for 10 months, from 2,740 to 5,480 tons of oil were released into the sea [Gourlay, 1988].

We can assume that similar accidents took place in some other regions of the world, including the former USSR. However, in Soviet times, the information about such events was very limited and tightly controlled. That explains why this book refers so often to the experience of other countries. Among the very few available data about drilling accidents in the former USSR are reports of long-term catastrophic blowouts of natural gas during the drilling in the Sea of Asov in 1982 and in 1985 (see Chapter 5). High levels of oil pollution in the areas of the Apsheronski Peninsula, Northern Caspian Sea, and some other places also resulted from accidental spills during exploratory and production drilling.

Broadly speaking, two major categories of drilling accidents should be distinguished. One of them covers catastrophic situations involving intense and prolonged hydrocarbon gushing. These occur when the pressure in the drilling zone is so high that usual technological methods of well muffling do not help. Lean holes have to be drilled to stop the blowout. The abnormally high pressure is most often encountered during exploratory drilling in new fields. The probability of such extreme situations is relatively low. Some oil experts estimate it at 1 incident for 10,000 wells [Sakhalin-1, 1994]. The need to drill lean holes emerges, on average, in 3% of accidental episodes.

The other group of accidental situations includes regular, routine episodes of hydrocarbon spills and blowouts during drilling operations. These accidents can be controlled rather effectively (in several hours or days) by shutting in the well with the help of the blowout preventers and by changing the density of the drilling fluid. Accidents of this kind are not so impressive as rare catastrophic blowouts. Usually, they do not attract any special attention. At the same time, their ecological hazard and associated environmental risk can be rather considerable, primarily due to their regularity leading, ultimately, to chronic impacts on the marine environment.

The statistics of drilling accidents, as mentioned previously, is very limited and controversial. For example, according to some data [VNIPImorneftegas, 1990], the probability of an accident during well drilling is 0.1–0.5% and during well repairing—1.0–2.5%. Other data [Sakhalin-1, 1994] indicate that the relative number of accidental episodes during exploratory drilling averages 1% a year, during the developmental drilling—0.4%, and during production and repair—0.03% each. The same source shows that the hydrocarbon blowouts happen only in 10% cases of total number of accidents. Accidents leading to large oil spills (of over 10,000 tons) account for less than 3% of them. Estimates also indicate that 2.3% of the total number of offshore drilling platforms are subjected to large accidents every year [Zalogin, Kuzminskaya, 1993]. The official statistics mostly take into account the accidental oil spills that, in contrast with gas blowouts, are easier to notice and control. However, about 33% of all known blowouts from offshore wells involve the emission of natural gas extracted on the shallow parts of the shelf. About half of these cases involve serious accidents and damage of drilling rigs [Moore, Hamilton, 1993].

Forecasting of hydrocarbon distribution resulting from the long-term well gushing and other catastrophic events (e.g., tanker accidents) is one of the most important directions of applied oceanography and marine ecology. Dozens of models have been developed in many countries to describe different scenarios of transferring and dispersing an oil spill. They give some idea about the possible consequences of accidental spills and sometimes provide real help in their prevention and liquidation [Duke, 1985; Monk, Cormack, 1992; GESAMP, 1993].

For example, such models were used to get estimates and prognoses for development of the Shtokmanovskoe gas condensate field in the Barents Sea [Borisov et al.,1994]. One such estimate admits the possibility that the mixture of gas, gas condensate, and produced water released during a well gushing can total from 8 to 38 million cubic meters a day. Volumes of gas condensate spills can reach from 10,000 to 100,000 m^3 in case of open well gushing for a month. Calculations of the drift and dispersion of such spills under different weather and oceanographic conditions of the Barents Sea show the possibility of hydrocarbon transfer over dozens and hundreds of kilometers from the place of blowout. This can result in a risk of lethal and sublethal impacts on marine organisms within areas of up to 1,400 km^2.

3.4.3. Transportation and Storage Accidents

Tanker transportation. Oil extracted on the continental shelf accounts for a considerable part (probably at least 50%) of annual volumes of oil transported by tankers (the latter constitute over 1 billion tons). This fact allows us to consider the tanker oil fleet as a part of the offshore oil and gas complex together with the systems of underwater pipelines. In fact, on some fields, the shuttle tankers are the main way of delivering hydrocarbons to the onshore terminals.

The ecological consequences of tanker oil transportation as well as the statistics and risk of tanker accidents have been thoroughly discussed in many publications [Nelson-Smith, 1977; Voloshin, 1989; IMO, 1990; GESAMP, 1993]. The accident probability for tankers with a carrying capacity of over 10,000 tons is estimated at 2.3% for every 10 million tons of dead weight. Other data show that at the end of the 1980s, the relative number of accidents of tankers with a carrying capacity of over 6,000 tons was about 2% [IMO, 1989].

The main causes of tanker accidents that lead to large oil spills include running aground and into shore reefs, collisions with other vessels, and fires and explosions of the cargo. According to official data [IMO, 1990], the amount of oil spilled during tanker accidents in 1989 and in 1990 were 114,000 and 45,000 tons, respectively. At the same time, the total volume of oil pollution caused by marine oil transportation was 500,000 tons a year.

The ratio of the oil volumes spilled during tanker accidents to the oil volumes discharged during routine tanker operations with ballast waters seems to be close to the analogous proportion of the oil volumes released into the sea during open well gushing to the oil volumes discharged during routine oilfield practices. Significantly, both large drilling accidents and large tanker catastrophes occur relatively rarely. The frequencies of such incidents as well as the oil volumes released in large spills differ from year to year.

The history of tanker accidents has been thoroughly described by both the scientific literature and the media. Analyzing the statistics and circumstances of such events indicates that they can hardly be avoided. Although the rate of tanker accidents has been declining over the past two decades, we should be prepared to deal with them in the future.

While speaking about the history of tanker transportation, we want to mention a sequence of large supertanker accidents starting with the catastrophic grounding of the tanker *Torrey Canyon* in the English Channel in 1967. The spill of 95,000 tons of oil caused heavy pollution of the French and British shores with serious ecological and fisheries consequences. This accident was followed by a number of other tanker accidents, including *Amoco Cadiz* (1978, 220,000 tons of oil spilled), *Exxon Valdez* (1989, 40,000 tons of oil spilled), and *Braer* (1993, 85,000 tons of oil spilled). Each of these episodes developed in accordance with its unique scenario. In all the situations, though, the levels of oil pollution reached lethal limits for marine fauna, mainly for birds and mammals. The consequences included much more damage than just ecological disturbances in the sea and on the shore. Chapter 7 will discuss this in more details.

In some cases, the tanker accidents occurred right in the zone of the oil field development. One of them happened in 1978 in the Shetland basin. The tanker *Esso Bernica* was holed during the mooring, and 1,100 tons of heavy oil fuel spilled into the coastal zone. The damage to nature and the local population was rather serious. Clean-up operations took over 6 months [Monk, Cormack, 1992].

One of the most dramatic situations developed in 1989 in the shallow waters of Prince William Sound near the Alaskan southern shore. The oil tanker *Exxon Valdez* ran aground and spilled over 40,000 tons of crude oil. As the oil spread along the coastline, it covered sea animals, birds, and plants. It turned hundreds of miles of this area (unique for its cleanness and biological resources) into an area of ecological disaster. The cleaning expenses totaled 2 billion dollars. The compensations for the damage to nature and the local population exceeded an already-paid 3.5 billion dollars and 15 billion dollars that are to be paid according to the court appellations [Anonymous, 1994a].

This relatively recent episode in the history of the offshore oil and gas industry causes an alarming association in the mind of a Russian reader. The *Exxon Valdez* catastrophe happened approximately at the same latitudes where the grand projects of the oil and gas developments on the Russian Arctic shelf have already been started (the shelves of the Barents and Kara Seas in vicinity of the

White Sea). The association gets even stronger if we take into account that considerable amounts of hydrocarbons extracted here are going to be transported by the tanker fleet. This will include tanker shuttles (including the ice types), large tankers with dead weight up to 120,000 tons, and supertankers. Each of these vessels is going to make dozens and hundreds of trips a year. This regular transportation activity is going to take place with the rest of the traffic in the area of the oil field developments and in addition to the general intense shipping and fishing in this Arctic basin. All of these factors considerably increase the probability of accidental situations occurring in the region. We must remember the high productivity and high vulnerability of the Arctic marine ecosystems. This region contains unique natural resources that are comparable to the rich resources of the Alaskan shelf.

This primary background information and general statistics about large tanker accidents (about 2% a year) allow us to conclude, without any calculations and modeling, that the risk of transportation accidents occurring on the Arctic shelves is going to be high. The consequences of these accidents can be catastrophic. Moreover, the environmental damage of possible accidents can exceed everything that has happened before in such cases, including the accidents on the Alaskan shelf. We wish this pessimistic prognosis would never become a reality. However, calculations conducted by some experts and given in the publication by Borisov et al. [1994] support these conclusions. These calculations and estimates show that in case of an accident of a tanker with a freight-carrying capacity from 5,000 to 50,000 tons (the type that is going to be used in the Barents Sea), the spilled oil and gas condensate can cover from 3,000 to 50,000 km^2. In case of a supertanker accident, up to 84,000 km^2 can be covered. This is enough, for example, to completely cover the Pechorskaya Sound with all its bays and estuaries. This amount could severely pollute adjoining vast fisheries regions, which is especially possible if a tanker accident happens near the Kolski Bay.

Very dangerous situations can emerge in case of a gas tanker accident. Gas carriers are going to be used together with oil tankers in the Barents Sea as well as on the eastern shelf of Sakhalin to transport liquefied natural gas. Gas tanker accidents, although less probable than the accidents with oil tankers, can cause so-called flameless explosions. It happens due to the rapid evaporation of the liquefied gas on the sea surface and formation of pieces of ice and gas clouds followed by combustion and explosions. Such explosions can destroy everything alive in areas of up to 400 km^2 [Voloshin, 1989]. In 1980, three out of seven accidents of the vessels transporting liquefied gas caused such kinds of explosions.

At last, the tragic apotheosis of possible outcomes is an accident involving a tanker that is transporting methanol—a rather toxic substance that is completely soluble in water. In case of an accident of such a vessel with a freight-carrying capacity of 35,000 tons, for example in the coastal zone of the Western Murman, the area of lethal impact to marine organisms will be from dozens and hundreds to thousands of square kilometers. In fact, it could cover the whole fisheries regions [Borisov et al., 1994]. We must wish once more that such a prognosis will not become a reality.

Storage.   Underwater reservoirs for storing liquid hydrocarbons (oil, oil-water mixtures, and gas condensate) are a necessary element of many oil and gas developments. They are often used when tankers instead of pipelines are the main means of hydrocarbon transportation. Underwater storage tanks with capacities of up to 50,000 m^3 either are built near the platform foundations or are anchored in the semisubmerged position in the area of developments and near the onshore terminals. Sometimes, the anchored tankers are used for this purpose as well.

Of course, a risk exists of damaging the underwater storage tanks and releasing their content, especially during tanker loading operations and under severe weather conditions. However, no summarizing quantitative assessments and statistics of such events are available. After the spill of 1,200 tons of crude oil in 1988 from an underwater storage tank during a storm in the North Sea, some countries introduced restrictions on installing such structures near the shore [Cairns, 1992]. The most dangerous are the accidents involving underwater storage tanks that contain toxic agents, for example methanol. Such accidents are possible in the area of Shtokmanovskoe field developments in the Barents Sea where over 3,000 tons of methanol products are planned to be stored underwater.

After the underwater storage tanks have completed their service, defining their future can be difficult. Thus, lately in England and Norway, the fate of one such obsolete oil storage tank has been discussed. This container has a rather impressive size (height—140 m, radius—25 m, and weight—14,500 tons). It contains 50,000 tons of seawater with oil residuals and 100 tons of oil-polluted slurry. The discussed alternatives include burying this huge structure or disassembling and transporting it to the shore [McIntyre, 1995].

Pipelines.   Complex and extensive systems of underwater pipelines have a total length of thousands of kilometers. They carry oil, gas, condensate, and their mixtures. These pipelines are among the main factors of environmental risk during offshore oil developments, along with tanker transportation and drilling operations. The causes of pipeline damage can differ greatly. They range from material defects and pipe corrosion to ground erosion, tectonic movements on the bottom, and encountering ship anchors and bottom trawls. Statistical data show that the average probability of accidents occurring on the underwater main pipelines of North America and Western Europe are, respectively, 9.3×10^{-4} and 6.4×10^{-4}. The main causes of these accidents are material and welding defects [Sakhalin-1, 1994].

Depending on the cause and nature of the damage (cracks, ruptures, and others), a pipeline can become a source of either a small and long-term leakage or an abrupt (even explosive) blowout of hydrocarbons near the bottom. The dissolution, dilution, and transferring of the liquid and gaseous products in the marine environment can be accompanied in some cases by ice and gas hydrates formation. The intensity and scale of toxic impacts on the marine biota in the accident zone can be, of course, very different, depending on a combination of many factors.

Modern technology of pipeline construction and exploitation under different natural conditions, including the extreme ones, achieved indisputable successes. However, there is no guarantee of nonaccidental hydrocarbon transportation along the pipelines either in the sea or on the land. More than enough evidence supports this conclusion, especially if we look at the Russian experience of many onland oil developments in Siberia and other places. There, oil pipeline leakage has become an everyday event. According to some data [Anonymous, 1995], the total length of obsolete pipelines posing a danger of accidents in Russia exceeds 35,000 km.

The situation for offshore oil developments is different, at least for those located on the shelves of North America and Western Europe. For example, in the North Sea, during more than a 20-year history of industrial oil and gas developments, only two cases of serious pipeline accidents have been reported [Haldane et al., 1992]. One of them happened in 1980. A pipeline ruptured because of a collision with an anchor. About 1,000 tons of crude oil spilled into the sea. The oil flow continued for 25 minutes until the automatic device blocked up the ruptured pipeline. The other episode happened in 1986. An abrupt pressure change in the system near the platform caused the damage of an underwater valve. The surface slick of about 1,500 tons of spilled oil happened to be rather stable in spite of the stormy weather. It drifted for ten days toward the Norwegian shore and then dispersed.

The relatively safe ecological situation during prolonged exploitation of 5,000 km of underwater pipelines in the North Sea certainly does not mean that this way of hydrocarbon transportation is totally reliable and safe. For instance, a spill of 18,000 tons of crude oil occurred on the shelf of Saudi Arabia [Nelson-Smith, 1977]. A spill of about 2,000 tons of heated crude oil into the coastal waters happened near the shore of New Jersey in the USA in 1990 [Burger, 1994]. In Russia in the Komi Republic alone, up to 1,000 pipeline accidents and leakages are recorded annually [Yablokov, 1995]. These examples indicate that pipeline oil and gas transportation does not eliminate the possibility of serious accidents and consequences. The approximate calculations show that in case of a rupture of an underwater main pipeline, the areas where the levels of hydrocarbons will be hazardous for marine biota can cover from 14 to 26 km^2 and spread up to 200 m deep [Borisov et al., 1994].

It is important to take into consideration that in a number of cases, the accidental oil and gas spills and blowouts on the onland main pipelines can pose danger to the coastal marine ecosystems. This can happen when onland pipeline accidents take place near big rivers or in locations of their crossing. Any pollution of river waters eventually affects the sea zone near the river mouth. Such a situation happened at the end of 1994 in the Usinsk area, Russia. An onland pipeline rupture led to the spill of more than 100,000 tons of oil with the danger of heavy pollution of the basin of Pechora River. The potential hazard of such situations can be even higher during oil and gas development on Sakhalin. The main pipelines are supposed to be laid along the entire eastern coast of the island, right across the main spawning rivers where reproduction of the unique populations of Sakhalin salmon takes place.

3.5. Impact on the Fishing Industry

While accidental situations occur periodically, physical interference with navigation and with fishing fleets is an immutable and obvious consequence of the oil and gas industry's activity on the shelf. The main factors causing interference with the fishing industry include:

- creating navigational safety zones with a radius from 500 to 1,000 m around every oil platform and underwater hydrocarbon storage tanks, where navigation is completely forbidden;
- laying unburied or untrenched pipelines on the bottom;

- leaving platforms, their fragments, and other structures in the area after termination of production;
- suspending wellheads for a period of time after completing exploratory drilling; and
- leaving various debris (steel framework, pipes, vessels, and so on) on the bottom in the areas of field development.

The data given below (UKOOA Factsheet, 1988 from [McIntyre, Turnbull, 1992]) about the number of offshore oil platforms and the length of pipelines in the British sector of the North Sea in 1987 give some basis for the rough estimations of the areas totally lost to the fishing industry:

Shallow water (Southern North Sea and Irish Sea):
steel platforms   108
concrete platforms   None
major pipelines (km)   1,120
Deepwater (Central/Northern North Sea):
steel platforms   37
concrete platforms   10
major pipelines (km)   3,587

At first, the sizes of these areas do not seem so large—not more than 0.1% of the total area of the North Sea, even if we add 100 more platforms located in the economic zones of other North Sea states. However, we should take into account the practice of combining several drilling platforms and satellite wells in one production system. These are connected by a network of pipelines, cables, storage tanks, and other equipment. In case of such organization in offshore oil production, the areas where fishing is impossible or complicated are certainly much larger. A similar situation results from using the anchor system for semi-submerged drilling platforms. In this case, safe navigation and fishing become impossible within a radius of up to 1,500 m and more.

Taking into account these facts and other considerations about the situation in the North Sea that developed after the beginning of offshore hydrocarbon exploitation, the British fishermen, in their appeal to the government, estimated the areas actually lost for fishing at 800 to 5,200 km^2 of sea on the UK continental shelf. The resulting annual losses of catch range from 580,000 to 3,000,000 pounds sterling [Buchan, Allan, 1992]. We must also note that the offshore oil and gas industry's activities also affect fisheries by causing physical interference, accidents, and losses or damage of fishing vessels and gear. Such incidents have become an everyday event in the North Sea. The number of claims to the oil and gas companies and special funds for compensation of damages has exceeded 1,200 cases over the past 15 years. In Norway, the compensation annually paid by the oil and gas operators for covering the fishermen's losses is estimated at 3.3 million pounds sterling.

Serious complications for fishing result from the suspending of exploration wells. A structure used to border a suspended well-head on the bottom of the sea takes an area of 3–4 m². It resembles an upside-down table with legs 3.5–4.5 m long and a central pipe 3.0–4.5 m long with a diameter of 0.5 m. At the end of the 1980s on the British shelf of the North Sea alone, 250 to 300 such structures existed [Buchan, Allan, 1992]. It is easy to predict the result of the bottom trawls encountering such structures and to imagine what British fishermen feel and say in such cases. Similar events take place during trawling in areas where anchors, ladders, oil and paint drums, and even tractors, trolleys, bulldozers, and other equipment left on the bottom of the oil development get caught by the trawls. In Norway, cleaning 6,300 km² of bottom area by removing 1,600 tons of such debris cost about 3 million pounds sterling [GESAMP, 1990]. It seems that the Russian fishing industry might encounter similar problems in the foreseeable future on the shelves of the northern, far-eastern, and other seas.

The pipeline system is an inseparable attribute of all offshore oil and gas developments. It presents one of the main obstacles for trawling fishing. This system grows as the field is being developed and, like blood vessels, covers extensive areas of the sea bottom. In the North Sea alone, such systems include at least 5,000 km of pipelines. These pipelines vary in size (diameter from 10 to 100 cm) and function. They include both temporary transportation lines located within the drilling area and main pipelines going all the way across the sea [Haldane et al., 1992]. Bottom trawling in such areas faces the risk of trawl loss or pipeline damage, especially when the latter lie openly on the bottom surface. Laying pipelines into a trench on the bottom and burying them is a large expense that oil companies try to avoid. Pipelines with a large diameter (40 cm and over) are often laid on the bottom surface without any deepening, although this practice causes the most interference and poses risks for trawling fishing.

Different publications, including the ones cited before, indicate that the relationships between the offshore petroleum and fishing industries are complicated by the absence of clear regulations that would take into account the fisheries interests in the zones covered by the network of underwater pipelines. The width of the areas where bottom fishing is dangerous or forbidden is often not defined by any official standard. Besides, no rules obligate the oil companies to compensate for fishing damages and losses. Thus, the fishermen often face a dilemma. If they trawl along the underwater pipelines they risk having an accident (that may be life threatening) and losing trawls without receiving any compensation. If they decide to avoid the risk, they have to leave the well-known areas and start looking for new places to fish.

The amount of possible fishing losses can be illustrated by calculating the potential catch in the zone that will be alienated from fishing because of the planned laying of a main pipeline in the Barents Sea. This zone covers an area of 5 miles on both sides of the 500-kilometer-long pipeline. This would cause the possible loss from 4,000 to 11,000 tons of bottom fish catch and from 12,000 to 45,000 tons of annual total catch (in case of a complete ban of any fishing). In other words, the fishing industry would lose 14 to 53 million dollars annually, depending on the way the pipeline is laid [Borisov et al., 1995].

Thus, the long-term experience in some countries indicates that the main concern of the fishing industry is not connected with pollution and environmental consequences of the offshore oil and gas industry's activity. Instead, the most serious impact involves displacement of traditional fishing areas and physical interference with trawl fishing (including vessel and gear damage). These occur because of construction of the offshore platforms, pipelines, and other structures of the field developments as well as from dumping debris and leaving various objects and materials on the bottom.

Before a similar situation emerges on the Russian shelves, the normative and legal base regulating relationships between the offshore petroleum and fishing industries should be prepared. In addition, a preventive approach during developing fishing strategy in the areas of future large-scale offshore activities should be applied. Such approach can stipulate, for example, replacement of trawling by longline fishing. These would avoid the difficulties and losses mentioned above and also ensure conservation of the bottom habitats that, to a considerable degree, define the productivity of the shelf ecosystems [Kokorin, 1994].

3.6. Decommissioning and Abandonment of Offshore Installations

Sooner or later (most often after 20–40 years) the hydrocarbon reservoir within any field becomes depleted. Issues that usually stay in the shade during the first stages of development then emerge. We are discussing the fate of thousands of enormous fixed platforms, underwater pipelines, and some other installations and structures after their decommissioning. If left on the bottom, they inevitably undergo degradation and dissipate over large areas. This poses obvious interference to navigation and fishing, as previously discussed.

3.6.1. Abandonment Options

First of all, it should be stressed that solving this problem faces big technical difficulties and involves impressive expenses. For example, in 1983, the cost of dismantling and removing all fixed platforms taken out of service on the world's shelves was estimated at about 10 billion dollars [GESAMP, 1990]. Other estimates [Lapshin et al., 1990] give a figure of 27 billion dollars.

The extremely high cost of dismantling and removal led to the need to revise some of the national and international regulations adopted about 40 years ago. Such a revision covered, in particular, the requirement set by the Convention on the Continental Shelf (Geneva, 1958) and the United Nations Convention on the Law of the Sea (Montego Bay, 1982) to remove abandoned offshore installations totally. At present, a more flexible and phased approach is used. It suggests immediate and total removal of offshore structures (mainly platforms) weighing up to 4,000 tons in the areas with depths less than 75 m and after 1998—at depths less than 100 m. In deeper waters, removing only the upper parts from above the sea surface to 55 m deep and leaving the remaining structure in place is allowed. The removed fragments can be either transported to the shore or buried in the sea. This approach, described in documents of the International Maritime Organization [IMO, 1989], considers the possibility of secondary use of abandoned offshore platforms for other purposes.

The scheme in Figure 21 shows possible dismantling and abandonment options for offshore installations depending on water depth and structure size. From the technical-economic perspective, the larger the structures are and the deeper they are located, the more appropriate it is to leave them totally or partially intact. In shallow waters, in contrast, total or partial structure removal makes more sense. The fragments can be taken to the shore, buried, or reused for some other purposes.

From the fisheries perspective, any options when the structures or their fragments are left on the bottom may cause physical inter-

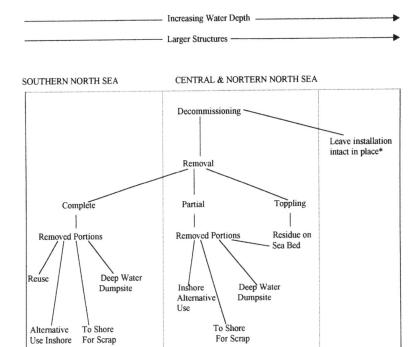

Figure 21 Dismantling and abandonment options for offshore installations and component parts [Side, 1992]. *—This option is for installations for which removal may prove either too hazardous or not technically feasible (e.g., some concrete gravity structures in the Northern North Sea).

ference with fishing activities. In these cases, the possibility of vessel and gear damages and corresponding losses does not disappear with termination of production activities in the area. Instead, abandoned structures pose the threat to fishing for many decades after the oil and gas operators leave the site. The obsolete pipelines left on the bottom are especially dangerous in this respect. Their degradation and uncontrolled dissipation over wide areas may lead to the most unexpected situations occurring during bottom trawling in the most unexpected places. At the same time, national and international agreements about the decommissioning and abandonment of offshore installations refer mostly to large, fixed structures like drilling platforms. The fate of underwater pipelines is still not affected by clear regulations.

3.6.2. Secondary Use of Offshore Fixed Platforms

The options of reusing abandoned platforms, their foundations, and other structures that are out of service have been actively discussed for the last 10 years. Rather extensive literature devoted to this issue includes many different projects and suggestions [Reggio, 1987; Side, Davies, 1989; Lapshin et al., 1990; Side, 1992; Dossena et al., 1995]. Usually, such publications discuss the possibility of reusing a platform either for research purposes or as an artificial reef. Both cases have interesting potentials for studying the ocean and increasing its bioproductivity.

An analysis of scientific potential of research stations permanently based on abandoned oil platforms in the Gulf of Mexico revealed several promising directions of marine research at such stations [Dokken, 1993; Gardner, Wiebe, 1993]. These include studying regulation of the marine populations and coral reproduction, making underwater observations, monitoring the sea level, and collecting oceanographic and meteorological information within the framework of international projects. Some other suggestions consider transformation of abandoned platforms into places for power generation using wind/wave and thermal energy [Rowe, 1993]. These platforms also could be used as bases for search and rescue operations or centers for waste processing and disposal [Side, 1992].

From the fisheries perspective, the most interesting projects are the ones aimed at converting the fixed marine structures into artificial reefs. Artificial reefs are known to be one of the most effective means of increasing the bioproductivity of coastal waters by providing additional habitats for marine life. They are widely and effectively used on the shelves of many countries. The possibilities to use abandoned oil platforms as reef structures and aquaculture farms were investigated in the former USSR in the Caspian Sea [Lapshin et al., 1990] where over 1,000 oil platforms are located [Bugrov et al., 1994]. However, due to different reasons, these attempts have never become anything but experimental developments. Similar research and even some large projects are in progress now on the shelves of many countries (Great Britain, Norway, Italy, Mexico, and so on). In these countries, fishing and marine aquaculture have to coexist with the offshore oil and gas industry and face its impacts even after hydrocarbon production is terminated.

The offshore structures can undoubtedly attract many species of migrating invertebrates and fish searching for food, shelter, and places to reproduce. In particular, observations in the Gulf of Mexico

revealed a strong positive correlation between the amount of oil platforms, growing since the 1950s, and commercial fish catches in the region. This tendency was supported by many observations [Side,
1992; Linton, 1994; Stanley et al., 1994]. It became one of the reasons
to suggest the positive impact of offshore oil and gas developments
on the fish populations and stock. Wide popularization of this fact led
to the mass movement using the slogan "From rigs—to reefs" in the
USA in the mid-1980s [Reggio, 1987].

However, further analyses of the fishing situation in the Gulf of
Mexico showed that the growth of the fish catch in this case was connected not with increasing the total stock and abundance of commercial species but with their redistribution due to the reef effect of
the platforms. A critical point here was the use of static gear methods
of fishing (e.g., lines and hooks) instead of trawl gears. Besides, the
areas around the platforms became very popular places of recreational and sport fishing. This also made a significant contribution to
the total catch volumes. Nothing similar was noted in the North Sea,
where the number of oil platforms has also been growing since the
1960s. However, the total catch did not correlate with this growth at
all and even decreased. This fact indicates the absence of any positive
impact of the reef effect of oil platforms on the commercial fish
catches in areas where the main way to fish is trawling.

We must add that the issue of using artificial reefs (including oil
drilling platforms and their parts) for increasing the bioproductive
potential of the sea is not as simple as it may seem. The optimal effect that exceeds the expenses can be reached only as a result of thoroughly analyzing particular ecological situation and possible
impacts of the reef structures on productivity of valuable species.
This analysis should take into account the available means to extract
the commercial product. Recently, a number of ecological and ichthyological studies have been conducted in the areas of the abandoned
oil platforms (for example, in Italy [Dossena et al., 1995]). The purpose of these studies is to find out if the structures just attract and
shelter the fish populations or radically change the populational
structure and promote fish abundance. The answer to this question
will define, to a certain extent, the scale and effectiveness of reusing
obsolete offshore oil and gas structures for increasing the biological
productivity of the sea.

At the same time, we should not forget about the previously discussed danger that abandoned offshore oil platforms and their fragments pose to navigation and trawling fishing. With an abundance
of such artificial reefs, this problem requires special regulations for

negotiating the inevitable conflict of interests. One such regulatory program has been developed and applied in the USA in the Gulf of Mexico on the shelf of Louisiana [Pope et al., 1993]. It requires mapping the area to indicate the locations of platforms, underwater pipelines, and other structures left on the bottom. The program also includes monitoring, collecting data, developing a warning system, and other activities necessary to control the situation and ensure safety in the region.

3.6.3. Explosive Activities

Complete or partial removal of steel or concrete fixed platforms that weigh thousands of tons is practically impossible without using explosive materials. Bulk explosive charges have been used in 90% of cases. This is very powerful, although short-term, impact on the marine environment and biota, which should not be neglected.

Unfortunately, information about the explosives' impact on water organisms is very incomplete. It covers mainly the impacts of relatively small explosive charges that are still used during seismic surveys and subsea trenching activities. One of the few reviews devoted to this issue [Side, 1992] includes a description of possible fish injury ranging from light hemorrhaging to ruptures of the body cavity depending on the power of the explosion. The author of this publication also notes the increased vulnerability of swimbladder fish to the impact of an underwater blast and the negative correlation between fish mortality and their weight.

It is extremely difficult to get any reliable estimates of possible mortality of marine organisms, especially fish, during an explosive activity even if the initial data, such as the type of explosive, depth of the water, bottom relief, and others, are known. This large uncertainty is connected, in particular, with the high heterogeneity of fish distribution that strongly depends on specific features of fish schooling behavior. Calculations show that with a 2.5-ton (TNT equivalent) charge, the mass of killed fish will be about 20 tons during each explosion. At the same time, if, for example, a school of herring happens to get into that zone, the fish kill figure may be much higher [Side, Davies, 1989].

One of the few known observations of fish damage in zones of explosive activity was done in 1992 in the Gulf of Mexico near the shore of Louisiana and Texas [Gitschlag, Herczeg, 1994]. In order to remove over 100 fixed platforms and other structures, more than 12,000 kg of plastic charges were exploded. The amount of dead fish floating on the surface was visually recorded after the explosions. It totaled to about 51,000 specimens. The actual number of killed fish was undoubtedly higher because many specimens could not float to the surface or did not get in the zone of visual observation.

Whatever number of adult fish actually died during the explosions, it will hardly influence the total abundance of commercial species. Much more hazardous for the fish stock are explosive impacts on fish larvae and juveniles. The threshold of lethal impacts for the younger organisms weighing up to several grams is tens of times lower than that for adult specimens [Yelverton et al., 1975; Side, 1992]. Thus, the zone of mortality of fish at the early stages of development is respectively wider. The quantitative estimates of possible effects at the populational level are even more complicated because of the absence of corresponding data and methods. Nevertheless, enough evidence exists to enforce strict regulations of explosive activities and to forbid them in areas and in seasons of spawning and fry development of commercial fish.

Removal of the offshore structures also decreases the number of habitats for structure-related fish. For example, in the mostly soft-bottom environment of the Gulf of Mexico, these structures provide hard substrates for marine organisms. The decline of stocks of reef fish observed in this region within the past decade can be connected, in particular, with elimination of over 400 oil-related structures that had served as an artificial habitat for marine life [MMS, 1995].

3.7. Environmental Hazard and Risk Assessment

We have discussed the extreme complexity and diversity of the offshore oil and gas industry's impacts on the marine environment and biota at all stages of its activity—from field exploration to production termination and decommissioning. Now it is quite logical to give a short description of general approaches and specific measures that are used to control, prevent, and manage the consequences of oil and gas production on the continental shelf.

Broadly speaking, all kinds of activities previously discussed can be divided into two groups, depending on the duration and intensity of their impact. The first group consists of a wide variety of different technological, transportation, and other operations that cause long-term (chronic) environmental effects. It includes regular discharges of liquid and solid wastes during drilling and hydrocarbon transportation, atmospheric emissions during flaring of associated gas, and other activities leading to persistent and constantly sustained disturbances in the environment. The second, also very large, group unites the short-term situations and impact factors that cause extreme (acute) consequences. Examples include large accidental hydrocarbon blowouts during drilling, explosive cutting activities, physical disturbances during structure installations, oil spills during tanker accidents, and so on.

This grouping is necessary since the choice of specific responses and ways of management (control, prevention, regulation, and others) depends, first of all, on the nature of the impact (long-term and controlled or short-term and accidental). Over the last 10–15 years, the search for optimal responses to both groups of impacts often uses the methodology based on the concept of hazard and risk [Ramsay, Grant, 1992; Anikiev et al., 1995]. This concept has been widely and successfully used for solving various theoretical and practical problems, including environmental protection in different areas [Izrael, 1984; Logofect, 1993]. Within its framework, the term *hazard* usually means the combinations of properties and characteristics of a material, process, object, or situation that are able to harm and cause a damage of ecological, economic, or any other nature. The term *risk* means the probability of realization of this hazard under specific conditions and within a certain period of time.

Practical application of this theory has a big advantage. In a number of cases, it helps to avoid the elements of uncertainty and subjectivity. Such uncertainty is often present in environmental literature (including this book) and even in official documents when they describe and assess the environmental situations using such expressions as serious danger, considerable hazard, heavy pollution, and so on. The methodology based on the hazard and risk concept provides quantitative estimates of actual risk based on relevant statistical data. These estimates then can be compared with risk-acceptance criteria. This opens the way for making better-grounded decisions regarding various areas of environmental control, regulation, and management of offshore oil field developments.

Practical application of the theory of environmental hazard and risk is limited in case of long-term impact assessments. In this situation, the quantitative exposure-effect assessment is either impossible or unreliable due to the complexity of many environmental cause-effect relationships and the background interference. In such cases, for example during regular technological discharges from the platforms and tankers, environmental control and management of the hazard is done with the help of international and national standards.

These are established both for the pollution sources and for the marine environment (see Chapter 8).

Legal, regulatory, and organizational measures are also used to reduce the hazard of situations associated with acute environmental risk, such as accidents, catastrophes, and other similar cases. The process of developing a system of regional and international environmental standards and requirements to oil tanker transportation and other offshore oil and gas industry's activities began as far back as the 1960s, after the dramatic oil spills described in Section 3.4 occurred. Special national and international services equipped with technical and chemical means were established to respond quickly in case of oil spills in the sea. The methods of remote control (including space control) and monitoring of oil pollution in the marine environment have become widely and effectively used. In some countries, the oil companies had to allocate part of their profits for financing the measures to detect and liquidate oil spills. For example, in the USA, such a fund contains 2 billion dollars [Neumann, 1993]. Besides, in 1990, the International Convention on Oil Pollution Preparedness, Response and Cooperation was developed and adopted by IMO.

For scientific substantiation of different measures of environmental protection (especially those aimed at increasing the safety of oil tanker transportation), the approaches and techniques of environmental hazard and risk analysis are used. The procedure of such analysis usually includes the following stages: identification of hazard, postulation of accidents, evaluation of their consequences, estimation of their frequencies, and calculations of the total risk (Figure 22). In order to reduce the risk and to find the optimal measures, this scheme is supplemented by analyzing the possibilities to control risk. These possibilities involve minimizing the frequency of emerging risk and reducing the severity of its consequences in all possible situations. The analysis takes into account available technical means and possibility of various (including extreme) natural conditions. The final result of all these analytical procedures, usually conducted with the help of special computer programs, is a set of optimal responses serving as a base for practical decision making [Ramsay, Grant, 1992].

The most important practical tasks solved with the use of just-described methodology are:

- calculating the stability and reliability of different modifications of the offshore platforms, depending on the intensity of natural impact under extreme situations;

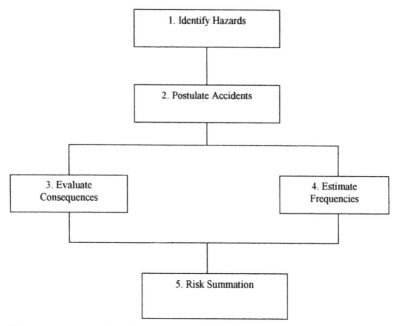

Figure 22 Risk analysis [Ramsay, Grant, 1992].

- calculating the stability and reliability of different modifications of underwater pipelines and other offshore structures;
- modeling and assessment of the probability of accidental situations and possible hydrocarbon discharges at different stages of oil and gas production;
- modeling and assessment of the probability of accidental situations and possible hydrocarbon discharges during tanker transportation of oil and liquefied natural gas;
- developing optimal strategies, specific contingency plans, and response techniques to control oil spills and mitigate their consequences; and
- updating environmental, technical, and legal measures and requirements to the oil and gas production industry.

Figure 23 illustrates one of the possible practical applications of the methodology based on environmental hazard and risk analysis. It gives so-called risk contours for the eastern coast of Scotland in the area of a tanker terminal servicing the largest gas-refining facilities in Great Britain. The risk contours calculated by two different methods show the distribution of probability of gas tanker accidents in the

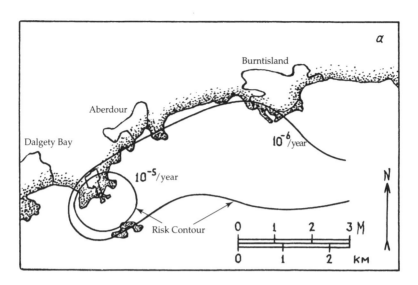

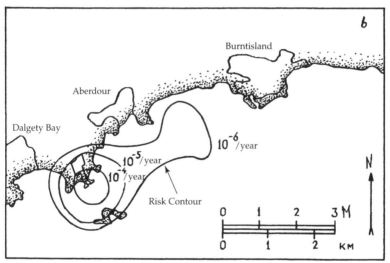

Figure 23 Risk contours: a—drawn using part computer, part manual calculation method; b—drawn using computer [Day, 1992].

vicinity of the terminal. They take into account the possibilities of ship collisions, running aground, fire, explosion, and other similar events modeled for specific conditions of the given region [Day, 1992]. This visual and quantitative illustration of the probability of possible accidents (which is, by the way, very low in this case) undoubtedly helps both to make practical decisions regarding such situations and to take the necessary measures to ensure environmental safety and protection of the local population.

Another example of using the methods based on risk analysis is given in Figure 24. It shows the probability of oil drift toward the seashore in case of an underwater pipeline oil spill. It is easy to see that in case of a pipeline oil spill at the distance of more than 60 km from the shore, the oil will not reach the shore and will disperse in the sea. Of course, this picture cannot be extrapolated to other areas. The scenario depends not only on the distance between the pollution source and the shore but also on the coastal currents, weather conditions, type of oil, pressure in the pipeline, and a number of other factors. Modeling and prognosis of such situations are essential for predicting the consequences of oil spills and preparing immediate response actions.

In conclusion, we want to stress once more the effectiveness of the methods based on the environmental hazard and risk analysis. They help to solve many practical safety and management problems at different stages of offshore oil and gas field development. These methods acknowledge the inevitability of environmental impacts

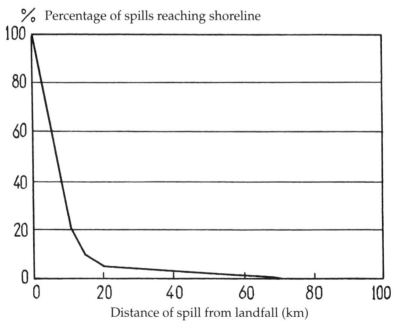

Figure 24 A pipeline oil spill beaching risk profile [Monk, Cormack, 1992].

during the offshore activities. They give the methodological base for quantitative assessments of these impacts and substantiate the measures to reduce their risk.

Conclusions

1. The large-scale and multistage activity of the offshore oil and gas industry imposes complex impacts on the environment. These impacts cause physical, chemical, and biological disturbances in the water column, on the bottom, and in the atmosphere. During the development of large oil and gas basins, local impacts can merge together and form zones of large-scale disturbances. The nature and intensity of these impacts depend on a combination of many natural and anthropogenic factors.

2. Physical impacts of the offshore oil and gas activity on marine biota include, in particular, seismic signals (sound waves) generated during seismic surveys. These can be hazardous for marine fauna, especially for the organisms at the early stages of development. The areas and periods of seismic surveys should not overlap with the areas and seasons of spawning and development of commercial species. Similarly, the explosive activities during the removal of abandoned platforms and other fixed structures should be strictly regulated and prohibited during the seasons and in the areas of spawning and mass migrations of commercial fish.

3. The main factor of environmental impact of offshore oil production is chemical pollution. It accompanies almost all kinds of activities at all stages of field development. Most liquid and solid waste discharges take place during offshore drilling. The volumes of these wastes reach up to 5,000 m^3 for each well in the form of drilling muds and cuttings in addition to thousands of cubic meters of everyday discharges of produced and ballast waters. The composition of drilling wastes is very changeable. They include, besides oil hydrocarbons, hundreds of chemicals and agents with different technological functions. The waste treatment methods presently used by the offshore oil and gas industry do not guarantee the full elimination of environmental disturbances in the areas of offshore oil and gas production.

4. A considerable part (up to 10–20%) of the total volume of offshore gas production is flared on the platforms. This forms one of the sources of atmospheric pollution. However, as of today, the quantitative estimates of the flows of the most dangerous products of hydrocarbon combustion are not available.

5. Accidents are inevitable companions of the offshore oil and gas production and transportation. The largest accidental spills introduce hundreds of thousands of tons of oil into the marine environment. They lead to serious ecological and fisheries consequences, especially in cases when oil reaches the shore and shallow coastal areas. Current national and international measures for accident prevention and mitigation of their consequences are not effective enough. The possibilities of their improvement are limited by the absence of a necessary database and extreme complexity, variability, and unpredictability of the spilled oil's behavior in the marine environment.

6. The coexistence of the fishing and the oil and gas industries on the continental shelf is complicated mainly by displacement of traditional fishing grounds and interference with fishing. This is caused by installing fixed oil platforms, laying underwater pipelines, suspending wellheads, and leaving various structures, equipment, and debris on the sea bottom. The long-term experience of many countries, primarily on the shelf of the North Sea, shows that trawling fishing in the areas of oil and gas field developments has become limited, complicated, and dangerous. It leads to considerable losses for the fishing industry that are not fully covered by compensation measures.

7. One of the global problems of the offshore oil and gas industry is connected with the fate of disused oil platforms and underwater pipelines. The number of abandoned offshore installations and structures all over the world totals many thousands, the length of the obsolete pipelines—tens of thousands of kilometers, and the cost of their dismantling and removal—billions of dollars. These structures left on the shelf dissipate and pose long-term threats (for hundreds of years) to navigation and fishing. The promising directions of the secondary use of offshore oil platforms include creating research stations, experimental energetic units, centers for waste processing and disposal, and bases for search and rescue operations. From the fisheries perspective, the most interesting ideas suggest using abandoned platforms (or their fragments) as artificial reefs that will attract marine organisms and possibly increase the bioproductive potential of the shelf waters. Conceptual, methodological, and economic aspects of these projects are presently being actively studied both at national and international levels.

8. During the last 10–15 years, environmental control and regulation of the offshore petroleum industry's impact have widely and effectively used methodology based on the probabilistic nature of environmental risk and the possibility of its quantitative assessment. The analysis of environmental hazard and risk is used to substantiate specific organizational, technical, economic, and legal measures. This approach is especially promising for solving problems regarding the environmental safety under conditions of acute (extreme) risk.

References

Alekperov, I. H., Kasimov, R. U. 1988. Use of infusoria for biotesting of thermally cleaned drilling cuttings. *Gidrobiologicheski zhurnal* 22(4):96–98. (Russian)

Anderson, J. W. 1985. Oil pollution: effect and retention in the coastal zone. In *Proceedings of the International Symposium on Utilization of Coastal Ecosystems: Planning, Pollution and Productivity.*—Rio Grande, Brazil, pp.187–214.

Anikeev, V. V., Mansurov, M. N., Fleishman, B. S. 1995. Environmental risk analysis for the offshore oil and gas field developments. *Docladi RAN* 340(4):566–568. (Russian)

Anonymous. 1994a. Exxon under attack again. *Mar.Pollut.Bull.* 28(5):272.

Anonymous. 1994b. Fisheries blast across bows of UK seismic industry. *First break* 12(11):547–548.

Anonymous. 1995. Russian pipeline spill. *Mar.Pollut.Bull.* 30(7):433–434.

Balashkand, M. I., Vekilov, E. H., Lovlya, S. A., Protasov, V. R., Rudakovski, L. G. 1980. *New sources of seismic explorations safe for ichthyofauna.* Moscow: Nauka, 80 pp. (Russian)

Bjoerke, H., Dallen, J., Bakkeplass, K., Hansen, K., Rey, L. 1991. Availability of seismic activities in relation to vulnerable fish resources. *Help Havforskningsinst. Egg-larveprogram* 38.

Borisov, V. P., Ponomarenko, V. P., Osetrova, N. V., Semenov, V. N. 1994. *Impact of the offshore oil and gas development on the bioresources of the Barents Sea.* Moscow: VNIRO, 250 pp. (Russian)

Borisov, V. P., Ponomarenko, V. P., Osetrova, N. V., Semenov, V. N. 1995. Bioresources of the Barents Sea and the project of development of Shtokmanovskoe gas condensate field. *Rybnoe khozyaistvo* 1:12–28. (Russian)

Buchan, G., Allan, R. 1992. The impact on fishing industry. In *North Sea oil and the environment: developing oil and gas resources, environmental impacts and responses.*—London and New York: Elsevier Applied Science, pp.459–480.

Bugrov, L. Y., Muraviev, W. B., Lapshin, O. M. 1994. Alternative using of petroleum-gas structure in the Caspian and Black Seas for fish-farming and fishing: real experience and rigs conversion prospects. *Bull.Mar.Sci.* 55(2–3):1331–1332.

Burger, J., ed. 1994. *Before and after an oil spill: the Arthur Kill.* New Brunswich, New Jersey: Rutgers University Press, 305 pp.

Burke, Ch. J., Veil, J. A. 1995. Synthetic-based drilling fluids have many environmental pluses. *Oil and Gas Journ.* (Nov.1995):59–64.

Cairns, W. J. 1992. Mitigation by design. In *North Sea oil and the environment: developing oil and gas resources, environmental impacts and responses.* —London and New York: Elsevier Applied Science, pp.281–332.

Chenard, P. G., Engelhardt, F. R., Blane, J., Hardie, D. 1992. Patterns of oil-based drilling fluid utilization and disposal of associated wastes on the Canadian offshore frontier lands. In *Drilling wastes.*—London: Elsevier Applied Science, pp.119–136.

Chevron Corporation Publication. 1993. *Oil and gas exploration and production offshore.* 4 pp.

Dalen, J. M., Knutsen, G. M. 1987. Scaring effects on fish and harmful effects on fish, larvae and fry by offshore seismic explorations. In *Proc. 12-th Intern.Congr.Accust. Halifax, July 16–18, 1986.*—New York, pp.93–102.

Dalen, J., Ona, E., Soldal, F. V., Saetre, R. 1996. Seismiske undersokelser til havs. *Fisken og havet.* 9:26.

Davies, J. M., Kingston, P. F. 1992. Sources of environmental disturbances associated with offshore oil and gas developments. In *North Sea oil and the environment: developing oil and gas resources, environmental impacts and responses.*—London and New York: Elsevier Applied Science, pp.417–440.

Day, P. E. 1992. Mossmorran case study. In *North Sea oil and the environment: developing oil and gas resources, environmental impacts and responses.* —London and New York: Elsevier Applied Science, pp.565–618.

Dokken, Q. 1993. Flower Gardens ocean research project: using offshore platforms as research stations. *Mar.Technol.Soc.J.* 27(2):45–50.

Dossena, G., Ratti, S., Cannavacciuolo, A., Sebastiani, G. 1995. Research on the impact of hydrocarbons. Exploration and production activities in the Adriatic Sea. *Les mers tributaires de Mediterrannee. Bulletin de 15, Institut oceanographique, Monaco.* Numero special 15. CIESM Science Series No.1: 203–214.

Duke, T. W. 1985. Potential impact of drilling fluids on estuarine productivity. In *Proceedings of the International Symposium on Utilization of Coastal Ecosystems: Planning, Pollution and Productivity.*—Rio Grande, Brazil, pp.215–240.

Gardner, B., Wiebe, P. 1993. Offshore platforms: an opportunity to contribute to global science. Introduction. *Mar.Technol.Soc.J.* 27(2):3–4.

GESAMP. 1990. *The state of the marine environment. GESAMP Reports and Studies. No.39.* UNEP, 112 pp.

GESAMP. 1992. *The state of the marine environment.* Oxford: Blackwell Scientific Publications Ltd, 128 pp.

GESAMP. 1993. *Impact of oil and related chemicals and wastes on the marine environment. GESAMP Reports and Studies. No.50.* London: IMO, 180 pp.

Gitschlag, G. R., Herczeg, B. A. 1994. Sea turtle observations at explosive removals of energy structures. *Marine Fisheries* 56(2):1–8.

Gourlay, K. A. 1988. Chapter 3. The black death: oil. In *Poisoners of the seas.* —London: Zed Books Ltd., pp.133–188.

Grahl-Nielson, O. 1987. Hydrocarbons and phenols in discharge water from offshore operations. Fate of the hydrocarbons in the recipient. *Sarsia* 72(3–4):375–382.

Gritzenko, A. I., Bosnyatski, G. P., Shilov, J. S. 1993. *Ecological problems of the gas industry.* Moscow: VNII prirodnikh gasov i gasovoi tekhnologii, 94 pp. (Russian)

Grogan, W. C., Blanchard, J. R. 1992. Environmental assessment. In *North Sea oil and the environment: developing oil and gas resources, environmental impacts and responses.*—London and New York: Elsevier Applied Science, pp.362–402.

Gusseinov, G. I., Alekperov, R. E. 1989. *Environmental protection during oil and gas fields development.* Moscow: Nedra, 230 pp. (Russian)

Haldane, D., Reuben, R. L., Side, J. C. 1992. Submarine pipelines and the North Sea environment. In: *North Sea oil and the environment: developing oil and gas resources, environmental impacts and responses.*—London and New York: Elsevier Applied Science, pp.481–522.

ICES. 1994. *International Council for the Exploration of the Sea. Report on the ICES Advisory Committee on the Marine Environment.* Copenhagen: ICES, 122 pp.

ICES. 1995. *International Council for the Exploration of the Sea. Report on the ICES Advisory Committee on the Marine Environment.* Copenhagen: ICES, 135 pp.

IMO. 1989. *International maritime organization. Analysis of serious casualties to sea-going tankers 1974–1988. IMO Publ.142/89.* London: IMO, 21 pp.

IMO. 1990. *International maritime organization. Guidelines and standards for the removal of offshore installations and structures on the continental shelf and the Exclusive Economic Zone.* London: IMO.

Izrael, J. A. 1984. *Ecology and control of the state of the environment.* Moscow: Gidrometeoizdat, 560 pp. (Russian)

Karamova, L. M. 1989. Natural gas and ecological problems. In *Ecology and impact of natural gas on the organism. Theses.*—Astrakhan. (Russian)

Kingston, P. F. 1991. The North Sea oil and gas industry and the environment. In *Proceedings Financial Times Conference on North Sea Oil and Gas.* —London: Financial Times, pp.19.1–19.6.

Kokorin, N. V. 1994. *Longline fishing.* Moscow: VNIRO, 423 pp. (Russian)

Kriksunov, E., Polonski, J., Vekilov, E. 1993. Seismic exploration with consideration of the nature protection requirements. *Neftyanik* 4:11–14. (Russian)

Krupinin, N. J. 1995. On the state of the environment in the Nizhnevartovski region. In *Ways and means of achieving the balanced eco-economical development in the oil regions of Western Siberia.*—Nizhnevartovsk: Uralski rabochi, pp.22–29. (Russian)

Kuznetsov, B., Samarskaya, T. 1993. Environmental protection during the offshore developments in the Barents Sea. *Neftyanik* 7:17–20. (Russian)

Lapshin, O. M., Karpenko, E. A., Nastjukov, A. B. 1990. *World experience of the secondary use of the offshore platforms for the aquaculture in the Caspian basin.* Moscow: VNIRO, 97 pp. (Russian)

Larson, R., Taylor, J., Scherer, B. 1991. Treatment of offshore produced water: an effective membrane process. In *Environment Northern Seas. Abstracts of Conference Papers.*—Stavanger (Norway): Industritrykk.

Lindberg, E. G. B., Roekke, N. E., Celius, H. K. 1990. *Greenhouse gas emissions from offshore oil and gas production on the Norwegian continental shelf.* Institut for Kotinenalsokkelundersoekelser og petroleumsteknol. A/S, Trondheim (Norway), 45 pp.

Linton, T. L. 1994. A comparison between the fish species harvested in the US Gulf of Mexico and the number of production platforms. *Bull.Mar.Sci.* 55(2–3):13–44.

Logofect, D. O. 1993. Ecology of risk and risk of ecology. *Izvestia RAN (Ser. biol.)* 6:883–896. (Russian)

Matishov, G. G. 1991. General causes of the crisis in the ecosystems of the North European seas. In *Ecological situation and protection of the flora and fauna of the Barents Sea.*—Apatiti: Izd-vo Kolskogo Filiala AN SSSR, pp.8–30. (Russian)

Mazur, I. I. 1993. *Ecology of oil and gas development.* Moscow: Nedra, 494 pp. (Russian)

McIntyre, A. D. 1995. The Brent Spar incident—a milestone event. *Mar.Pollut.Bull.* 30(9):578.

McIntyre, A. D., Turnbull, R. G. H. 1992. Environment. In *North Sea oil and the environment: developing oil and gas resources, environmental impacts and responses.*—London and New York: Elsevier Applied Science, pp.27–56.

Menzie, C. A. 1982. The environmental implications of offshore oil and gas activities. *Environ.Sci.Technol.* 16:454–472.

Middleditch, B. S. 1981. *Environmental effects of offshore oil production—the Buccaneer gas and oil field study.* New York—London: Plenum Press.

Mikhaylova, L. V. 1995. Chemical pollution—one of the main environmental problems of the Ob-Irtish region. In *Ways and means of achieving the balanced eco-economical development in the oil regions of Western Siberia.* —Nizhnevartovsk: Uralski rabochi, pp.43–46. (Russian)

Minton, R. C., Begbie, R., Tyldsey, D., Park, D. 1991. Oily cuttings cleaning system ready for onshore testing. *Ocean Industry* 26(9):54–56.

MLA. 1993. *Marine Laboratory Aberdeen. Annual Review (1991–1992).* Aberdeen: Agriculture and Fisheries Department. The Scottish Office, 68 pp.

MMS. 1995. *Mineral Management Service. Outer continental shelf natural gas and oil resource management program: cumulative effects 1987–1991. OCS report MMS 95-0007.* Herdon, Va.: U.S. Department of the Interior (USDOI).

Monk, D. C., Cormack, D. 1992. The management of acute risks. Oil spill contingency planning and response. In *North Sea oil and the environment:*

developing oil and gas resources, environmental impacts and responses. —London and New York: Elsevier Applied Science, pp.619–642.

Moore, B., Hamilton, T. 1993. Shallow gas hazard—the HSE perspective. *Petroleum Review* 47(560):403–407.

Muraveyko, V. M. 1991. *Animals' electrosensory systems.* Apatiti: Izd-vo Kolskogo Filiala AN SSSR, 106 pp. (Russian)

Muraveyko, V. M., Belyaev, A. V., Shparkovski, I. A. 1991. Impact of the geophysical methods of marine exploration on pelagic and benthic biocenoses. In *Ecological situation and protection of the flora and fauna of the Barents Sea.*—Apatiti: Izd-vo Kolskogo Filiala AN SSSR, pp.78–84. (Russian)

Neff, J. M. 1998. Fate and effects of drilling mud and produced water discharged in the marine environment. In *U.S.-Russian Government Workshop on Management of Waste from Offshore Oil and Gas Operation (April 1998, Moscow).*

Neff, J. M., Rabalais, N. N., Boesch, B. F. 1987. Offshore oil and gas development activities potentially causing long-term environmental effects. In *Long-term environmental effects of offshore oil and gas development.*—London: Elsevier Applied Science, pp.149–173.

Nelson-Smith, A., trans. 1977. *Oil pollution and marine ecology.* Moscow: Progress, 301 pp. (Russian)

NERI. 1996. *National Environmental Research Institute. Oil exploration in the Fylla Area. An initial assessment of potential environmental impact. NERI Technical Report No.156.* 75 pp.

Neumann, R. 1993. Legislation passed to bill ocean polluters. *Search* 24(10):284.

NRC, 1983. *National Research Council (U.S.). Drilling discharges in the marine environment.* Washington D.C.: National Academy Press.

NSTF. 1994. *North Sea Task Force. The Quality Status Report on the North Sea.* Fredensborg: Olsen and Olsen, 250 pp.

O'Keeffe, D. 1985. A computer program for predicting the effects of underwater explosions on the swimbladder of fish. In *Workshop on the effects of explosive use in the marine environment. COGLA EPB Technical Report No5.*—Ottawa, Canada, pp.324–353.

Panov, G. E., Petryashin, L. F., Lisyani, G. N. 1988. Environmental protection in the oil and gas industry. In *Industry and environment.*—Moscow: TSMN GKNT, 77 pp. (Russian)

Pavlov, D. S., Savvaitova, K. A., Sokolov, L. I., Alekseev, S. S. 1994. *Rare and endangered animals. Fish: Reference guidebook.* Moscow: Visshaya shkola, 334 pp. (Russian)

PARCOM. 1991. *Triennial Report of discharges from offshore exploration and exploitation installation in 1989. Thirteenth meeting of the Paris Commission 13/4/7. E.*

Pope, D. L., Moslow, T. F., Wagner, J. B. 1993. Geological and technological assessment of artificial reef sites, Louisiana outer continental shelf. *Ocean Coast.Manage.* 20(2):121–145.

Protasov, V. R., Bogatirev, P. B., Vekilov, E. H. 1982. *Ways to preserve the ichthyofauna during different kinds of the underwater activities.* Moscow: Legkaya i pishchevaya promyshlennost, 88 pp. (Russian)

Ramsay, C. G., Grant, S. 1992. Hazard and risk. In: *North Sea oil and the envi-ronment: developing oil and gas resources, environmental impacts and re-sponses.*—London–New York: Elsevier Applied Science, pp.559–584.

Ray, J. P. 1985. The role of environmental science in the utilization of coastal and offshore waters for petroleum development by the US industry. In *Proceedings of the International Symposium on Utilization of Coastal Ecosystems: Planning, Pollution and Productivity.*—Rio Grande, Brazil, pp.127–149.

Reggio, V. C. 1987. *Rigs to reefs: the use of obsolete petroleum structures as artifi-cial reefs. OCS Report MMS 87-0015.* Minerals Management Service, Gulf of Mexico Regional Office. Louisiana.

Rowe, D. M. 1993. Possible offshore application of thermoelectric conver-sion. *Mar.Technol.Soc.J.* 27(3):43–48.

Sakhalin-1. 1994. *Technico-economical calculations for the project Sakhalin-1. Vol.11: Preliminary environmental impact assessment and suggestions on en-vironmental protection.* EXXON, Sodeko, SMNG, 180 pp. (Russian)

Shears, J. R. 1989. *An analysis of the frequency, location and magnitude of oil spills around the British Isles, and an evaluation of their environmental damage.* Department of Geography, University of Southhampton, 47 pp.

Side, J. S. 1992. Decommissioning and abandonment of offshore installa-tions. In *North Sea oil and the environment: developing oil and gas resources, environmental impacts and responses.*—London–New York: Elsevier Applied Science, pp.523–552.

Side, J., Davies G. 1989. *Fisheries and environmental criteria in abandonment strategies. Report of the SERC funded study.* Edinburgh: Institute of Off-shore Installations.

Side, J. S., Johnston, C. S. 1985. *Alternative uses of offshore installations. Report of the SERC funded study.* Edinburgh: Institute of Offshore Installations.

Smith, J. P., Ayers, R. C., Tait, R. D. 1997. *Perspectives from research on the envi-ronmental effects of offshore discharges of drilling fluids and cuttings.* Exxon Publications. 20 pp.

Somerville, H. J., Bennett, D., Davenport, J. N., Holt, M. S., Lynes, A., Mahie, A., McCourt, B., Parker, J. G., Stephenson, J. G., Watkinson, T. G. 1987. Environmental effect of produced water from North Sea oil operations. *Mar.Pollut.Bull.* 18(10):549–558.

Somerville, H. J., Shirley, D. 1992. Managing chronic environmental risks. In *North Sea oil and the environment: developing oil and gas resources, envi-ronmental impacts and responses.*—London–New York: Elsevier Applied Science, pp.643–664.

Stanley, D. R., Wilson, C. A., Cain, C. 1994. Hydroacoustic assessment of abundance and behavior of fishes associated with an oil and gas plat-form of the Louisiana Coast. *Bull.Mar.Sci.* 55(2–3):2–3.

Stradomskaya, A. G., Semenov, A. D. 1991. Oil pollution levels in the water and bottom sediments in the shallow areas of the Caspian Sea. In *The-ses of the second All-Union Conference on Water Toxicology.*—Spb., Vol.2, pp.194–195. (Russian)

Swan, J. M., Neff, J. M., Young, P. C., eds. 1994. *Environmental implications of offshore oil and gas development in Australia—the findings of an independent scientific review.* Sydney: Australian Petroleum Exploration Association Limited, 696 pp.

UKOOA. 1988. *The abandonment of offshore installations: fact sheet on oil and gas activities.* London: UKOOA.

UNEP. 1988. Environmental impact of the discharges of the water-based drilling muds. In *Industry and environment.*—Moscow: TSMN GKNT, 77 pp. (Russian)

Vekilov, E. H., Geranin, M. P. 1989. Environmental protection measures during the offshore oil and gas production activities. In *Complex studies of the pollution of the World Ocean in connection with development of its mineral resources.*—Leningrad: Sevmorgeologia, pp.89–95. (Russian)

VNIIP. 1994. *The state of the environment and environmental protection activities in the former USSR.* Moscow: VNIIP, 111 pp. (Russian)

VNIPImorneftegas. 1990. *The analysis of the accidents on the shelf and suggestions on reducing the frequency of such accidents during oil and gas production.* Moscow, 159 pp. (Russian)

Voloshin, V. I. 1989. *Pollution of the marine environment by ships.* Leningrad: Gidrometeoizdat, 272 pp. (Russian)

Voronin, V. M., Lukashina, V. A., Muraveyko, V. M. 1989. *Impact of electrical fields on the water organisms.* Apatiti: Izd-vo Kolskogo Filiala AN SSSR, 39 pp. (Russian)

Yablokov, A. 1995. Accident near Ussinsk. *Ekologicheski monitoring* 1:8–9. (Russian)

Yelverton, J. T., Richmond, D. K., Hicks, W., Saunders, K., Fletcher, E. R. 1975. *The relationship between fish size and their response to underwater blast.* Washington DC: Defense Nuclear Agency.

Zalogin, B. S., Kuzminskaya, K. S. 1993. Change trends in the nature of the World Ocean under anthropogenic impact. In *Life of the Earth: nature and society.*—Moscow, pp.66–72. (Russian)

Zick, V., Dimov, G., Malinovski, J., Berchia, I., Gramberg, I., eds. 1990. *Geology and mineral resources of the World Ocean.* Warsaw: INTERMORGEO, 756 pp. (Russian)

Chapter 4

Biogeochemical and Ecotoxicological Characteristics of Crude Oil and Oil Hydrocarbons in the Marine Environment

An unprecedented variety of the world's scientific literature has been devoted to the fate and environmental effects of crude oil and oil hydrocarbons in the sea. According to the most modest estimates, at least 10,000 studies on these topics have been published over the past 15–20 years. In Russia, research of the properties and behavior of oil and its products in the water environment started as far back as the end of the last century. Interest on these issues was associated with concern about the possible negative impact of oil on the fish stock in the basin of the Volga River and the Caspian Sea [Grimm, 1891; Arnold, 1897].

Such persistent attention to oil hydrocarbons in the water environment is justified. Abundant evidence clearly demonstrates the global distribution of oil contamination. Worldwide public response occurs at every large accidental oil spill in the sea. Concerns about the scale and consequences of oil pollution have grown over the last decades, especially after the expansion of the oil industry on the continental shelf. Undoubtedly, the oil input is one of the leading factors of the industry's impact on the marine environment and biota (see Chapter 3). This chapter will briefly discuss the biogeochemical and toxicological characteristics of oil and its main components. At present, many data on the oil levels in the sea are subject to a profound revision due to new developments in analytical and ecotoxicological methods [GESAMP, 1993; ICES, 1994a]. Therefore, this chapter will use the latest and most reliable studies conducted using a relatively similar methodological basis.

4.1. Composition and Main Properties of Crude Oil and Its Fractions

The signs of oil presence in the sea include oil slicks on the surface and oil deposits on the bottom, tar balls and lumps drifting in the water column, oil odor in seawater, tainting of seafood organisms, and so on. They became a common attribute of practically all marine regions a long time ago. However, the origin and prehistory of each of these manifestations of oil pollution are rather complex and not always obvious. It is primarily connected with the variety of oil hydrocarbon sources. Unlike xenobiotics (e.g., DDT, PCBs, or strontium-90) that are present in the water only due to human activity, hydrocarbons can enter the marine environment from natural sources (e.g., marine seeps, erosion of sediments, etc.) as well.

Regardless of the origin, a number of chemical and physical mechanisms define the behavior of all crude oils in the marine environment. The composition and properties of crude oils from different reservoirs differ so much that it is difficult to speak about typical oil. Figure 25 shows the main fractions of crude oil.

From the chemical perspective, crude oil (petroleum) is a complex mixture of several thousand different hydrocarbons (liquid hydrocarbons constitute 80–90% of its mass). Crude oil also contains traces of organic sulfur (mercaptans, thiophenes, disulfides, thiophans), nitrogen (homologues of pyridine, acridine, hydroquinoline), and oxygen (naphthenic acids, asphaltenes, resins). Besides, crude oil contains some water (up to 10%), dissolved hydrocarbon gases (up to 4%), mineral salts (mainly chlorides—up to 4 g/l), and many trace elements. The trace element ratio (most often between vanadium and nickel) may serve as an additional characteristic of oil's origin and properties.

Oil hydrocarbons are usually divided into four main groups:

- aliphatic saturated hydrocarbons with straight or branched chains of carbon atoms (40–50% of volume); they are called alkanes or paraffins;
- saturated cyclic and polycyclic compounds where alkyl groups can substitute for hydrogen atoms (25–75%); they are called naphthenes or cycloparaffins;
- aromatic unsaturated cyclic compounds where alkyl groups can also substitute for hydrogen atoms (usually up to 10–20%, more rarely—up to 35 %); and

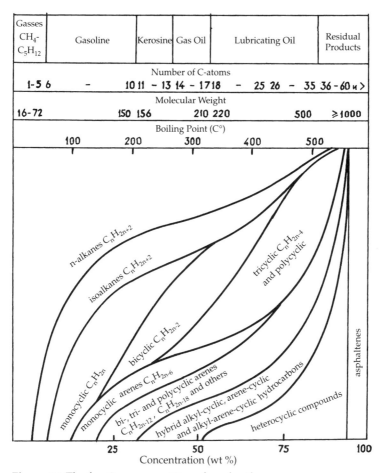

Gasses CH_4-C_5H_{12}	Gasoline	Kerosine	Gas Oil	Lubricating Oil	Residual Products
			Number of C-atoms		
1-5 6	–	10 11 – 13	14 – 17 18 –	25 26 – 35	36 – 60 и >
			Molecular Weight		
16-72		150 156	210 220	500	≥1000
			Boiling Point (C°)		
100	200		300	400	500

Figure 25 The fraction composition of crude oil.

- aliphatic unsaturated noncyclic hydrocarbons with a straight or branched chain and a double-bonded carbon link; they are called alkenes or olefins. Compounds of this group are not a part of crude oil's composition but are the main product of its cracking.

    Crude oil contains some hydrocarbons with mixed (hybrid) composition, for example paraffin–naphthenic, naphthene–aromatic. Figure 26 gives some examples of the chemical structure of oil compounds.

    The behavior and biological impacts of oil and oil products in water ecosystems are ultimately defined by their physical and

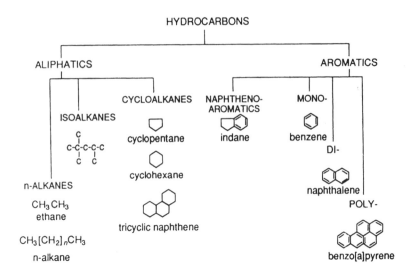

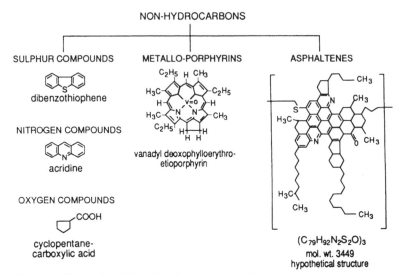

Figure 26 Examples of the chemical structure of some common components of crude oil [GESAMP, 1992].

physicochemical properties, primarily by specific gravity, volatility, and water solubility. Tables 17 and 18 give some characteristics of crude oil. Crude oils are classified as heavy, average, and light according to their density. Most crude oils usually have a specific gravity less than 1.0, which means that they are lighter than water. On average, about 1–3% (sometimes up to 15%) of crude oil dissolves in the water, and between 10% and 40% of its original volume evaporates. The data in Table 18 show that especially rapid (within minutes and hours) evaporation is typical for low-molecular-weight alkanes, cycloalkanes, and benzenes. Polycyclic aromatic hydrocarbons (PAHs), such as anthracene and pyrene, do not transform into the gaseous phase. They stay in the marine environment and undergo complex reactions such as oxidation, biodegradation, and photochemical processes. These usually lead to the formation of more polar, and hence more soluble, substances.

The solubility of oil hydrocarbons depends, first, on their molecular structure and molecular mass. Aromatic hydrocarbons are the most water soluble. Naphthenes do not dissolve in the water. Increase in molecular mass usually leads to lower water solubility. These general properties of oil ultimately explain the relative enrichment of dissolved oil fractions by the most soluble low-molecular-weight aromatic and aliphatic hydrocarbons right after oil is released on the water surface. The subsequent rather rapid evaporation of these hydrocarbons (see Table 18) leads to increasing contribution of less volatile (although not so well soluble) fractions of aromatic hydrocarbons.

The differences in properties of oil components also lead to physical fractionalizing of crude oil in the water environment. This causes oil to be present simultaneously in several physical states, including:

- surface films (slicks);
- dissolved forms;
- emulsion (oil-in-water, water-in-oil);
- suspended forms (oil aggregates floating on the surface and in the water column, oil fractions adsorbed on suspended particles);
- solid and viscous components deposited on the bottom; and
- compounds accumulated in water organisms.

In conditions of chronic pollution, oil's dominant form in the water environment is usually emulsified oil. The presence of high-molecular-weight compounds in oil composition and the impact

Table 17

Main Physicochemical Characteristics of Hydrocarbons of Crude
Oil [Nelson-Smith, 1977]

Compound	C_n*	Boiling Point (°C)	Melting Point (°C)**	Gravity (g/cm³)	Water Solubility (mg/kg)
Paraffins:					
Methane	1	–161.5	–	0.424	90 ml/l (20°C)
Ethane	2	–88.5	–	0.546	47 ml/l (20°C)
Propane	3	–42.2	–	0.542	65 ml/l (18°C)
Butane	4	–0.5	–	0.579	150 ml/l (17°C)
Pentane	5	36.2	–	0.626	360 (17°C)
Hexane	6	69.0	–	0.660	138 (15°C)
Heptane	7	98.0	–	0.684	52
Octane	8	125.7	–	0.703	15
Nonane	9	150.8	–	0.718	10
Decane	10	174.1		0.730	3
Undecane	11	195.9	–	0.741	–
Dodecane	12	216.3	–	0.766	–
Tridecane	13	235.6	–5.5	0.756	–
Tetradecane	14	253.6	6.0	0.763	–
Pentadecane	15	270.7	10.0	0.769	–
Hexadecane	16	287.1	18.0	0.773	–
Heptadecane	17	302.6	22.0	0.778	–
Naphthenes:					
Cyclopropane	3	–33.0	–	–	insignificant
Cyclobutane	4	13.0	–	–	–
Cyclopentane	5	49.3	–	0.751	–
Methylcyclo-pentane	6	71.8	–	0.749	–
Cyclohexane	6	80.7	–	0.779	–
Methylcyclo-hexane	7	100.9	–	0.769	–
Ethylcyclopentane	7	103.5	–	0.763	–
Ethylcyclohexane	8	131.8	–	0.788	–
Trimethyl-cyclohexane	9	141.2	–	0.777	–
Aromatic:					
Benzene	6	80.1	–	0.879	820 (22°C)
Toluene	7	110.6	–	0.866	470 (16°C)
Ethylbenzene	8	136.2	–	0.867	140 (15°C)
p-xylene	8	138.4	–	0.861	–
m-xylene	8	139.1	–	0.864	80
o-xylene	8	144.4	–	0.874	–
Isopropylbenzene	9	152.4	–	0.864	–

Table 17 (*continued*)

Compound	C_n*	Boiling Point (°C)	Melting Point (°C)**	Gravity (g/cm³)	Water Solubility (mg/kg)
n-propylbenzene	9	159.2	–	0.862	60 (15°C)
Naphthalene	10	217.9	80.2	1.145	20
2-methylnaphthalene	11	241.1	37.0	1.029	–
1-methylnaphthalene	11	244.8	–22.0	1.029	–
Dimethylnaphthalene	12	262.0	–18.0	1.016	–
Trimethylnaphthalene	13	285.0	92.0	1.010	–
Anthracene	14	354.0	216.0	1.250	–

Notes: * C_n—number of carbon atoms in the molecule; **—melting points are given mainly for compounds which are solid under marine conditions.

Table 18
Evaporation and Dissolution of Petroleum Hydrocarbons in the Water Environment at 25°C [Bogdashkina, Petrosyan, 1988]

Hydrocarbon	Partial Pressure (Pa)	Solubility (mg/l)	T_e* (hour)	T_d* (hour)
n-pentane	68,400	40	0.012	2,000
n-heptane	6,100	2.5	0.14	3.2×10^4
n-decane	175	0.05	4.7	1.6×10^6
n-dodecane	16	0.003	520	2.6×10^8
Benzene	12,700	1,780	0.065	45
n-xylene	1,170	180	0.71	40
Naphthalene	11	32	750	2,500
Phenanthrene	0.2	1.2	4.2×10^4	6.7×10^4
Anthracene	0.001	0.04	8.3×10^5	2.0×10^6
Pyrene	0.001	0.14	8.3×10^5	5.7×10^5

Note: *—The time of evaporation (T_e) or dissolution (T_d) of 50% of the substance released into the water environment.

of hydrodynamic factors (wind mixing and others) promote self-emulsification. The prevailing form of oil entering the water environment is also emulsified oil.

The described patterns of physicochemical behavior are typical not only for crude oil but also for oil products of any composition and origin. All of them are present in the water environment in one or several of the states listed previously. The content and distribution of these forms and fractions in the marine environment are studied within the framework of different programs of control and monitoring of oil pollution. For example, long-term observations in the Baltic Sea showed that 3.6% of the total amount of oil exists in the form of slicks, 0.4% is adsorbed on suspended particles, 15% is accumulated in bottom sediments, 64% is emulsified, and 17% is present in dissolved state [Nesterova et al., 1986].

Hydrocarbons of crude oil formed under natural conditions are usually saturated with alkyl groups. In contrast, anthropogenic aromatic compounds formed as a result of incomplete combustion of organic materials and fuels (e.g., oil, coal, peat, wood, garbage) have considerably fewer alkyl radicals. Microbial degradation and photo-oxidation of aromatic compounds of crude oil in the marine environment also lead to the reduced numbers of alkylated derivatives. To a certain extent, the number of alkyl radicals helps to distinguish relatively fresh fractions of crude oil and hydrocarbons formed either as a result of incomplete combustion or during environmental degradation of aromatic compounds.

One of the main difficulties in interpreting data on marine pollution by oil is connected with the existence of hydrocarbons similar to the ones of crude oil but produced by living organisms. The most complicated situations occur when oil is present in low, background concentrations. The hydrocarbons of biogenic origin mainly include olefins in contrast with aromatic substances that dominate in crude oil. Besides, the molecules of biogenic aliphatic hydrocarbons mostly contain an odd number of carbon atoms [GESAMP, 1993].

Numerous studies have revealed the ability of marine organisms (e.g., bacteria, phytoplankton, zooplankton, benthic algae) to enrich the water environment with organic substances [Khailov, 1971; Mironov, 1972; Romankevich, 1977; Neff, 1979; NRC, 1985; GESAMP, 1993]. Some organisms synthesize and release saturated and unsaturated hydrocarbons (mainly olefins, paraffins, and sometimes aromatic compounds) in the marine environment; others produce complex mixtures of dissolved and suspended organic matter by decomposing biogenic materials.

Biological production of hydrocarbons in the World Ocean occurs on a global scale. Most likely, the total flow of biogenic hydrocarbons considerably exceeds the intensity of all anthropogenic inputs of oil hydrocarbons with analogous composition into the marine environment. According to approximate estimates [Gurvich, 1993; Minas, Gunkel, 1995], due to photosynthesis alone, about 10–12 million tons of hydrocarbons (mostly aliphatic) are produced annually in the World Ocean. However, the dispersion of these processes

over enormous areas of the World Ocean and the natural balance between the speed of biogenic production and degradation prevent adverse environmental impacts of these constantly present biogenic
hydrocarbons. Most likely, these hydrocarbons even help to maintain
the stability of marine communities due to participation in the complex processes of regulating the ecological metabolism in the sea. The
same cannot be said about oil hydrocarbons of anthropogenic origin
that often enter the marine environment in short periods of time, cir-
culate within limited areas, and may be present in concentrations that
cause toxic effects on marine organisms.

4.2. Sources and Volumes of Input into the World Ocean

Table 19 illustrates the variety of oil hydrocarbon sources and gives
expert estimates of the scales of distribution and impact of each of
these sources on the marine environment. Tables 20–22 present more
detailed quantitative information.

First, note the approximate nature of the given estimates. They
can vary up to 1–2 orders of magnitude, especially in cases of natural
oil sources, atmospheric input, and river runoff. In particular, the absolute figures in Table 21 raise some doubts. The authors of these estimates stress the difficulty to get credible figures because of the
absence of reliable methods and statistics [GESAMP, 1993]. Nevertheless, the data in Tables 20–22 seem to give an adequate reflection
of the relative significance of different oil hydrocarbon inputs in
the marine environment. Many experts believe that on a global scale,
the main anthropogenic flows of oil hydrocarbons into the marine
environment come from land-based sources (refineries, municipal
wastes, river runoff, and so on) and transportation activity (tanker oil
transportation and shipping). Enough evidence exists to agree with
this opinion [NRC, 1985; GESAMP, 1990; 1993]. Polycyclic aromatic
hydrocarbons (PAHs), especially benzo(a)pyrene, as Table 22 shows,
enter the marine environment mostly due to atmospheric deposition.

Table 20 illustrates the general trend of declining total input of
oil hydrocarbons into the World Ocean over the years. The global
situation reflected in this table certainly may differ at the regional
level. This depends on natural conditions, degree of coastal urbanization, density of population, industrial development, navigation,
oil and gas production, and other activities. For example, in the
North Sea, offshore production input reached up to 28% of the total
input of oil hydrocarbons in 1987 (see Table 21) instead of a "modest"

Table 19

Sources and Scale of Oil Hydrocarbon Input into the
Marine Environment

Type and Source of Input	Environment		Scale of Distribution and Impact		
	Hydrosphere	Atmosphere	Local	Regional	Global
Natural:					
Natural seeps and erosion of bottom sediments	+	–	+	?	–
Biosynthesis by marine organisms	+	–	+	+	+
Anthropogenic:					
Marine oil transportation (accidents, operational discharges from tankers, etc.)	+	–	+	+	?
Marine non-tanker shipping (operational, accidental, and illegal discharges)	+	–	+	?	–
Offshore oil production (drilling discharges, accidents, etc.)	+	+	+	?	–
Onland sources:					
sewage waters	+	–	+	+	?
oil terminals	+	–	+	–	–
rivers, land runoff	+	–	+	+	?
Incomplete fuel combustion	–	+	+	+	?

Note: +, –, and ? mean, respectively, presence, absence, and uncertainty of corresponding parameters.

2% on the world scale shown in Table 20. This equaled the annual input of more than 23,000 tons of oil products at the background of their general changeable flow of 120,000–200,000 tons a year in the North Sea [Bruns et al., 1993]. One can expect similar situations in other regions of intensive offshore oil and gas developments, for example, in the Gulf of Mexico, Red Sea, Persian Gulf, or Caspian Sea. Just remember the persistent pollution in oil production areas in the Caspian Sea or the amounts of annual discharges (about 40 million tons of produced waters polluted by oil products) during offshore

Table 20

Estimates of Global Inputs (thousand tons/year) of Petroleum Hydrocarbons into the Marine Environment

Source:	1973*	1979**	1981*	1985***	1990***
Land-Based sources:				34%	1,175 (50%)
Urban runoff and discharges	2,500	2,100	1,080 (500–1,250)	–	–
Coastal refineries	200	60	100 (60-600)	–	–
Other coastal effluents	–	150	50 (50-200)	–	–
Oil Transportation and Shipping:				45%	564 (24%)
Operational discharges from tankers	1,080	600	700 (400–1,500)		
Tanker accidents	300	300	400 (300–400)		
Losses from non-tanker shipping	750	200	320 (200–600)		
Offshore production discharges	80	60	50 (40–60)	2%	47 (2%)
Atmospheric fallout	600	600	300 (50–500)	10%	306 (13%)
Natural seeps	600	600	200 (20–2,000)	8%	259 (11%)
Total discharges	**6,110**	**4,670**	**3,200**	**100%**	**2,351**

Notes: *—[NRC, 1985]; **—[Kornberg, 1981]; ***—[GESAMP, 1993].

drilling in the Gulf of Mexico [Anonymous, 1993]. At the same time, no reliable balance estimates exist for these regions.

The continental shelf of the Gulf of Mexico is also distinctive for intense seepage of natural liquid and gaseous hydrocarbons. Some authors [Kennicutt et al., 1992] believe that this can lead to formation of oil slicks and tar balls on the sea surface, which makes assessing and identifying anthropogenic pollution more difficult. In any case, the input of oil hydrocarbons from natural sources into the Gulf of Mexico is larger than in many other areas.

In the Baltic Sea (Table 21), the Sea of Asov, and the Black Sea, the leading role in oil input most likely belongs to land-based

Table 21
Relative Contributions (%) of Petroleum Hydrocarbon Inputs into the World's Oceans, the North Sea, the Danish Sea Area, and the Baltic Sea [GESAMP, 1993]

Sources	Global 1985	North Sea 1987	Danish Sea Area 1983	Baltic Sea 1984
Natural sources	8	2	–	–
Offshore production	2	28	15.7	?
Maritime transportation	45	3	22.8	7.8
Atmosphere	10	12	10.5	1.3
Land-based discharges and runoff:	34	43	51.8	90.6
- *refineries*			1.1	2.4
- *municipal waste water*			17.5	15.4
- *industrial waste water*			5.2	0.9
- *urban runoff*			6.1	13.1
- *rivers*			21.9	58.8
Dumping at sea	1	12	–	

Table 22
Estimates of the Inputs (tons/year) of Polycyclic Aromatic Hydrocarbons (PAHs) and Benzo(a)pyrene [Neff, 1979]

Source	Total PAHs	Benzo(a)pyrene
Biosynthesis	2,700	25
Petroleum	170,000	20–23
Domestic and industrial wastes	4,000	29
Surface land runoff	2,940	118
Fallout and rainout	50,000	500
Total	230,040	697

sources, which are dominated by river inflow. The Danube River alone annually brings to the Black Sea about 50,000 tons of oil, half of the total oil input of about 100,000 tons [Konovalov, 1995].

Traditional shipping and oil transportation routes are more exposed to the impacts of oil-containing discharges from tankers and other vessels than other areas. For example, observations in the Caribbean basin [Atwood et al., 1987; Jones, Bacon, 1990; Corbin, 1993], where annually up to 1 million tons of oil enter the marine environment, showed that about 50% of this amount came from tankers and other ships [Hinrichsen, 1990]. In the Bay of Bengal and the Arabian Sea, inputs from tanker and other ship discharges equal, respectively, 400,000 tons and 5 million tons of oil a year [Hinrichsen, 1990]. The most intense tanker traffic exists in the Atlantic Ocean and its seas, which accounts for 38% of international maritime oil transportation. In the Indian and Pacific Oceans, this portion is, respectively, 34% and 28% [Monina, 1991].

Enforcing stricter requirements to activities accompanied by oil discharges led to global declining of oil inputs in the marine environment mentioned above [GESAMP, 1993]. In 1981, oil transportation and shipping in general were responsible for discharging about 1.4 million tons of oil products. This amount was reduced to 0.56 million tons in 1990 (see Table 20). The reduction mainly occurred as a result of adopting stricter international regulations concerning transportation operations in the sea (International Convention for Prevention of Pollution from Ships and others). The total oil input into the sea during the same period, according to the estimates given in Table 20, dropped from 3.20 to 2.35 million tons.

Although this tendency of decreasing oil pollution caused by tanker transportation and shipping gives some reason for environmental optimism, two alarming circumstances should not be neglected. One of them has already been mentioned. Strikingly high volumes of oil's input are reported for some regions (e.g., the Caribbean basin, northern part of the Indian Ocean, Mediterranean Sea). These volumes may total hundreds of thousands or even millions of tons of oil [Hinrichsen, 1990]. They are directly connected with highly intensive shipping and tanker transportation in these areas. Some estimates indicate that annual oil input into the marine environment may reach 7.3 million tons [Panov et al., 1986; GESAMP, 1994]. Other researches give even higher figures. For example, data summarized by S.M. Konovalov [Konovalov, 1995] suggest that global oil input into the World Ocean reaches 20 million tons a year, and pollution caused by tankers accounts for 50% of it. Note that annually about 6,500 large tankers transport more than 1.2 billion tons of oil and oil products. In spite of the fact that the latter estimates are considerably higher than the ones based on official statistics (Table 20), they have not been refuted thus far. This raises serious concerns about the actual situation in different marine regions and in the World Ocean in general.

The other circumstance that can affect the tendency of decreasing oil pollution from tankers involves accidental spills. Accidental situations during oil tanker transportation repeatedly happened in the past. Remember, for example, two relatively recent tanker accidents, the *Exxon Valdez* and the *Braer*, that spilled 40,000 tons of oil into Alaskan waters in 1989 and 85,000 tons near the shore of the Shetland Islands in 1993, respectively. The probabilistic nature of accidental situations and highly variable volumes of spilled oil (see Figure 27) do not allow definite conclusions to be made. Although the level of oil pollution has tended to decrease, large volumes of spilled oil could change this situation.

A number of dramatic events show the vulnerability of making optimistic prognosis about decreasing oil pollution at the regional and global levels. For instance, catastrophic large-scale events took place in the Persian Gulf during and after the 1991 Gulf War. Between 0.5 and 1 million tons of oil were released into the coastal waters. Besides, products of combustion of over 70 million tons of oil and oil products were emitted into the atmosphere [Fowler, 1993]. Another large-scale accident occurred in Russia in September–November 1994. About 100,000 tons of oil were spilled on the territory of the Komi Republic. This threatened to cause severe oil pollution for the basin of Pechora River and, possibly, the Pechora Bay.

It must be remembered that catastrophes, in spite of the obvious evidence of their consequences and all the attention they attract, are inferior to other sources of oil pollution in their scales and degree

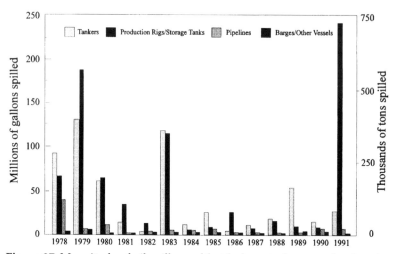

Figure 27 Magnitude of oil spills worldwide from tankers, production rigs and storage tanks, pipelines and barges and other vessels from 1978 to 1991 [Swan et al., 1994].

of environmental hazard. Land-based oil-containing discharges and atmospheric deposition of products of incomplete combustion can accordingly give 50% and 13% of the total volume of oil hydrocarbon input into the World Ocean (see Table 20). These diffuse sources continuously create relatively low but persistent chronic contamination over huge areas. Many aspects of chemical composition and biological impacts of these contaminants remain unknown.

4.3. Fate and Behavior in the Marine Environment

Complex processes of oil transformation in the marine environment start developing from the first seconds of oil's contact with seawater. The progression, duration, and result of these transformations depend on the properties and composition of the oil itself, parameters of the actual oil spill, and environmental conditions. The main characteristics of oil transformations are their dynamism, especially at the first stages, and the close interaction of physical, chemical, and biological mechanisms of dispersion and degradation of oil components up to their complete disappearance as original substances. Similar to an intoxicated living organism, a marine ecosystem destroys, metabolizes, and deposits the excessive amounts of hydrocarbons, transforming them into more common and safer substances.

Hundreds, if not thousands, of publications give detailed description of oil's behavior in the marine environment in different situations of oil spills and discharges of oil-containing wastes. Figures 28–30 and Table 23 give the general idea about oil dispersion and transformation in the sea.

Physical transport. The distribution of oil spilled on the sea surface occurs under the influence of gravitation forces. It is controlled by oil viscosity and the surface tension of water. Only ten minutes after a spill of 1 ton of oil, the oil can disperse over a radius of 50 m, forming a slick 10-mm thick. The slick gets thinner (less than 1 mm) as oil continues to spread, covering an area of up to 12 km^2 [Ramade, 1978]. During the first several days after the spill, a considerable part of oil transforms into the gaseous phase. This portion can reach up to 75%, 40%, and 5–10% for light, medium, and heavy oils, respectively [Wang, Fingas, 1994]. Besides volatile components, the slick rapidly loses water-soluble hydrocarbons. The rest—the more viscous fractions—slow down the slick spreading.

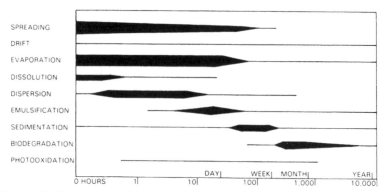

Figure 28 Time span and relative magnitude of processes acting on spilled oil: line length—probable time span of any process; line width—relative magnitude of the process both through time and in relation to other contemporary processes [Baker et al., 1990].

Further changes take place under the combined impact of meteorological and hydrological factors and depend mainly on the power and direction of wind, waves, and currents. An oil slick usually drifts in the same direction as the wind. The speed equals 3–4% of the wind speed, often exceeding the speed of the water [Mironov, 1989]. While the slick thins, especially after the critical thickness of about 0.1 mm, it disintegrates into separate fragments that spread over larger and more distant areas. Storms and active turbulence speed up the dispersion of the slick and its fragments. A considerable part of oil disperses in the water as fine droplets that can be transported over large distances away from the place of the spill. This happened after the accident of the tanker *Braer* near the shore of the Shetland Islands in January 1993 [MLA, 1993].

By now, over 20 models have been developed to predict oil slick trajectories. Their effectiveness for solving practical tasks of predicting and preventing the consequences of oil spills ultimately depends on the availability of the data needed for calculations. These data must include characteristics of the oil (type and properties), spill (volume and rate), and environmental parameters (wind and current velocity and direction, depth, temperature, and others). Unfortunately, this information is seldom available in the needed time and in the needed amounts. Creating banks of such data for the areas where accidental situations are especially possible certainly will increase the reliability and effectiveness of using these models for predicting oil slick behavior.

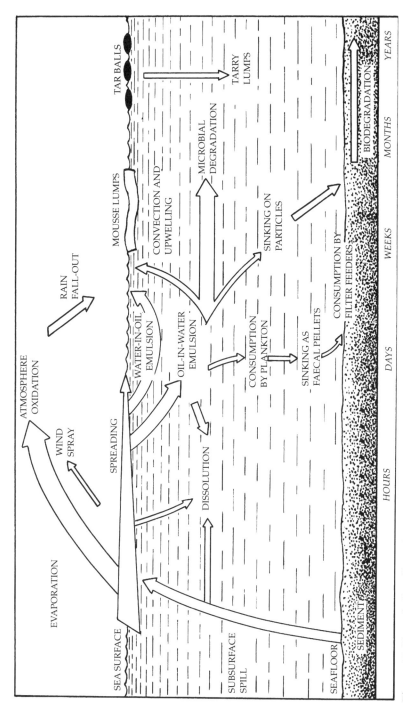

Figure 29 Fate of oil spills in the ocean [Rasmussen, 1985; Emerson, 1994].

135

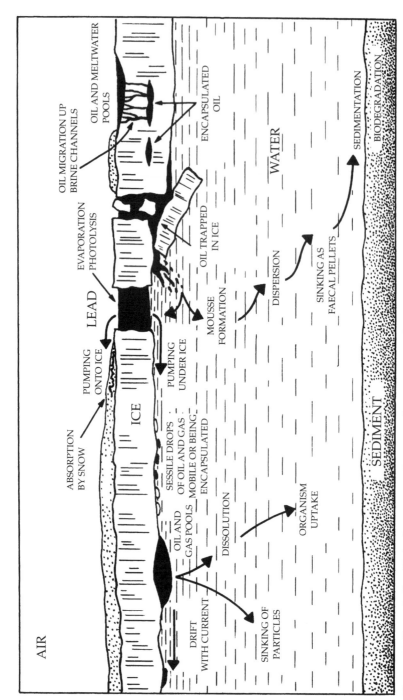

Figure 30 Fate of oil spills in the ocean in the winter [Emerson, 1994].

136

Table 23
Biogeochemical Processes of Transformation and Transport of Oil and Oil Products in the Marine Environment

Process	Zone Where the Process Takes Place			
	Water Column	Sea–Atmosphere	Sea–Seashore	Water–Bottom Sediments
Transformation of chemical composition and physical state	Dissolution emulsification; chemical and microbial degradation	Evaporation; chemical and photochemical oxidation; microbial degradation; emulsification	Sorption on solids; bioaccumulation; separation into fractions under salinity gradient	Microbial and chemical degradation; biological accumulation
Distribution and transport	Distribution in the water column; sedimentation of oil aggregates	Slick spreading; wind and current transport	Sedimentation; biosedimentation; washing ashore	Accumulation in sediments; resuspension; transport of solids

137

Dissolution. Most oil components are water-soluble to a certain degree (see Tables 17 and 18), especially low-molecular-weight aliphatic and aromatic hydrocarbons. Polar compounds formed as a result of oxidation of some oil fractions in the marine environment also dissolve in seawater. Compared with evaporation, dissolution takes more time. Hydrodynamic and physicochemical conditions in the surface waters strongly affect the rate of the process.

Some authors believe that, on average, 1–3% of crude oil dissolves in seawater, and the concentration of these dissolved fractions under an oil slick does not exceed 0.1 mg/l [Baker et al., 1990]. Data indicate that maximum concentrations of dissolved oil hydrocarbons stable in seawater are 0.3–0.4 mg/l [Grahl-Nielsen, 1987]. Exceeding these levels may be followed by creating unstable oil-in-water emulsions and forming oil slicks on the sea surface (see Emulsification).

At the same time, numerous studies show much higher levels of dissolved oil in the marine environment [Mironov, 1972; Nelson-Smith, 1977; Bogdashkina, Petrosyan, 1988; Baker et al., 1990]. For example, in the Baltic Sea, 17% of oil was reported to be present in the dissolved state [Nesterova et al., 1986]. After the accidental grounding of the *Amoco Cadiz* near the French shore in 1978, about 14% of the spilled oil dissolved in the water, which increased the rate of its microbial degradation (see Microbial degradation) [GESAMP, 1993].

Emulsification. Oil emulsification in the marine environment depends, first of all, on oil composition and the turbulent regime of the water mass. The most stable emulsions such as water-in-oil contain from 30% to 80% water. They usually appear after strong storms in the zones of spills of heavy oils with an increased content of non-volatile fractions (especially asphaltenes). They can exist in the marine environment for over 100 days in the form of peculiar "chocolate mousses". Stability of these emulsions usually increases with decreasing temperature. The reverse emulsions, such as oil-in-water (droplets of oil suspended in water), are much less stable because surface-tension forces quickly decrease the dispersion of oil. This process can be slowed with the help of emulsifiers—surface-active substances with strong hydrophilic properties used to eliminate oil spills (see Chapter 6). Emulsifiers help to stabilize oil emulsions and promote dispersing oil to form microscopic (invisible) droplets. This accelerates the decomposition of oil products in the water column.

Oxidation and destruction. Chemical transformations of oil on the water surface and in the water column start to reveal themselves no earlier than a day after the oil enters the marine environment. They mainly have an oxidative nature and often involve photochemical reactions under the influence of ultraviolet waves of the solar spectrum. These processes are catalyzed by some trace elements (e.g., vanadium) and inhibited (slowed) by compounds of sulfur. The final products of oxidation (hydroperoxides, phenols, carboxylic acids, ketones, aldehydes, and others) usually have increased water solubility. An experimental research showed that they have increased toxicity as well [Izrael, Tsiban, 1988]. The reactions of photooxidation, photolysis in particular, initiate the polymerization and decomposition of the most complex molecules in oil composition. This increases the oil's viscosity and promotes the formation of solid oil aggregates [GESAMP, 1977; 1993].

Sedimentation. Some of the oil (up to 10–30%) is adsorbed on the suspended material and deposited to the bottom. This mainly happens in the narrow coastal zone and shallow waters where particulates are abundant and water is subjected to intense mixing. In deeper areas remote from the shore, sedimentation of oil (except for the heavy fractions) is an extremely slow process.

Simultaneously, the process of biosedimentation happens. Plankton filtrators and other organisms absorb the emulsified oil. They sediment it to the bottom with their metabolites and remainders. The suspended forms of oil and its components undergo intense chemical and biological (microbial in particular) decomposition in the water column. However, this situation radically changes when the suspended oil reaches the sea bottom. Numerous experimental and field studies show that the decomposition rate of the oil buried on the bottom abruptly drops. The oxidation processes slow down, especially under anaerobic conditions in the bottom environment. The heavy oil fractions accumulated inside the sediments can be preserved for many months and even years. For example, 20 years after an accidental oil spill near the Atlantic shore of the United States, bottom sediments still had a considerable amount of oil residuals [Teal, 1993].

The ratio between dissolved and suspended forms of oil and its components in the marine environment varies within a very wide range. This depends upon the combination of environmental factors and the composition, properties, and origin of the oil hydrocarbons [Nemirovskaya, 1994]. For example, for the Baltic Sea, this ratio varies from 0.2 to 2.1 [Nesterova et al., 1986].

Microbial degradation. The fate of most petroleum substances in the marine environment is ultimately defined by their transformation and degradation due to microbial activity. About a hundred known species of bacteria and fungi are able to use oil components to sustain their growth and metabolism. In pristine areas, their proportions usually do not exceed 0.1–1.0% of the total abundance of heterotrophic bacterial communities. In areas polluted by oil, however, this portion increases to 1–10% [Atlas, 1993].

Biochemical processes of oil degradation with microorganism participation include several types of enzyme reactions based on oxygenases, dehydrogenases, and hydrolases. These cause aromatic and aliphatic hydrooxidation, oxidative deamination, hydrolysis, and other biochemical transformations of the original oil substances and the intermediate products of their degradation.

The degree and rates of hydrocarbon biodegradation depend, first of all, upon the structure of their molecules. The paraffin compounds (alkanes) biodegrade faster than aromatic and naphthenic substances. With increasing complexity of molecular structure (increasing the number of carbon atoms and degree of chain branching) as well as with increasing molecular weight, the rate of microbial decomposition usually decreases. For example, for anthracene and benzo(a)pyrene, this rate is tens to hundreds of times lower than for benzene [Tinsli, 1982; Bogdashkina, Petrosyan, 1988]. Besides, this rate depends on the physical state of the oil, including the degree of its dispersion. The most important environmental factors that influence hydrocarbon biodegradation include temperature, concentration of nutrients and oxygen, and, of course, species composition and abundance of oil-degrading microorganisms. These complex and interconnected factors influencing biodegradation and the variability of oil composition make interpreting and comparing available data about the rates and scale of oil biodegradation in the marine environment extremely difficult. This explains why, in spite of vast research, answers to many urgent questions are not available yet. These include extremely important issues concerning the effect of low temperatures on the rates of oil degradation in the marine environment. Meanwhile, the answer to this question would help to elucidate the fate of oil both in the high latitude regions and in the deep ocean areas where the water temperature is usually below 5°C.

Many studies, including the ones recently conducted in the Arctic areas [Koronelli et al., 1994], demonstrate that microbial degradation of oil hydrocarbons slows under low temperatures. At the same time, some experimental data and field observations conducted during the winter in the North Sea [Bruns et al., 1993] indicate that oil-degrading microorganisms can adapt to the low temperatures. It was shown that they can intensively degrade oil hydrocarbons even at the temperatures as low as 2°C. This discrepancy of results seems to be connected with the complex and multiple impacts of many factors. These include the content of nutrients and oxygen, species composition of the microorganisms, availability of substrate for microorganism development, oil level, and others. For example, according to some estimates [Atlas, 1994], rates of hydrocarbon degradation in pristine marine waters are typically less than $0.03 \text{ g}/\text{m}^3$ per day, while in polluted areas these values could reach $0.5–50 \text{ g}/\text{m}^3$ per day.

Aggregation. Oil aggregates in the form of petroleum lumps, tar balls, or pelagic tar can be presently found both in the open and coastal waters as well as on the beaches. They derive from crude oil after the evaporation and dissolution of its relatively light fractions, emulsification of oil residuals, and chemical and microbial transformation. Around 5–10% of spilled crude oil and up to 20–50% of weathered oil and oil products of ballast waters and washing waters from tankers form these aggregates [Benzhitski, 1980]. The chemical composition of oil aggregates is rather changeable. However, most often, its base includes asphaltenes (up to 50%) and high-molecular-weight compounds of the heavy fractions of the oil.

Oil aggregates look like light gray, brown, dark brown, or black sticky lumps. They have an uneven shape and vary from 1 mm to 10 cm in size (sometimes reaching up to 50 cm). Their surface serves as a substrate for developing bacteria, unicellular algae, and other microorganisms. Besides, many invertebrates (e.g., gastropods, polychaetes, and crustaceans) resistant to oil's impacts often use them as a shelter. These organisms can create peculiar fouling communities [Benzhitski, 1980].

Oil aggregates can exist from a month to a year in the enclosed seas and up to several years in the open ocean [Benzhitski, 1980]. They complete their cycle by slowly degrading in the water column, on the shore (if they are washed there by currents), or on the sea bottom (if they lose their floating ability).

Self-purification. As a result of the processes previously discussed, oil in the marine environment rapidly loses its original properties and disintegrates into hydrocarbon fractions. These fractions have different chemical composition and structure and exist in different migrational forms. They undergo radical transformations that slow after reaching thermodynamic equilibrium with the environmental parameters. Their content gradually drops as a result of dispersion and degradation. Eventually, the original and intermediate compounds disappear, and carbon dioxide and water form. Such self-purification of the marine environment inevitably happens in water ecosystems if, of course, the toxic load does not exceed acceptable limits.

The basic condition of maintaining the ecosystem's ability to process the exceeding amounts of oil under conditions of chronic pollution is the balance between the rates of input and destruction of oil products. Theoretical and experimental approaches have been developed to assess the assimilation ability of marine ecosystems [Tziban et al., 1985; GESAMP, 1977; Nemirovskaya, 1994]. Researchers have suggested balance equations to calculate oil pollution of the marine environment [Simonov, Zubakina, 1979]. However, the application of these methods to assess regional situations has not become common and widespread. The reason seems to be the lack of adequate information about the oil's behavior under natural conditions. The extreme complexity of the process, which depends on the combined impacts of physical, chemical, and biological factors, makes accurate quantitative estimates of the self-purifying ability of the marine environment difficult.

Assessing oil's behavior, dispersion, and transformation presents a special difficulty when spills occur in areas covered with ice. In such situations, the following events typically occur (see Figure 30):

- increasing viscosity of crude oil under low temperatures;
- limited distribution due to oil's adsorption on the ice surface and accumulation in the porous layers and cavities of the ice cover;
- drifting of the oil slick together with the ice mainly under influence of the wind; and
- reduced rates of microbial and photochemical degradation under conditions of low temperature and limited oxygen input.

These processes are still poorly studied from the quantitative perspective. However, to a considerable extent, they will define the fate and consequences of possible oil spills in areas of future developments in the Russian northern and far-eastern seas.

Further research of oil's behavior in the marine environment, including oil inputs and the self-purification ability of different ecosystems in different situations of oil pollution, is critically important. Such studies should substantiate the national and international strategies regarding pollution prevention and the choice of practical measures to eliminate accidental oil spills (see Chapter 6, Section 6.3).

4.4. Environmental Levels and Distribution in the Marine Ecosystems

By now, an extensive database exists on the content and distribution of oil and its components for practically all areas of the World Ocean. National, international, regional, and global programs of oil pollution observations are in progress and numerous publications describe their results. At the same time, the attempts to interpret and summarize the large volumes of accumulated information, undertaken from time to time [GESAMP, 1977; 1993; UNEP, 1990; NSTF, 1994], face a number of difficulties. They derive from the complexity of oil's composition, the changeability of the forms of its existence, and the variability of the ways it transforms in the sea under the multiple impacts of environmental factors. Besides, insufficient quality assurance procedures and methodological inconsistencies of many studies can result in inadequate conclusions made on their basis [Topping, 1992]. This is especially possible while identifying individual oil hydrocarbons when they occur at low levels comparable within the background content of natural organic substances with a similar or identical composition.

In this review we will try to overcome, at least partially, the limitations and difficulties mentioned above. To do so, it seems proper to use the results of latest and most reliable studies conducted based on widely accepted and relatively similar methodologies and quality assurance procedures.

4.4.1. Background Characteristics

The question about background levels of oil hydrocarbons is the first issue that emerges when measuring the content of these substances in the water and other components of the marine ecosystems. Strictly speaking, background concentrations are those concentrations that had existed in nature before they changed as a result of human activity (see Figure 31). For xenobiotics, such as chlorinated synthetic compounds and artificial radionuclides, these concentrations equal zero (in practice—the detection limit). However, in case of oil and its components, the situation is more complicated. Oil contamination became a factor of a global scale a long time ago. As noted previously, it can come from both natural and anthropogenic sources. At present, finding the area in the World Ocean where the traces of natural and anthropogenic oil hydrocarbons and the products of their

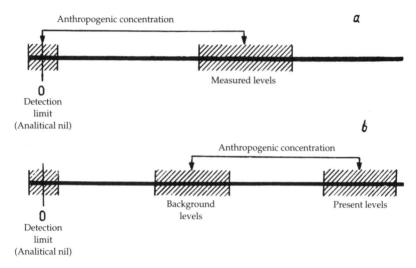

Figure 31 Anthropogenic and background concentrations in the marine environment (from [Laane, 1992] with changes): a—for xenobiotics (e.g., DDT, PCBs, strontium-90); b—for oil and other substances of both natural and anthropogenic origin.

transformation cannot be detected with the use of modern analytical methods is practically impossible. In most cases, distinguishing the natural background levels from anthropogenic concentrations of hydrocarbons is rather difficult. These difficulties considerably increase with decreasing measured levels.

This is true, first of all, for polycyclic aromatic hydrocarbons (PAHs). PAHs are present in crude oil (up to 30% of its volume). They have increased water solubility as compared with hydrocarbons of aliphatic structure. PAHs are released during burning of all kinds of combustible materials. They can be transported in the atmosphere over large distances. Eventually they can fall out on the water surface forming global hydrocarbon contamination of the marine environment [Rovinski et al., 1988]. PAHs are always present (sometimes dominating) in the organic fraction of sea aerosols. Their levels, as shown in detailed observations in the North Sea [Preston et al., 1992], highly correlate with the content of lead. Lead also enters the atmosphere with products of combustion of different kinds of fuel, especially as a part of automobile exhaust.

Figure 32 shows the scale of anthropogenic contribution of PAHs in bottom sediments from one of the fjords in Canada. It illustrates the process of increasing PAH levels due to increased anthro-pogenic input since the 1920s.

These and some other data make it possible to consider PAHs as indicators of the anthropogenic pollution of the marine environment by hydrocarbons. It must be also remembered that the previously widely accepted opinion about the considerable scale of the natural biosynthesis of PAHs in the marine environment [Nelson-Smith, 1977] is in doubt at present [Laane, 1992]. If such biosynthesis happens at all, it most likely occurs with low intensity and gives mainly pyrene as the final product. Hypothetical total phytoplankton production of PAHs in the World Ocean is estimated at about 2,700 tons a year, which includes 26 tons of benzo(a)pyrene. These volumes constitute less than 1% of the total anthropogenic input of these hydrocarbons into the marine environment [Neff, 1979].

Revealing global background levels of oil hydrocarbons (first of all PAHs) in the World Ocean is a difficult task. It is complicated by the uneven pattern of their distribution as well as by the difficulties of measuring the natural background of many hydrocarbons. Taking into account the global nature of oil contamination, it would

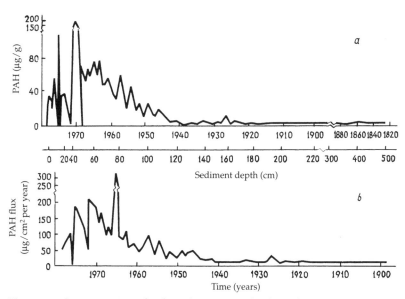

Figure 32 Concentration of polycyclic aromatic hydrocarbons (PAHs) (a) and PAH flux (b) as a function of sediment core in the Saguenay Fjord, Quebec [Smith, Levy, 1990].

be reasonable to consider as background levels those concentrations that can be found in the most remote areas not subjected to direct anthropogenic impact. These areas should include some pelagic ocean areas (especially in the Southern Hemisphere) and primarily Antarctic waters. Table 24 summarizes the latest data on the content of dissolved alkanes and aromatic hydrocarbons in the surface waters and other components of the Antarctic ecosystem.

These data show relatively high variability of hydrocarbon concentrations in all components of the marine ecosystem of Antarctica. Such a situation is partly explained by the fact that, lately, the consequences of human activity, including shipping, fishing, and the supplying of polar research stations, can be found even here. Oil spill incidents have already occurred, including the *Bahia Paraiso* oil spill [Kennicutt, 1992] which caused increasing concentrations of oil hydrocarbons in the bottom sediments in this area (see Table 24).

Although the data for seawater in the clean Antarctic areas vary within considerable limits, they help to estimate the minimum background levels of oil hydrocarbons. For PAHs, according to all available Antarctic data [Cripps, 1992], the median concentration of 0.02 μg/l can be considered as such minimum level. To compare, the

Table 24

Concentrations of Aliphatic and Polyaromatic Hydrocarbons (PAHs) in Different Components of Marine Ecosystem of Antarctica

Studied Samples	n-alkanes	PAHs	Reference
Seawater, surface layer (μg/l)	0.1–8.9	0.007–0.072	Cripps, 1992
Terrigenic suspense (μg/l)	0.003–0.010	0.01–0.060	Cripps, 1992
–" –	0.013–0.103	0.002–0.010	Desedery et al., 1991
Biogenic suspense (μg/l)	450–1,380	–	Matsueda, Handa, 1986
Sea ice (mg/m²)	0.13–6.20	–	Nicols et al., 1988
Phytoplankton (μg/l)*	0.02–0.09	0.03	Cripps, 1992
Bottom sediments (μg/kg):			
Upper layer (2.5 cm)	64–2,900	14–280	Cripps, 1992
–" –	1,000–9,300	84–2,500	Danishevskaya et al., 1989
Upper layer (10 cm)	728	195	Venkatesan, Kaplan, 1987
*Polluted samples***	1,200–77,000	1,000–60,000	Cripps, 1992; Kennicutt et al., 1992
*Polluted samples****	100–1,723	200–1,653	Kennicutt et al., 1992

Notes: *—Calculated for 20 mg of phytoplankton (dry weight) in 1 liter of seawater. **—Collected in the areas of oil spills. ***—Collected a year after the *Bahia Paraiso* incident.

background PAH level for the Pacific Ocean is estimated at 0.04 μg/l [Tkalin, 1993].

The background levels of the total content of all extracted non-polar (mainly aromatic) hydrocarbons in the surface waters of the Antarctic (in the area of the Antarctic peninsula) vary within 0.15–4.65 μg/l, with the average level less than 1 μg/l. This is considered the upper limit of the natural content of the sum of aromatic hydrocarbons in seawater [Weber, Bicego, 1990]. For normal alkanes, such estimation is complicated because of relatively high natural bioproduction of these hydrocarbons. In earlier publications [Saliot, 1981; Grahl-Nielsen, 1987], the background (natural) levels of oil hydrocarbons in seawater was thought to be less than 1 μg/l for volatile compounds (with 5–12 carbon atoms in the molecule) and 1–10 μg/l for nonvolatile hydrocarbons (with more than 12 carbon atoms).

It is important to note that one study conducted in the Eastern Antarctic showed considerably lower levels of PAHs in Antarctic waters (0.003–0.005 μg/l) [Green et al., 1992]. The authors of this research did not make any special comments about the discrepancy between their results and other available data. However, these differences may reflect the fundamental methodological issues typical for studies of low background levels. In this case, discrepancies with the rest of available data could derive from the use of different procedures of sample filtration. Sample filtration usually precedes the chemical analyses of trace amounts of any substances in natural waters. The lack of coordination of relatively simple pretreatment procedures (e.g., filtration), as well as the use of different reference materials, seems to be a major cause of incomparability of the results of many expensive research and monitoring programs. No doubt, similar reasons explain to some extent the high variability of many data on the oil levels cited in this book.

As to the minimum background level of PAHs in bottom sediments, it most likely varies below 5 μg/kg. Long-term research showed that these levels are typical for sediments on the northwestern shelf of Australia. This region has not been exposed to anthropogenic impact thus far, although oil and gas exploratory activities are planned to begin in the near future [Pendoley, 1992].

From a broader perspective, the data on the background levels of oil contamination are highly uncertain and inconclusive. Hence, wrong interpretations of monitoring results are possible. The lower the measured levels, the higher the probability of inaccurate data. Most likely, one of the ways to solve this problem is accumulation and summarizing of data for each region using modern approaches.

These approaches should take into account the latest requirements for quality assurance of marine measurements [Topping, 1992; ICES, 1995], capabilities of modern analytical equipment (e.g., devices for gas and liquid chromatography, fluorescent spectrometry, mass spectrometry), and methods of identifying hydrocarbons of natural and anthropogenic origin [Oradovski, 1986; GESAMP, 1993; Nemirovskaya, 1994].

4.4.2. Content and Distribution in Seawater

Available data on the seawater levels of oil hydrocarbons in different regions are so variable that finding any reliable average figures is difficult. Such situation is explained both by the complexity of the biogeochemical behavior of oil (variety of forms; dynamism of physical, chemical, and biological processes of dispersion and transformation; existence of natural sources; and so on) and by analytical difficulties and inconsistencies mentioned above. High variability is typical even for the latest available data on oil pollution of the surface seawater, as summarized in Table 25.

This circumstance makes revealing statistically valid average figures even for individual regions, not to mention the total area of the World Ocean, practically impossible. Distinguishing the ranges of hydrocarbon concentrations for the main ecological zones of the hydrosphere seems to be a more reasonable approach. This book makes such an attempt based on the most reliable analytical results recently published. Some of these results are presented in Table 25. Chapter 2 gave typical ranges of oil hydrocarbon concentrations in the surface waters (see Figure 8 and Table 5).

In spite of the wide variability of concentration ranges for each of the distinguished zones, the tendency of oil levels to increase from the ocean pelagic areas to the enclosed seas, coastal waters, and estuaries is still quite obvious. A more detailed analysis of these materials shows that, generally, the distribution of oil hydrocarbons in the World Ocean agrees with other characteristics of the field of global pollution of the marine environment [Patin, 1979, 1995]. Such characteristics include:

- maximum contamination of the euphotic layer;
- patchy distribution of contaminants;
- localization in the upper microlayer;
- depositing in bottom sediments;

Table 25

Concentrations of Oil Hydrocarbons in the Surface Waters of the Seas and Oceans

Area	Concentration (μg/l)		Reference
	PAHs*	TOHs*	
Surface waters of different regions (summarized data)	0.05–0.5	–	Neff, 1993
Oceanic waters (worldwide)	–	1–10	Swan et al., 1994
White Sea	–	20–300	Sapozhnikov, Sokolova, 1994
– " –	–	70	VNIIP, 1994
Baltic Sea	–	1.07–2.95	Bruns et al., 1993
Open waters	–	50	VNIIP, 1994
Open waters	0.004–0.021	–	Witt, 1995
Gulf of Riga and Gulf of Finland	0.02–0.08	–	Nemirovskaya, 1994
Straits	0.01–0.41	1.3–70.0	NSTF, 1994
North Sea	–	1.4–5.4	Bruns et al., 1993
North Atlantic			
(British shelf)	–	0.7–12.0	Law et al., 1992
Coastal waters	–	0.7	
Estuaries	–	12	
North-East Atlantic	–	0.1–0.3	Theobald et al., 1992
Shetland Islands (heavy pollution during an oil spill)	–	100–4,295	MLA, 1993
Barents Sea	–	28–203	Matishov, 1991
South-eastern part	–	100–150	Borisov et al., 1994
Bays of Kola Peninsula	–	140–610	Semenov et al., 1991
Caribbean Sea	–	0.4–93.6	Corbin, 1993
Coastal waters	–	0.4–339.8	Bernard et al., 1995
Atlantic Ocean (coastal waters of Brazil)	–	0.19–43.3	Weber, Bicego, 1991
Sea of Asov	–	100–850	Semenov et al., 1991
– " –	–	150	VNIIP, 1994
Black Sea	–	50–150	Eletski, Hosroev, 1992
– " –	–	50–100	Savin, Podplyotnaya, 1991
Mediterranean Sea	–	10–40	GESAMP, 1993
– " –	0.085	4–15	Nemirovskaya, 1991
Rhone Delta (NW Mediterranean)	0.03–0.05	–	Bouloulassi, Saliot, 1991
Shelf of Corsica	–	1.36–5.97	De Domenico et al., 1994
Coastal waters of Libya	–	4.1–77.6	Hinshery et al., 1991
Caspian Sea	–	70–150	VNIIP, 1994
Northern part	–	50–200	Stradomskaya, Semenov, 1991
Middle and southern parts	–	300–2,000	Kostrov et al., 1991

Table 25 (*continued*)

Area	Concentration (μg/l)		Reference
	PAHs*	TOHs*	
Aegean Sea	–	0.1–10	Balci, 1993
Indian Ocean	0.002–0.046	4–14	Nemirovskaya, 1991
Coastal waters of India	–	3.9–18.1	Kadam, Bhangale, 1993
Persian Gulf	–	16.1–88.5	Madany et al., 1994
Persian Gulf and Gulf of Oman	–	4.4–63.0	Emara, 1990
Arabian Sea (northern part)	–	2–17	Gupta, 1991
Red Sea	0.008–0.070	3–12	Nemirovskaya, 1991
Pacific Ocean			
North-western part	0.04	15	Tkalin, 1991
Shelf of Sakhalin	0.04	4–15	Tkalin, 1993
Coastal waters of China	–	13	Zhijie, 1990
Coastal zone of Patagonia, Argentina	–	0.6–40.9	Esteres, Comendatore, 1993
Strait of Malacca (west coast of Peninsula Malaysia)	–	5–386	Abdullan et al., 1994
– " –	–	29–284	Law et al., 1995
South China Sea	0.05	–	Tkalin, 1991
– " –	–	37–980	Law et al., 1995
Philippine Sea	–	10	Tkalin, 1991
Sea of Japan	0.06	18	Tkalin, 1991
– " –	–	10–50	VNIIP, 1994
Coastal waters of Australia	–	0.2–22.6	Phillips et al., 1992
– " –	0.05–0.41	0.1–180	Swan et al., 1994
Weddell Sea, Antarctic Peninsula	–	0.15–4.65	Weber, Bicego, 1990

Notes: *—PAHs—sum of polyaromatic hydrocarbons, individually identified and measured by the methods of chromatography and mass-spectrometry; TOHs—total oil hydrocarbons extracted by organic solvents, measured by the methods of UV-fluorometry or UV- and IR-spectrophotometry.

- increased levels in the contact zones (water–air, water–sediment) and marginal areas of the World Ocean;
- overlapping the fields of maximum pollution and bioproduction; and
- relative stability of the oil flows and typical levels in the marine environment.

These are the most general characteristics of the global picture of oil pollution. In concrete situations (especially in areas influenced by local input of oil and oil products), the levels and patterns of oil hydrocarbon distribution can differ considerably. At the same time, practically always and everywhere, the concentrations of oil substances in the surface microlayer are considerably (up to 10 times and more) higher than their levels in the water column.

At the end of the 1970s, a large international project of global monitoring of marine oil pollution was conducted. It was based on about 100,000 visual observations and several thousands chemical analyses of water samples from all regions of the World Ocean. The results revealed global distribution of dissolved and emulsified oil hydrocarbons in the surface waters in concentrations of up to several micrograms per liter (in areas with heavy pollution—up to several milligrams per liter). The observations found a high frequency of floating oil aggregates and oil slicks, which covered up to 0.5% of the studied surface of the ocean [Izrael, Tziban, 1988]. Their distribution clearly correlated with the locations of shipping routes, especially with areas of oil tanker transportation and their accidents, as reflected in Figure 33.

This situation, as seen from the data in Table 25, remains generally the same at present, although the levels of oil and oil products have significantly decreased in many regions during the last 5–10 years [Kadam, Bhangale, 1993; GESAMP, 1993]. Oil pollution in all its manifestations is gravitated toward the coastal and shelf waters, estuaries, zones of industrial discharges, land runoffs, and areas of regular shipping and marine oil transportation. The discharges of ballast waters and other oil-containing ship wastes still remain a considerable factor of oil pollution. They sometimes lead to situations when the levels of oil hydrocarbons in the open waters are higher than that in the coastal zone [Corbin, 1993].

Data in Table 25 show that the northern and southern Russian seas (the Barents, Black and Caspian Seas, and the Sea of Asov) are among the most polluted areas of the World Ocean. The levels of chronic oil pollution here reach hundreds and thousands of micrograms per liter. These levels are often an order and more higher than the maximum permissible concentration (MPC) of oil in the water (50 $\mu g/l$) accepted in Russia. Note that in the near future, these areas, primarily the shelves of the Barents and Caspian Seas, can become arenas of large-scale and intensive development of oil and gas resources (see Chapter 1).

Nonaromatic hydrocarbons (paraffins and others) are usually found in seawater in concentrations considerably (an order and more) higher than PAH levels. Nevertheless, the PAHs are typically the object

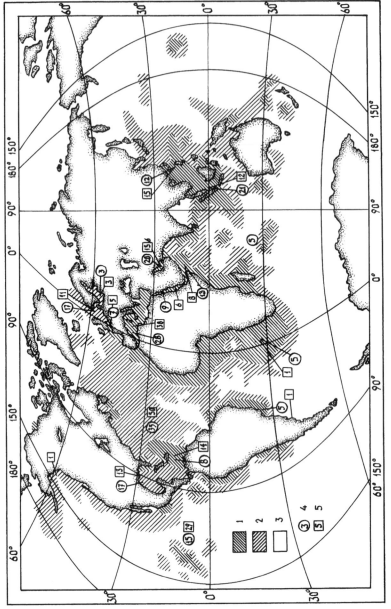

Figure 33 Zones of oil pollution of the World Ocean and areas of tanker accidents (capacity over 6 thousand tons) in 1973–1990: 1—catastrophic oil pollution; 2—critical; 3—not determined; 4 and 5—number of accidents with and without oil spills, respectively [Monina, 1991].

of marine monitoring because they have an increased toxicity and their presence may indicate the anthropogenic origin of oil products.

Especially high concentrations of PAHs are observed in the thin surface microlayer of seawater. For example, in the coastal waters in the Plymouth area (Great Britain), this thin layer contained 100–100,000 $\mu g/l$ of the sum of PAHs. This exceeded the content of the same hydrocarbons in the water column thousands of times [NERC, 1994]. In the offshore Baltic areas enrichment factor (the ratio of PAH concentration in the surface microlayer of 0–2 mm to concentration in the surface water layer of 5–10 m) was greater than 10 [Witt, 1995].

Among all PAHs, benzo(a)pyrene, which displays carcinogenic properties and comes mostly from anthropogenic sources, attracts special attention. The content of this especially hazardous toxicant can reach up to 10% of the sum of all other PAHs. Typical concentrations of benzo(a)pyrene in the open ocean waters and enclosed seas are 0.001–0.01 $\mu g/l$ and 0.01–0.1 $\mu g/l$, respectively. In the coastal waters and in areas of chronic pollution, they can reach up to 0.1–10 $\mu g/l$ [Tziban et al., 1985; Ilnitski et al., 1993].

Besides the analyses of dissolved hydrocarbons, many programs of marine monitoring include observations of the content of floating oil aggregates (petroleum lumps, tar balls, and others) in seawater. Such observations conducted in the 1970s [Polikarpov, Benzhitski, 1975; Nesterova et al., 1979; Benzhitski, 1980] showed the wide distribution of oil aggregates in the marine environment. Their average content in different seas and oceans varied from 0.1 mg/m^2 to 10 mg/m^2. The total amount of solid oil residuals in the open waters of the World Ocean reached about 0.5 million tons by 1980 [Benzhitski, 1980].

These results were generally supported by later observations. In particular, relatively high levels of oil aggregates (up to 10–100 mg/m^2) were found in the Mediterranean Sea [Golik et al., 1988], Indian Ocean [Gupta et al., 1989], Caribbean Sea [Atwood et al., 1987], Gulf of Mexico [Zheng, Van Fleet, 1988], and some other regions [GESAMP, 1993]. After the military conflict in the Persian Gulf, tar residuals and oil slicks covered at least 500 km of the coastal areas of Saudi Arabia. These caused severe damage to marine flora and fauna, especially to populations of sea birds and ecosystems of mangroves and salt marshes [Canby, 1991; McKinnon, Vine, 1991].

The oil aggregates are often found on the seashore, primarily on beaches adjoining areas of marine oil transportation. The results of recent studies [Asuaquo, 1991; Badawy et al., 1993; Corbin et al., 1993; GESAMP, 1993] indicate that typical levels of oil aggregates on the sea beaches vary from 0.04–100 g/m². If the content of oil residuals exceeds 100 g/m², the beach becomes unusable for people according to sanitary-hygienic norms. In some cases, oil pollution of beaches is connected not only with marine oil transportation but with offshore oil and gas production operations as well [Atwood et al., 1987; Corbin et al., 1993].

The last 10 to 15 years have been characterized by some decreasing of global oil pollution of the World Ocean. Most likely, this is resulted from implementation of more rigid international requirements for discharges of oil-containing wastes from ships and oil tankers. At the same time, the presence of increased concentrations of oil hydrocarbons in seawater and bottom sediments of some regions, particularly in coastal zones with slow water circulation, still remains a problem raising serious concerns.

4.4.3. Content and Distribution in Bottom Sediments

Many features of oil hydrocarbon behavior in the water environment, described previously, appear to be typical for their distribution in bottom sediments as well. Table 26 summarizes the data on concentrations of oil hydrocarbons in the upper (0–2 cm) layer of bottom sediments. The table shows that the ranges of hydrocarbon levels in sediments from different regions are rather wide and variable. Similar to the data of seawater analyses, the variability of sediment levels can be partially explained by the analytical inconsistency. Neglecting the standard procedures of data normalization increases the differences caused by diversity in the physical and chemical composition of bottom sediments [ICES, 1994b].

In spite of these limitations, many programs of monitoring marine pollution (including the ones in areas of oil and gas production) give preference to the studying of samples of bottom sediment over samples of seawater. One of the advantages of such approach is connected with the fact that the rate of chemical and microbial degradation of suspended oil hydrocarbons radically drops when they settle on the bottom. A considerable part of these hydrocarbons accumulate in sediments in concentrations of up to hundreds (sometimes thousands and tens of thousands) of times higher than their levels in seawater. The highest levels of oil in bottom sediments are typical in river mouths, deltas, estuaries, bays, gulfs, and ports as well as in areas of regular shipping, oil production, and transportation (see Table 26). This tendency, as previously noted, can be seen at

Table 26
Concentrations of Oil Hydrocarbons in the Surface Layer of Bottom Sediments

Area	Concentration (mg/kg of dry matter)		Reference
	PAHs*	TOHs*	
Barents Sea	0.2–0.4	1.0–10	Ilyin, Petrov, 1994
Kola Bay	–	20–1,950	Kalitovich, 1991
White Sea	–	0.5–2.0	Sapozhnikov, Sokolova, 1994
Baltic Sea (Gulf of Riga and Gulf of Finland)	0.019–0.230	–	Nemirovskaya, 1994
Baltic Sea (muddy sediments)	0.70–1.88	–	Witt, 1995
Baltic Sea (sandy sediments)	0.01–0.97	–	Witt, 1995
North Sea	0.7–2.7	–	Glamer, Fomsgaard, 1993
North Sea (southern part)	0.1–10.0	1–100	Stebbing, Dethlefsen, 1991
– " –	5–90**	100–900**	Stebbing, Dethlefsen, 1991
Beaufort Sea (coastal waters)	0.85	–	Yunker, MacDonald, 1995
North Atlantic	0.002	–	Laane, 1992
Shelf of Newfoundland	0.075–0.5	–	Hellou et al., 1994
Shelf of Canada	0.063–0.936	–	Pelletier et al., 1991
Bay of Sydney on the shelf of Canada	1–100	–	Addison, 1992
Bays, harbors, and ports of Canada and the United States	1–76	–	GESAMP, 1993 (summarized data)
Gulf of Mexico (shelf of southern Florida)	–	1.26–3.44	Snedaker et al., 1995
Gulf of Mexico (Sarasota Bay)	0.02–26	–	Sherblom et al., 1995
Bays, harbors, and ports (Europe, North America)	0.5–700	20–700	Baker et al., 1990 (summarized data)
South Atlantic			
Near Montevideo	0.1–5.0	–	Moyano et al., 1993
Near Rio de Janeiro	–	4.9–9.0	Knoppers et al., 1990
Caribbean Sea (coastal waters of Puerto Rica)	0.5–58.8	5.7–1,891.6	Klekowski et al., 1994
Mediterranean Sea			
Open waters	0.16–0.86	7.1–30.3	Liplatou, Saliot, 1991
Rhone Delta (NW Mediterranean)	1.2–2.4	4.6–350.4	Liplatou, Saliot, 1991

Table 26 (*continued*)

Area	Concentration (mg/kg of dry matter)		Reference
	PAHs*	TOHs*	
Adriatic Sea	0.06–2.58	–	Cariochia et al., 1993
– " –	0.027–0.527	–	Guzella, Paolis, 1994
Ligurian Sea	–	20–214	Danovaro et al., 1995
Arabian Sea (northern part)	–	2–6	Gupta, 1992
Black Sea (near Odessa)	–	3,500	Savin, Podplyetnaya, 1991
Sea of Asov	–	610–1,010	Semenov et al., 1991
Caspian Sea (northern part)		1,000	Stradomskaya, Semenov, 1991
– " –	–	1,000–10,000	Makarov et al., 1995
Indian Ocean (coastal waters of India)	–	40–300	Ram, Kadam, 1991
Gulf of Oman	0.4–0.8	0.2–9.7	Badawey et al., 1993
South China Sea	–	3–1,332	Law et al., 1995
Strait of Malacca	–	52–276	Abdullah et al., 1994
– " –	–	5–47	Law et al., 1995
Pacific Ocean (shelf of Patagonia)	–	0–126	Esteres, Commendatore, 1993
Bering Sea	0.002–0.004	9–32	Sokolova et al., 1995
West Australia	<0.005	–	Pendoley, 1992
East Antarctica	7	–	Green et al., 1992
Antarctica	–	1–171	Cripps, 1992

Notes: *—see *Notes* for Table 25. **—For grain sizes <6 μm.

present even in such remote regions as Antarctica. Local levels of PAHs in sediments of some areas of Antarctica were reported to be much higher than the ones given in Table 26 [Cripps, 1992; Green et al., 1992].

Similar to the situation for surface waters, Table 26 reveals relatively high levels of oil hydrocarbons in bottom sediments of Russian seas, including the Barents, Black, and Caspian Seas and the Sea of Asov. These are exactly the regions where large projects of oil and gas field developments are going to be realized in the near future. For example, in the Northern Caspian Sea in 1991, the content of oil products in bottom sediments reached 5,000 mg/kg in the areas adjoining the Tengiz field. In the areas of oil terminals these levels were as high as 60,000 mg/kg [Stradomskaya, Semenov, 1991].

The most typical and large-scale mechanism of forming oil pollution on the bottom involves river runoff transporting large volumes of suspended material. This suspense is deposited at the river mouth and estuarine areas of the sea creating large alluviums of finely dispersed materials. These serve as peculiar sedimentational traps and accumulate the main mass of river suspense, including the suspended fractions of oil in concentrations usually higher than the levels of dissolved oil compounds.

Practically always and everywhere, hydrocarbon concentrations gradually decrease when moving from the sediment surface to the deeper layers, as shown in Figure 32. Sediments with a higher degree of dispersion and higher levels of organic matter usually trap more oil substances, including both PAHs and aliphatic hydrocarbons. An example of positive linear relationship between the concentrations of PAHs and the percentage of total organic carbon (TOC) is given in Figure 34. According to a rough estimation, PAHs constitute about 10% of the total amount of oil hydrocarbons (mainly aromatic) in bottom sediments. Approximately the same proportions are often found for the relative content of benzo(a)pyrene in the sediments at the background of total PAHs. Sometimes this content reaches up to 30% of PAHs and exceeds the maximum permissible concentration (MPC) for benzo(a)pyrene ($20 \, \mu g/kg$) adopted in Russia, as found in the Riga and Finland Gulfs in the Baltic Sea [Nemirovskaya, 1994].

Benzo(a)pyrene is an indisputable leader among many toxicants due to its carcinogenic properties. The control of its content in the environment, especially in marine sediments, is a necessary

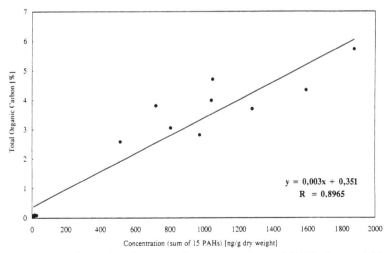

Figure 34 Relationship between the concentration of PAHs (sum of 15 individuals) and the percentage of total organic carbon (TOC) in surface sediments [Witt, 1995].

element of many programs of monitoring [Izrael, Tziban, 1988; Rovinski et al., 1988; Ilnitski et al., 1993; ICES, 1994b]. The background levels of benzo(a)pyrene in bottom sediments in regions distant from pollution sources vary from 0.01 μg/kg to 10 μg/kg of the dry matter [Laane, 1992; Law, 1986; Ilnitski et al., 1993]. Some publications [Law, 1986; Ilnitski et al., 1993] clearly note a drastic increase in PAH levels (up to 100 mg/kg), especially benzo(a)pyrene (up to 8 mg/kg), in zones of regular discharges of oil-containing waste. This is accompanied by obvious symptoms of metabolic disturbances in benthic organisms as well as by the structure and function changes in benthic communities. The latter deserves special attention when assessing the ecological situation in areas of intensive offshore oil and gas developments. Section 4.5 and Chapter 7 will discuss this in more detail.

4.4.4. Accumulation in Marine Organisms

Summarized data in Tables 27 and 28 show great variability of levels of oil hydrocarbons and benzo(a)pyrene in the organs and tissues of marine organisms from different regions and groups. In living organisms, besides the obvious reasons for such variability (diversity of pollution situations, complexity of oil's composition, dynamism of its transformation in the sea, and analytical discrepancy), one more group of factors increase the chaos of results. These include the particular features of biological accumulation and transformation of hydrocarbons in living organisms. Regardless of input sources oil hydrocarbons into living organism (biosorption, assimilation with food, and so on), these substances are immediately involved in a complex chain of biochemical transformations. Enzyme reactions radically change the chemical structure of original oil compounds, converting them into metabolites that do not resemble the parent substances. Such transformations happen on the background of the biosynthesis and biochemical transformations of natural hydrocarbons in the organs and tissues of living organisms.

   This is why the measurement and identification of oil hydrocarbons in biological samples remain a difficult analytical task. To accomplish it, both sophisticated analytical methods (gas and liquid chromatography with high resolution; infrared and fluorescent spectroscopy; mass spectrometry, and others) and sensitive biochemical techniques are used. The latter include methods of measuring the induced enzyme activity (e.g., mixed function oxidase, cytochrome P-450) in the liver of bottom fish as a selective response to PAH presence in the organism. Numerous studies [McDonald et al., 1992; Hellou et al., 1994; ICES, 1994b; 1995] show that these enzyme systems can serve as reliable biochemical indicators of accumulation of PAHs and other hydrocarbons with carcinogenic properties in marine organisms. Methods of assessing the cytochemical disturbances in the digestive cells of mussels are used with the same purpose [Pelletier et al., 1991].

In spite of discouragingly wide ranges of data given in Tables 27 and 28 and the contradictory nature of many publications on the accumulation of petroleum substances in the marine biota, some general conclusions and statements are possible, including:

1. A positive correlation exists between the content of oil hydrocarbons in pelagic and benthic organisms and their levels in seawater and bottom sediments, respectively. The PAH concentrations in the marine organisms are at least 2–3 orders of magnitude higher than their levels in seawater. The ratios between PAH levels in marine organisms and bottom sediments are considerably lower and can be close to 1.

2. The levels of aliphatic hydrocarbons in living organisms (as well as in abiotic components of marine ecosystems) are practically always higher than the PAH content.

3. Water organisms accumulate oil and its fractions due to extraction of suspended and emulsified forms during filtration, biosorption in the organs and tissues contacting with seawater (gills, skin), and assimilation with food. The relative contribution of these channels in actual uptake of oil compounds in different water organisms certainly depends on many factors. These include the systematic characteristics of the organism, its habitat, the type of feeding, the pollution situation, and others.

4. The levels of PAHs and other oil components in organisms ultimately depend upon the balance and ratios between the rate of their input into the organism, the efficiency of biochemical transformation in organs and tissues, and the rate of excretion. Benthic invertebrates (especially bivalves) usually have an increased ability to accumulate oil due to their high filtrational activity, contact with bottom sediments, and less-developed and less-active enzyme and metabolic systems as compared with fish systems. This is why sessile and low active benthic organisms (mussels and others) are often used as standard objects during the marine monitoring [Tavares et al., 1988; Murray et al.,

Table 27

Concentrations of Oil Hydrocarbons in Marine Organisms
(AHs—Aliphatic Hydrocarbons; PAHs—Polycyclic Aromatic Hydrocarbons, BaP—Benzo(a)pyrene)

Organism and Area of Sampling	Concentrations (Wet Biomass)			Reference
	Sum of AHs (mg/kg)	Sum of PAHs (mg/kg)	BaP (μg/kg)	
Coastal waters				
Fish (muscles)	–	n × 0.01	–	GESAMP, 1991 (summarized data)
Mollusks (soft tissues)	–	–	0.6–2,770	–"–
Oysters (soft tissues)	–	n × 10	–	–"–
Mussels (soft tissues)	3–3,500	0.1–1.1*	0–28.5	Baker et al., 1990 (summarized data)
Baltic Sea				
Fish (muscles)	–	–	0.09–1.14	Veldre, 1991
Zooplankton (biomass)	–	–	0.26–21	–"–
Mussels (soft tissues)	–	0.01–0.11	0.1–3.0	Granby, Spliid, 1995
North Sea				
Zooplankton (biomass)	–	35.6–64.6	–	Peters et al., 1992
Ob Bay				
Fish (liver)	–	<10	–	Knyazeva et al., 1991
Caspian Sea				
Sturgeon (muscles)	2.8–4.0	0.3–2.1	–	Kostrov et al., 1991
Sturgeon (gills)	0.73	2.0–2.3	–	–"–
Sturgeon (liver)	0.9–2.7	0.5–0.9	–	–"–

Table 27 (*continued*)

Organism and Area of Sampling	Concentrations (Wet Biomass)			Reference
	Sum of AHs (mg/kg)	Sum of PAHs (mg/kg)	BaP (μg/kg)	
Black Sea				
Fish (muscles)	–	0.4–2.0	–	Mironov et al., 1991
Mussels (soft tissues)	–	0.1–0.2	–	–"–
Mussels (from clean areas)	–	0.02	0.6	Shchekaturina et al., 1995
Mussels (near Sevastopol)	–	0.75	24	–"–
Mediterranean Sea (near Naples)				
Fish (muscles)	–	0.09–1.93	0.44	Cocchieri et al., 1990
Bivalves (soft tissues)	–	0.19–0.30	0–21	–"–
Atlantic Ocean (shelf of Newfoundland)				
Crabs (soft tissues)	–	0.04–0.06	–	Hellou et al., 1994
Crabs (pancreas)	–	0.27–1.28	–	–"–
Atlantic Ocean (shelf of Newfoundland and Labrador)				
Seals and whales (muscles)	–	0.1–5.5*	–	Hellou et al., 1990
Shelf of Canada				
Mussels (soft tissues)	10–440	–	–	Pelletier et al., 1991
Shelf of the Great Britain				
Dolphins (muscles)	–	0.11–2.40	–	Law, Whinnett, 1992
Shetland Islands				
Fish (muscles)**	–	<2.5	–	MLA, 1993
Bivalves (tissues)**	–	>10	–	–"–
Shelf of Spain				
Mussels (soft tissues)	1.57–1.72	–	–	Amejjeires et al., 1994
Zooplankton (biomass)	–	16–245	–	Peters et al., 1994

Table 27 (*continued*)

Organism and Area of Sampling	Concentrations (Wet Biomass)			Reference
	Sum of AHs (mg/kg)	Sum of PAHs (mg/kg)	BaP (µg/kg)	
Shelf of Brazil				
Mollusks (*soft tissues*)	0.2–8.0	0.2–0.7	–	Tavares et al., 1988
North-East Atlantic				
Seals (*muscles*)	–	0.02–0.45	–	Hellou et al., 1991
Canary Islands				
Fish (*muscles*)	8.87–12.81	–	–	Quintero, Diaz, 1994
Gulf of Mexico (shelf of South Florida)				
Fish (*muscles*)	0.095–1.427*	–	–	Snedaker et al., 1995
Shellfish (*tissues*)	0.68–3.20*	–	–	–"–
Corrals	11.1–101.6*	–	–	–"–
Caribbean Sea				
Fish (*muscle*)	–	0.23–1.79	–	Singh et al., 1992
Fish (*skin*)	–	0.28–22.19	–	–"–
Crabs (*muscles*)	–	0.42–0.53	–	–"–
Crabs (*gills*)	–	0.6–16.2	–	–"–
Coastal waters of the USA				
Lobsters (*muscles*)	–	–	0.5–1.6	GESAMP, 1991 (summarized data)
Lobsters (*pancreas*)	–	–	1.6–8.0	–"–
Lobsters (*muscles*)**	–	–	30–40	–"–
Lobsters (*pancreas*)**	–	–	700–1,400	–"–

162

Table 27 (continued)

Organism and Area of Sampling	Concentrations (Wet Biomass)			Reference
	Sum of AHs (mg/kg)	Sum of PAHs (mg/kg)	BaP (μg/kg)	
Persian Gulf				
Fish (muscles)	6.4–32.6*	–	–	Al–Saad, 1990
Bering Sea				
Zooplankton	67–1,016	0.05–1.8	4–150	Sokolova et al., 1995
South China Sea				
Fish (muscles)	–	8.62	–	Lin et al., 1991
Oysters (soft tissues)	–	21.4	–	–"–
West Australia (shelf)				
Oysters (soft tissues)	1.0–4.9	0.01–0.15	<10	Pendoley, 1992
South Australia (shelf)				
Bottom fish (muscle lipids)	7–5,440	0.0004–0.056	–	Nicholson et al., 1994
Bottom fish (liver lipids)	666–2,141	0.0005–0.093	–	–"–
Australia (shelf)				
Mussels (soft tissues)	–	0.04–0.07	0.43–0.96	Murray et al., 1991
Mussels (soft tissues)**	–	0.07–0.37	1–29	–"–
Antarctica (shelf)				
Fish (muscles)	–	0.04–0.08	–	McDonald et al., 1992
Fish (liver)	–	0.15–0.32	–	–"–
Fish (bile)	–	19–69	–	–"–

Notes: *—Data for dry weight. **—From the areas of local pollution.

163

Table 28
Concentrations of Benzo(a)pyrene in Organisms from Different
Seas (from [Ilnitski et al., 1993] with changes)

Area	Marine Organisms	Concentration (μg/kg of dry matter)
Baltic Sea		
Lahepere Bay	algae	1.7
Pärnu Bay	algae	7.3
	zooplankton	0.26–5.1
	fish	0.06–8.50
Gulf of Finland	fish	0.02–4.48
Black Sea		
Near Yevpatoriya	algae	3.2
	bivalves (shell)	traces
	bivalves (body)	15.0
Near Gurzuf	algae	6.0
	jellyfish	71.7
	fish	1.3–3.7
Kalamitski Bay	plankton	107.3
	jellyfish	71.7
	bivalves (shell)	0.4
	bivalves (body)	15.4
Near Karadag	plankton	84.5
	jellyfish	58.6
	bivalves (shell)	traces
	bivalves (body)	4.1
South-eastern part	fish*	0.137–4.492
	mollusks*	0.25–0.60
Barents Sea	micro- and mezoplankton	0.053 ± 0.033
	macroplankton	0.015 ± 0.006
	shrimps	0.005
	fish	0.16
	zoobenthos	0.008 ± 0.003
	mussels (body)	0.72
	mussels (shell)	0.002
Sea of Japan		
Ussuriysk Bay	plankton	131.8
Shamore Bay	jellyfish	81.1
	mussels (body)	33.9
	mussels (shell)	10.5
	scallops (body)	25.8
	scallops (shell)	8.2

Note: *—Concentrations of benzo(a)pyrene are given in μg/kg of wet biomass.

1991; Farrington, Tripp, 1993; Ameijeres et al., 1994], in particular, in areas of oil and gas production [Somerville et al., 1987].

5. The distribution of oil hydrocarbons within marine organisms is extremely heterogeneous. Typically, the concentrations are higher in the organs and tissues that are exposed to contaminants through direct contact with the water and bottom environment as well as in the organs and systems of assimilation and excretion of hazardous substances. Besides, due to the lipophilic properties of oil hydrocarbons, they accumulate in the organs and tissues with an increased content of fats and lipids, including membrane structures. In particular, in fish, high PAH levels are most often found in their liver and bile as well as in gills, gonads, fat deposits, and tissues with high lipid content (see Table 27). In mussels and other invertebrates, high PAH concentrations are typical for digestive glands and reproductive organs enriched by lipids. Marine fish, invertebrates, and mammals often contain oil aggregates in their stomach and digestive tract [GESAMP, 1977; Ambrose, 1994].

6. Most likely, the effect of increasing the toxicant concentrations in marine organisms with increasing trophic level, shown, for example, for heavy metals and chlorinated hydrocarbons, is not typical for most oil hydrocarbons and other substances of oil origin. At the same time, the possibility of such effects should not be excluded for some PAHs, in particular, benzo(a)pyrene. The organisms of the upper trophic levels might accumulate higher amounts of this contaminant due to its stability and affinity to lipids.

Thus, the general picture of interaction of oil hydrocarbons with marine biota includes extremely complex, multifactorial, and variable processes that control the uptake, accumulation, transformation, and removal of these substances from the organisms. The same processes ultimately predetermine the biological effects and consequences of oil pollution in the marine environment, including areas of offshore oil and gas production.

4.5. Effects of Oil on Marine Organisms, Populations, and Ecosystems

The results of even the most advanced and precise chemical analyses *per se* cannot be used as a base for assessing biological effects and ecological hazards of any pollution of the marine environment. Studying such issues requires using methodology of water toxicology, ecotoxicology, applied hydrobiology, and ecology. The research in this direction started over a hundred years ago. At present, every year hundreds of publications are devoted to the problem of biological effects of oil and its components in water environment. This undoubtedly reflects the universal concern about the ecological situation in the hydrosphere.

At the same time, in spite of the relatively long history and the abundance of studies, many key issues of this problem still are the subjects for discussions. Certainly, the complexity of oil composition and behavior in the water environment increases the uncertainty of data on the biological consequences of oil pollution. However, the main problems that complicate reaching a commonly accepted interpretation of results arise from critical importance of possible conclusions. These conclusions ultimately define the ecological standards and requirements for the oil and gas industry, including economical sanctions for the harm caused by pollution.

Clearly, all of the above directly relate to the oil and gas production on the continental shelf. Chapter 7 will discuss in detail the eco-fisheries consequences of offshore oil development. However, the assessment of these issues cannot be complete and objective without consideration and analysis of abundant toxicological information about the oil impact on marine organisms, populations, and ecosystems.

4.5.1. General Characteristics and Mechanisms of Biological Effects

The previous sections described oil and its components in the marine environment using mainly the terms and concepts of analytical chemistry (levels, concentrations, and so on). At the same time, from the ecological perspective, the primary significance should be attached not to the levels of oil pollution *per se* but to the biological effects and consequences caused by these levels.

Extensive scientific literature on marine oil toxicology provides a broad range of answers regarding the degree of biological hazard of oil pollution. Some studies reveal no adverse effects under oil concentrations of up to several milligrams per liter. Others, in contrast, give evidence of metabolic disturbances in water organisms caused by the presence of trace amounts (hundredths and thousandths of a milligram per liter) of dissolved oil hydrocarbons (see, for example, Mironov, 1972; Nelson-Smith, 1977; Patin, 1979; Mikhailova, 1991; Stegeman et al., 1993; GESAMP, 1993). A similar discrepancy in opinions can be found about the environmental risk of oil spills and the hazards of low levels of chronic oil contamination [GESAMP, 1977; Hyland, Schneider, 1978; Simonov, Zubakina, 1979; Izrael, Tziban, 1988; GESAMP, 1989; Howarth, 1991].

Such broad diversity of opinions and results is not surprising if we take into account:

- extremely wide (practically unlimited) range of possible responses of marine biota to the oil exposure (see Table 29);
- complex dynamics and interaction of all biological parameters studied in the laboratory and field conditions (see Figures 35 and 36);
- endless variations of such conditions (toxicant levels, exposure duration, temperature, test organisms, stages of their development, and many others);

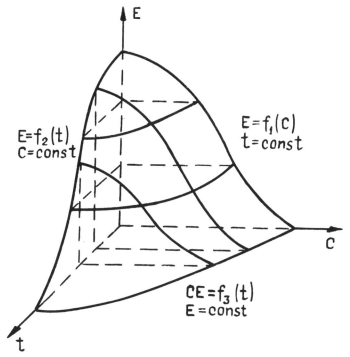

Figure 35 General outline of toxic effect *E* (survival, reproduction, physiological parameters, and other measured end points as compared with control) depending on toxicant concentration (*C*) and the duration of exposure (*t*).

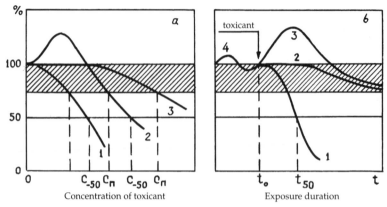

Figure 36 Types of responses of biological systems (in per cent of change of biological, physiological, and biochemical parameters) to toxic exposure and stress impact: a—depending on toxicant concentration in the environment (1—acute effects; 2—stimulation followed by inhibition; 3—slow changes; C_t—threshold concentrations during different types of responses); b—depending on duration of toxic (stress) exposure (1—dynamic of acute toxicity; 2, 3—chronic effect of threshold concentrations; 4—natural variability of biosystem before t_0; t_0—start of toxic exposure; t_{50}—time when 50% decrease of tested parameter is observed; shaded area shows conditionally acceptable (reversible) effects with corresponding C_t) [Patin, 1982].

- variety of properties and toxicity of different oils and their fractions; and
- diversity of bioassay methods and other procedures used for studying biological responses and toxic effects in the water environment.

    After considering these factors, the previously mentioned discrepancies in data and conclusions about the oil levels that cause adverse biological effects will not be so surprising. Instead, this should raise concerns about the ability of modern science to assess the ecotoxicological situation and suggest proper managerial regulations.
    Some of the described difficulties can be overcome by using quantitative criteria to measure the toxicity. These include:

- concentration–response curve describing the relationships between different exposure concentrations of an agent and percentage response of the exposed test population;

- median lethal concentration (LC_{50}), i.e., the concentration that causes the death of 50% of test organisms in acute experiments with a duration from 24 to 96 hours; and
- median effective concentration (EC_{50}), i.e., the concentration that causes some sublethal response in 50% of test organisms.

The data in Figure 37 give some idea about the values of LC_{50} of oil hydrocarbons for marine organisms and about their correlation with the body size. Figure 38 shows the correlation between the survival time and the size of developing eggs of marine invertebrates in acute experiments with diesel fuel [Patin, 1979]. In both cases, the resistance of tested organisms and eggs declined with decreasing size. These results possibly reflect the general tendency of increased sensibility and vulnerability of smaller organisms (including embryos and larval stages) to toxic impacts, in this case—to the presence of oil hydrocarbons. At the same time, some data do not agree with this tendency [Anderson, 1985].

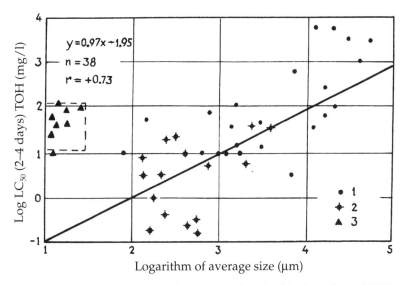

Figure 37 Correlation between the toxicity of total oil hydrocarbons (TOH) for marine organisms and organisms' body size: 1—adult specimens; 2—early stages of development; 3—EC_{50} (effective median concentrations) for unicellular algae; n—amount of data; r—coefficient of correlation (dotted line marks data not included in regression equation) [Patin, 1982].

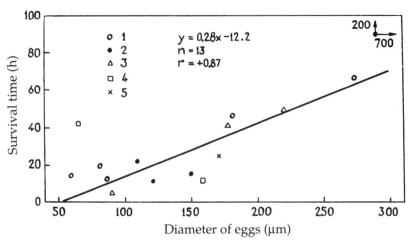

Figure 38 Correlation between the survival time of developing eggs of marine invertebrates and the eggs' size under exposure in 0.5% solution of diesel fuel in seawater: 1—Mollusca; 2—Echinodermata; 3—Annelida; 4—Urochordata; 5—Arthropoda (*Balanus cariosus*) [Patin, 1978].

4.5.2. *Toxic and Threshold Concentrations in Seawater*

The diagram in Figure 39 presents experimental data on the effects of dissolved oil on marine organisms of different systematic and eco-logical groups. This diagram is based on summarized publications [Patin, 1979; Patin, 1982; 1995] and the data in Table 29.

The wide range of toxic and threshold concentrations (i.e., low-est observed effect concentrations) of dissolved oil hydrocarbons for the studied groups of marine biota is quite evident. It probably re-flects not only the variety of conditions and methods of toxicological studies but also the differences in responses of different species of water organisms. These have been repeatedly observed during field studies in areas of heavy oil pollution [Nelson-Smith, 1977; Hyland, Schneider, 1978; Mazmanidi, 1991; Yankyavichus et al., 1992]. The most radical and chronic changes of species composition and structure are found in benthic communities dwelling on oil-polluted sediments. In particular, such changes are typical for some areas of long-term offshore oil and gas productions. Chapter 7 will discuss this further.

The data in Figure 39 and Table 29 (as well as Figures 37 and 38) clearly show the increased vulnerability (sensitivity) of early

Table 29

Effects of Oil Hydrocarbons on Marine Organisms (Laboratory Studies)
(TOH—Total Oil Hydrocarbons; PAH—Polycyclic Aromatic Hydrocarbons; BaP—Benzo(a)pyrene)

Marine Organisms	Concentration (μg/l), Oil Toxicant, Exposure (days)	Effects	Reference
Unicellular algae:			
Cultures of mass species and natural phytoplankton (Baltic, Black, Caspian, Mediterranean, and Red Seas)	50–500 TOH 1–5	Inhibition of photosynthesis and photoluminescence, reduced division rate	Patin, 1982
Natural phytoplankton (Barents Sea)	100 TOH 1–5	Reduced primary production, biomass, and number of cells	Khromov, 1977
Natural phytoplankton	100 TOH 1–5	Changes in species composition, reduced primary production	Jankyavichus et al., 1992
Macrophytes:			
Polysiphonia breviarticula	100 Crude oil 5	Dying off sprouts	Mironov, 1972
Crustaceans:			
Acartia tonsa (nauplius)	100–400 TOH 1–7	Reduced survival	Patin, 1982

Table 29 (*continued*)

Marine Organisms	Concentration (μg/l), Oil Toxicant, Exposure (days)	Effects	Reference
Niphargoides maeotcus (nauplius)	100–400 TOH 8–40	Reduced survival	Patin, 1982
Homarus americanus	100–1,000 Crude oil 15	Swollen chromatophores	SEPA, 1990
Corophium bonelli	30–40 Chronic exposure	Reduced larval survival and density of settling on the bottom *in situ*	Bonsdorff et al., 1990
Crustaceans (8 species, Arctic)	100–500 Naphthalene 4	50% mortality in acute tests	Anderson, 1985
Zooplankton (Arctic)	10–1,000 TOH	Sublethal responses	Perey, Wells, 1985
Zooplankton (Narragansett Bay, Rhode Island, USA)	40 PAH 180	Changes in abundance and species structure	NRC, 1985
Zooplankton (Baltic Sea)	1–10 TOH Chronic exposure	Changes in species structure	Jankyavichus et al., 1992
Zooplankton (North Sea)	TOH in the platform discharges 100	–"–	Davies et al., 1981

Table 29 *(continued)*

Marine Organisms	Concentration (μg/l), Oil Toxicant, Exposure (days)	Effects	Reference
Bivalves:			
Mass species of the Caspian Sea (*Monodacna caspia*, *Didacna trigonoides*, *Cerastoderma lamarcki*)	50–100 TOH 60	Reduced survival	Patin, 1982
Mytilus edulis	30 PAH 38–182	Altered structure and function of lysosomes	SEPA, 1990
Mytilus edulis	10–100 TOH	Reduced filtrational activity	Hyland, Schneider, 1978
Mytilus edulis	125 TOH 120	50% mortality	Widdows, 1992
Mytilus edulis (larvae)	25–35 Diesel fuel 10	Reduced growth rate, 50% mortality	Stroemgren, Nielson, 1991
Nassarius obsoletus	1–4 Kerosene	Abnormal feeding behavior	Hyland, Schneider, 1978
Mercenaria mercinaria	100 Phenol 1	Cellular damage to gills and digestive gland	SEPA, 1990
Mussels, oysters	1–100 Crude oil	Oil flavor and odor	Hyland, Schneider, 1978

Table 29 (*continued*)

Marine Organisms	Concentration (μg/l), Oil Toxicant, Exposure (days)	Effects	Reference
Annelids:			
Ophryotrocha labronica (larvae)	100 TOH 15	Reduced survival	Patin, 1982
Nereis diversicolor	100–500 TOH 8–26	– " –	Patin, 1982
Echinoderms:			
Paracentrotus lividus	4.5 BaP 3 hours	Abnormal cell division in embryos	SEPA, 1990
Benthic communities of invertebrates:			
Zoobenthos	190 TOH 175	Reduced abundance and species diversity, damages of larvae	Grassle et al., 1981
Zoobenthos	30–40 TOH Chronic exposure	Suppression of larvae ability to settling on the bottom	Bonsdorff et al., 1990
Fish:			
Salmo trutta caspicus (developing eggs and larvae)	10–20 TOH 60–70	Reduced survival, abnormal development	Patin, 1982
Rhombus maeoticus (developing eggs)	10–100 TOH 2–3	Morphological and physiological disturbances	Patin, 1982
Platichthus flesus (developing eggs)	50–100 TOH	Reduced survival, abnormal development	Mazmanidi, 1991

Table 29 (*continued*)

Marine Organisms	Concentration (μg/l), Oil Toxicant, Exposure (days)	Effects	Reference
Clupea harengus pallas (developing eggs)	0.9 Benzene 1	Delays of embryonic development	SEPA, 1990
Clupea harengus (larvae)	26–62 TOH 1–7	Reduced respiration rate under 15°C	Galleogo et al., 1995
Fundus het.roclitus	200 Naphthalene 15	Cell pathology in sense organ of lateral line	SEPA, 1990
Hypomesus pretiosus (developing egg)	54–113 Crude oil 3–21	Nervous cell death in embryo	SEPA, 1990
Menidia menidia	140 Crude oil 7	Enlargement of olfactory organ	SEPA, 1990
Scophthalmus maximus (larvae)	1–10 BaP 9	Reduced survival, disturbances of enzyme systems	Peters et al., 1992
Pollachius virens, *Gadus morhua* (developing eggs)	50 TOH 2–14 hours	Reduced survival	Foeyn, Serigstad, 1989
Salmo salar (juveniles)	200 Crude oil 30	50% mortality	Borisov et al., 1994

175

Table 29 (*continued*)

Marine Organisms	Concentration (μg/l), Oil Toxicant, Exposure (days)	Effects	Reference
Gadus morhua	50–300 Crude oil 84–91	Delays of gametogenesis, histological changes in gonads	Karataeva, 1993
Gadus morhua	50–100 TOH	Avoiding behavior	Bohle, 1982
Gadus morhua (developing eggs and larvae)	50–100 TOH 14	Reduced growth rate, morphological and behavioral anomalies	Tilseth et al., 1981
Gadus morhua	150–300 TOH 91	Reduced growth rate, gametological and histological anomalies	Khan et al., 1981
Gadus morhua (developing eggs and larvae)	30–200 TOH (heavy fraction) 10–28	Reduced growth rate, abnormal feeding behavior	Solberg et al., 1981
Gadus morhua (larvae)	50–280 TOH 1	50% decrease in oxygen consumption	Saragstad, 1987
Gadus morhua (larvae)	120–340 TOH	Disturbances of fatty acids metabolism	Barnung, Grahl-Nielson, 1987
Fundulus hetroclitus (larvae)	<100 TOH 30	Morphological anomalies	Linden et al., 1980

176

Table 29 (*continued*)

Marine Organisms	Concentration ($\mu g/l$), Oil Toxicant, Exposure (days)	Effects	Reference
Herring (larvae)	25 TOH Acute tests	Reduced survival	Law, Hudson, 1986
Acipenser baeri Brandt	10 TOH	Avoiding behavior	Palatnikov, Mamedov, 1988
Marine fish (larvae, fry)	10–100 TOH	Sublethal responses	Borisov et al., 1994
Fish of the Barents and White Seas (cod, salmon, haddock)	25–35 TOH	Threshold responses of respiratory system	Shparkovski, 1993
Commercial species of fish (developing eggs, larvae)	<1 PAH Hours–days	Development anomalies	Rice, 1985
Commercial species of fish	1–100 Crude oil	Oil odor and flavor	Mironov, 1972; Hyland, Schneider, 1978; GESAMP, 1993

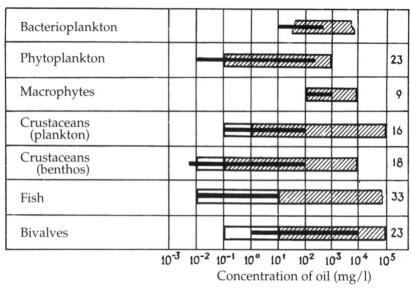

Figure 39 Ranges of toxic (rectangles) and threshold (bold lines) concentrations of dissolved fractions of oil for main groups of marine organisms (not-shaded sections—ranges of toxic concentrations for early life stages; figures at right—number of studies species). Toxic concentrations—limits of concentrations which cause more than 50% decrease of tested parameters as compared with control in the tests with duration no less than 2–4 days. Threshold concentrations—minimum levels in seawater which cause up to 50% decrease of tested parameters in the tests comparable in duration with the life span of tested organisms [Patin, 1982].

developmental stages of most studied species of fish and invertebrates. Toxic concentrations of oil (causing lethal effects or irreversible disturbances of vitally important functions) for eggs (embryos), larvae, and young marine animals are usually considerably lower than for adult organisms. Minimum levels equal approximately 10^{-2}–10^{-1} mg/l. Similar figures are given in some other summarized publications [Capuzzo, 1987; SEPA, 1990; Neff, 1993]. These concentrations, as Figure 39 shows, overlap with threshold levels that cause physiological, behavioral, and some other reversible anomalies but do not usually cause the death of the organism.

The early life stages are not the only sensitive phases of ontogenesis. Critical stages also include the periods of metamorphoses in marine organisms. For example, the sensitivity of copepods and other crustaceans to oil's toxicity increases several times during exoskeleton formation [Anderson, 1985].

Among the environmental factors that affect the nature and consequences of oil's impact on water organisms, the leading role belongs to temperature. The intensity of physiological processes in the organism usually increases with the temperature. As a rule, this results in a more rapid adverse effect of most toxicants under high temperature. At the same time, with oil hydrocarbons, the possibility of a reverse correlation between the hazardous effects and temperature exists. For example, eight species of invertebrates (mainly crustaceans) from Alaska were less resistant to the naphthalene impact at 4–8°C than the same species from a warmer region tested at 22°C [Anderson, 1985]. The likelihood of increased sensitivity of the Arctic specimens is questionable and requires additional studies. More probable, the negative correlation between temperature and adverse effects in this study can be explained by higher rate of decomposition and evaporation of the most toxic oil fractions under higher temperature.

4.5.3. Toxic and Threshold Concentrations in Bottom Sediments

The review of biogeochemical properties and behavior of oil in the marine ecosystems showed that oil substances concentrate at the water-atmosphere and water-bottom sediment boundaries (in the biotopes of hyponeuston and benthos, respectively). This fact suggests that the main events ultimately defining the biological consequences of marine oil pollution take place at these boundary habitats. The results of experimental and field observations describing the responses of benthic organisms to the presence of oil hydrocarbons in bottom sediments are especially valuable.

Unfortunately, information on this issue is rather limited. Incorporating the methodology of bottom sediment toxicity tests into marine monitoring via some national and international programs [Bewers, Karau, 1993; Matthiessen, 1994; ICES, 1994b] gives some hope of filling this gap in marine ecotoxicological knowledge.

The results of rather scanty (compared with the data for seawater) experimental and field observations, given in Table 30, indicate that lethal effect on benthic organisms occurs when the oil level in bottom sediments ranges within 1,000–7,000 mg/kg. Sublethal and threshold effects (impairment of feeding, behavior, physiological functions, as well as pathological changes in organs and tissues, including carcinogenic effects) usually appear at concentrations of 100–1,000 mg/kg. For the most toxic components and fractions of oil, especially for PAHs, these effects are possible under even lower levels—from 10–100 mg/kg.

Table 30
Toxic Effects of Oil Pollution of Bottom Sediments on Benthic Marine Organisms

Marine Organisms	Concentrations (μg/kg of dry matter), Oil Toxicant, Exposure (days)	Effects	Reference
Invertebrates:			
Uca pugnas (crab)	1,000–7,000 Hydrocarbons Chronic exposure	Death, sublethal effects	Anderson, 1985 (summarized data)
Melita nitida (amphipod, juveniles)	100 Crankcase oil <2	Death	Borowsky et al., 1993
Ampelisca (amphipod, ostracod)	23 Fuel oil 14	86% decrease in abundance	Widbom, Oviatt, 1994
Schizopera khabeni (copepod)	10 PAH mixture 4–14	Sublethal effects	Lotufo, Fleeger, 1995
Macoma baltica (bivalve)	640–3,890 Oil Chronic exposure	Death	Anderson, 1985 (summarized data)
_"_	0.2–0.4 Naphthalene From 1 hour to 3 days	Death, impaired burying ability	_"_

180

Table 30 (continued)

Marine Organisms	Concentrations (μg/kg of dry matter), Oil Toxicant, Exposure (days)	Effects	Reference
Macoma inquinata (bivalve)	1,200 Oil Chronic exposure	Death, impaired burying ability	Anderson, 1985 (summarized data)
Protothaca staminea (bivalve)	1,200 Oil Chronic exposure	Physiological and biochemical disturbances	-"-
Bivalves	100–150 Oil hydrocarbons Chronic exposure	Tumors	Ilnitski et al., 1993
Oysters	>0.010 Oil	Oil odor and flavor	Laevastu, Fukuhara, 1984
Abarenicola pacifica (polychaete)	500–1,000 Oil Chronic exposure	Disturbances of feeding, reduced glycogen content	Augenfeld, 1980
Nereis diversicolor (polychaete)	100 Oil Chronic exposure	Death	Anderson, 1985
Nereis virens	74–5,200 Oil	Behavioral disturbances, impaired burying ability, death	Olla et al., 1984

Table 30 (*continued*)

Marine Organisms	Concentrations (μg/kg of dry matter), Oil Toxicant, Exposure (days)	Effects	Reference
Benthic Organisms and Communities	50–700 Oil and its fractions	Reduced survival and reproduction, disturbances of community structure	Law, Blackman, 1981 (summarized data)
	2 Naphthalene	-"-	-"-
	>2 Naphthalene Chronic exposure	Stress effects in zoobenthos	Armstrong et al., 1979
	<2 Naphthalene Chronic exposure	-"-	Moore, 1983
	20–400 Oil Chronic exposure	Reduced abundance and species deversity	Stebbing, Dethlefsen, 1991
	10–100 Oil hydrocarbons Chronic exposure	Decrease in species diversity	Reiersen et al., 1989
	50–60 Oil Chronic exposure	-"-	Kingston, 1992
	0.01–0.1 Naphthalene, benzo(a)pyrene Chronic exposure	No effects observed	ICES, 1994

Table 30 (continued)

Marine Organisms	Concentrations (μg/kg of dry matter), Oil Toxicant, Exposure (days)	Effects	Reference
Bottom Fish	10–100 Aromatic hydrocarbons Chronic exposure	Increased frequency of tumors and diseases	GESAMP, 1993 (summarized data)
	500 Oil Chronic exposure	No effect of oil avoiding observed	Moles et al., 1994
	4–300 Oil Chronic exposure	Oil odor and flavor	Laevastu, Fukuhara, 1984
Flounder	1–100 PAH Chronic exposure	Specific induced activity of liver enzyme systems	Addison, 1990
Pseudopleuronectes americanus	<1,000 PAH Chronic exposure	Biochemical changes in fish liver	Pyane et al., 1988
Limanda limanda	210 Oil hydrocarbons 144	Biochemical and cytological changes in fish liver	Moore et al., 1991
Mangroves *Rhizophora mangle*	10–60 PAH Chronic exposure	Mutagenic changes in photosynthetic system	Klekowski, 1994

This was shown, in particular, in a detailed study of the oil pollution of bottom sediments in some bays, ports, and harbors on the western coast of North America [GESAMP, 1991]. This research found that the concentrations of PAHs (mainly transformed high-molecular-weight products of incomplete combustion of heavy fractions of oil) in bottom sediments varied within a wide range (1–76 mg/kg), depending on the anthropogenic press on the studied area. The frequency of tumor formations and other symptoms of carcinogenic pathologies in bottom fish in polluted areas directly correlated with the levels of PAHs.

Mutagenic disturbances, induced by the presence of PAHs in bottom sediments (0.1–58.8 mg/kg), were also found in the mangroves of the coastal zone of Puerto-Rico [Klelowski et al., 1994]. Figure 40 shows a direct and strong correlation (correlation coefficient

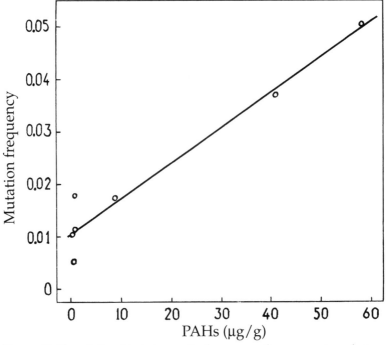

Figure 40 The relationship between the frequency of mangrove trees heterozygous for recessive (chlorophyll-deficient) mutations in a population and the concentration of polycyclic aromatic hydrocarbons (PAHs) in the underlying sediment [Klekowski et al., 1994].

0.94) between the frequency of recessive mutations in the coastal mangroves and the level of oil pollution in bottom sediments.

Species and populational responses to oil pollution of bottom sediments are revealed ultimately on the level of benthic communities. Numerous studies show that their structural and functional characteristics can be considerably disturbed under conditions of chronic oil pollution. Similar disturbances in benthos occur in areas of discharges from the offshore oil and gas activities, as Chapter 7 will discuss.

4.5.4. Oil as a Multicomponent Toxicant

The main ecotoxicological characteristic of oil is the extreme complexity and variability of its composition in both abiotic components and biota. From the first seconds of contact between oil and the marine environment or living organism, complex chains of physical, chemical, and biochemical transformations begin. The progress, rate, and biological effects of these transformations differ for different oil fractions and compounds. Oils from natural and man-made sources, just-extracted crude oils, weathered oils, and oils from different natural fields, all have considerably different characteristics, including toxic properties.

From the ecotoxicological perspective, oil in the marine environment can be considered as a multicomponent complex toxicant that consists of hundreds of substances (mainly hydrocarbons) with changeable composition and properties. The most important feature of oil's toxic impact is its integrated (non-selective) nature. Every vital function, process, mechanism, or system within a living organism is affected in one or another way by oil in the marine environment. On the other hand, it is practically impossible to distinguish any specific function or system within an organism that shows a selective response to the oil exposure (as it happens, for example, in the case of organophosphate exposure).

At the same time, we should remember the general dependence of toxic properties of all substances on their structure [Tinsly, 1982]. In particular, oil hydrocarbons with complex and branched molecules, especially those with a cyclic structure, are usually more toxic than the ones with simpler molecules and straight chain of carbon atoms. Similarly, increasing molecular weight within a homologous series of hydrocarbons is associated with a proportional increase in toxicity. These specifics of molecular structure explain an increased toxicity of aromatic hydrocarbons, especially of high-molecular-weight PAHs. These substances are among the main objects of marine ecotoxicological studies. Most other hydrocarbon compounds either do not dissolve in water well (e.g., paraffin hydrocarbons) or evaporate quickly (e.g., monoaromatic hydrocarbons.)

Table 31 gives a general idea about the acute toxicity of some oil fractions and hydrocarbons. Figure 41 illustrates the phenomenon of increasing the toxic properties of aromatic hydrocarbons with increasing molecular weight and structural complexity. Usually, the simultaneous presence of several oil hydrocarbons does not cause a synergetic effect, i.e., the toxicity of their mixture does not exceed the toxicity of each component [Neff, 1993].

Table 31
Toxicity of Oil and Oil Hydrocarbons for
Developing Eggs of Cod (96-hour LC_{50})
[Falk-Petersen, Kjorsvik, 1987]

Hydrocarbon	LC_{50} (mg/l)
Oil	
Crude	about 24
Weathered	1–5
Photoxidized:	
by sun light	>4
by artificial light	about 4
1,3-Dimethylnaphthalene	0.1–0.3
2-Methyl-1-naphthol	<0.3
1-Methylnaphthalene	0.3–1.0
2-Methylnaphthalene	0.3–1.0
1,2-Dimethylnaphthalene	0.3–1.0
1-Naphthol	1.25
Naphthalene	1–3
1,4-Dimethylnaphthalene	1–3
1,8-Dimethylnaphthalene	1–3
2-Naphthol	>3
2,4-Dimethylphenol	3.7
2,6-Dimethylphenol	4
Xylenes	4
4-Methylphenol	5
2-Methylphenol	12
2,3-Dimethylphenol	13
2,5-Dimethylphenol	13
3,4-Dimethylphenol	13
Phenol	>30
3-Methylphenol	>30
3,5-Dimethylphenol	>30

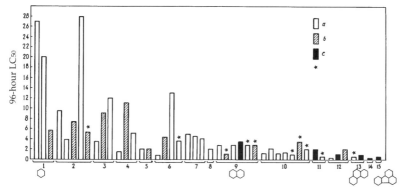

Figure 41 Toxicities of aromatic hydrocarbons (96-hour LC_{50}) depending on their structure for marine organisms of different groups: a—crustaceans (shrimps, copepods, amphipods); b—fish (salmon fry, striped bass); c—polychaetes; *—24-hour LC_{50}; 1—benzene; 2—toluene; 3—n-xylene; 4—o-xylene; 5—p-xylene; 6—ethylbenzene; 7—trimethylbenzene; 8—tetramethylbenzene; 9—naphthalene; 10—methylnaphthalene; 11—trimethylnaphthalene; 12—fluorene; 13—phenanthrene; 14—methyl-phenenthrene; 15—fluorantene (from [Anderson, 1985] with changes).

Within a large group of PAHs, such high-molecular-weight hydrocarbons as benzo(a)pyrene, benzo(a)anthracene, and analogous substances should be distinguished. Although their carcinogenic and mutagenic properties are rather well studied in mammals, water toxicological research of these pollutants is very limited. Evidence indicates that they can cause mutagenic and carcinogenic effects in a number of marine organisms [Bogdashkina, Petrosyan, 1988; GESAMP, 1991; 1993; Ilnitski et al., 1993]. Benthic and demersal forms, including many species of fish, living in contact with polluted bottom sediments, are especially sensitive to these substances. In particular, laboratory and field observations [Anderson, Gossett, 1986] found that at PAH concentrations of 3–5 mg/kg and higher, bottom fish showed obvious symptoms of tumor damage and developed other carcinogenic and mutagenic changes in the organs and tissues.

4.5.5. Methodological Aspects of Marine Ecotoxicology of the Oil

The data given in Figures 37–39 and 41 as well as in Tables 29–31 reflect the variety of responses of marine organisms to oil exposure under experimental (mainly laboratory) conditions. Such experiments help to reveal the mechanism of oil intoxication, screen individual substances according to their toxicity, find the species-specific responses to oil presence, and study many other important problems. Nevertheless, this approach is a target of certain criticism because it often involves an extrapolation of laboratory results to the natural conditions. The long-term experience of the Russian Federal Research Institute of Fisheries and Oceanography [Patin, 1979; 1988] indicates that such extrapolations can be quite valid when used to estimate threshold levels of environmental contamination. In particular, in Russia, laboratory experiments are widely used to establish the maximum permissible concentrations (MPC) for different toxicants in the water environment. At the same time, the extrapolations of laboratory results are questionable when used to explain the mechanisms of cumulative toxic effects in populations and communities exposed to chronic pollution.

A number of new approaches in marine ecotoxicology developed over last 20 years successfully combine the advantages of laboratory and field studies. In particular, they provide an opportunity to conduct controlled field studies, including ecotoxicological research *in situ* under conditions that are very close to the natural situations. These studies involve exposing mesocosms in the water column or on the bottom. Mesocosms represent natural communities and their biotopes (water, sediments) isolated from the environment with the help of containers made from transparent materials. The volumes of these containers can be up to 2,000 m^3 [Patin, 1979; Grassle et al., 1981; NRC, 1985; Gray, 1987; Howarth, 1991; GESAMP, 1993]. The results of such experiments, partially included in Table 29, give unique information on the structural and functional responses of marine populations and communities to long-term (up to 6 months) exposure under low levels of oil. Such studies showed, in particular, the increased vulnerability of benthic and plankton crustaceans (especially small forms) to the oil impact. At the same time, bivalve mollusks and some other benthic invertebrates (e.g., polychaetes) were relatively resistant. The domination of these resistant organisms in benthic communities can serve as an indicator of critical levels of oil pollution.

Recently, the methodology of marine ecotoxicological studies has been enhanced by a group of sensitive and selective methods. These methods assess sublethal effects by measuring biochemical changes in enzyme and immune systems (biomarker techniques). In particular, these involve registration of mixed function oxidase induction in the microsomal fractions of the fish liver [Addison, 1992; Peters et al., 1992; Stegeman et al., 1993; Sergeev, Bogovski, 1993].

Laboratory experiments and field observations revealed a clear correlation between some biochemical parameters of fish and invertebrates and PAH concentration in seawater and bottom sediments. Most often, these parameters included the level of cytochrome P-450, activity of enzyme systems, and the integrity of lysosomes.

Biomarker methodology is becoming an important element in a number of national and international programs of marine monitoring [Addison, 1992; Stegeman et al., 1993; ICES, 1994b; 1995]. Possibly, it will provide new data and ideas for solving complex and controversial problem of assessing cumulative biological effects under chronic oil contamination of the sea.

Biochemical transformations of some xenobiotics in living organism can lead to the formation of substances with higher toxicity than that of the original compounds. Experiments have proven that biochemical decomposition of some oil fractions can also form metabolites more toxic than the original hydrocarbons. This happens in the cytochrome P-450 system in the liver of some animals [Archakov et al., 1988]. Evidence indicates that these biochemical mechanisms explain the high toxicity and carcinogenic properties of substances such as benzo(a)pyrene. These mechanisms can also increase the toxicity of lower-molecular-weight oil hydrocarbons. Something similar, that is increasing oil toxicity, also happens in the marine environment as a result of photochemical degradation of oil under the impact of solar radiation [Bogdashkina, Petrosyan, 1988; GESAMP, 1993].

To neutralize the harmful effects of a toxicant, organisms develop a complex chain of compensatory processes, beginning from primary biochemical responses at the subcellular (molecular) level. At each level of biological hierarchy (subcellular, cellular, organismal, populational, and ecosystem), one of two scenarios may be realized. First, toxic effects can be compensated and eliminated. Second, they can be transferred to the next level where the counteraction between the hazardous effects and protective response mechanisms repeats but in a different form specific for this level.

Under natural conditions, all these responses (metabolic, trophic, behavioral, etc.) are tangled together in a complex mosaic of organismal, populational, and community effects. It may even lead to integral response of the whole ecosystem. The final result of these complicated processes is a structural and functional changes in the marine ecosystem, usually described by indexes of species diversity and dominance, abundance, biomass, and production. The analysis and assessment of these changes at the background of natural dynamics of ecological processes are extremely difficult. Usually, reliable results are possible only if the studies are conducted in areas with relatively high levels of oil pollution. Chapter 7 will discuss such studies in the areas of offshore oil and gas developments.

4.6. Levels, Thresholds, and Zones of Manifestation of Biological Effects

The abundance and diversity of toxicological and biogeochemical data discussed above complicate their integral comprehension. Graphic generalization of these materials can help to overcome this difficulty. Figures 42 and 43 show the outcome of such an attempt in

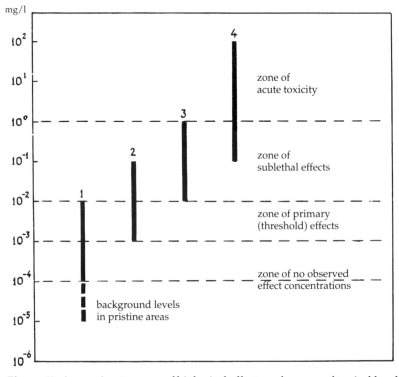

Figure 42 Approximate zones of biological effects and ranges of typical levels of dissolved (mainly aromatic) oil hydrocarbons in seawater (concentrations above 1 mg/l include both dissolved and emulsified forms of oil): 1—pelagic areas (open waters); 2—coastal areas; 3—bays, estuaries, ports, and other shallow semi-closed areas; 4—areas of heavy local pollution (oil spills, discharges, etc.).

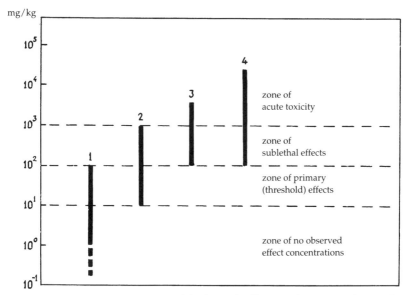

Figure 43 Approximate zones of biological effects and ranges of typical levels of oil hydrocarbons in bottom sediments: 1—pelagic areas (open waters); 2—coastal areas; 3—bays, estuaries, ports, and other shallow semi-closed areas; 4—areas of heavy local pollution (oil spills, discharges, etc.).

relation to seawater and bottom sediments. These graphs make it possible to compare the levels of oil hydrocarbons causing particular biological effects with actual levels of oil pollution found in different areas of the World Ocean. Certainly, such data presentation has the approximate nature and may be supplemented and improved by new materials.

The results of many studies, including the ones previously discussed, indicate that the upper limit of the zone of no observed effect (safe) concentrations of dissolved oil hydrocarbons equals approximately 10^{-3} mg/l. Most likely, the hazardous effects under lower concentrations, noted by some researchers [Rice, 1985], could be explained by the presence of individual fractions with higher toxicity. In any case, this possibility does not agree with the bulk of available toxicological data. The zone of no observed effect concentrations distinguished on the graph covers pelagic areas (open waters) of the seas and oceans located far from the shore. At present, these areas hardly experience any hazardous consequences of background oil contamination.

The next zone on the concentration scale (10^{-3}–10^{-2} mg/l) can be conventionally called the zone of primary threshold effects. It covers relatively clean pelagic and coastal waters where toxicological studies indicate the possibility of reversible (mainly physiological and biochemical) responses to the presence of oil hydrocarbons. The effects caused by these levels are usually compensated at the cellular and organismal levels. As observations under natural conditions show, they do not lead to any obvious biological consequences in the sea and, as a rule, cannot be detected in actual situations of oil contamination [GESAMP, 1993; Neff, 1993].

This zone of primary threshold effects is an intermediate one between the zone of no observe effect levels and the zone of obvious sublethal levels. It is here—within the range of 10^{-3}–10^{-2} mg/l—where one should look for values of the maximum permissible concentrations (MPC) of oil hydrocarbons in seawater. Determining levels of permissible pollution is a complex, debatable, and important area of applied water toxicology. In the former USSR and now in Russia, the environmental MPC for the over 1,500 different pollutants have been established [Register of MPC, 1995]. These MPC are based on the results of acute and chronic experiments with organisms of different trophic levels conducted according to a standard procedure [Patin, 1988; Patin, Lesnikov, 1988]. The official standard of MPC for dissolved hydrocarbons (0.05 mg/l) adopted about 30 years ago was established based on the results of experiments with freshwater organisms and then extrapolated to the marine environment. The research data, including the ones given in Table 28, suggest that this standard could be changed to 0.01 mg/l of dissolved oil hydrocarbons in seawater.

Next on the concentrational scale are the zones of sublethal and lethal effects, 10^{-2}–1.0 mg/l and above 1.0 mg/l, respectively. These are typical for gulfs, estuaries, ports, harbors, bays, and other shallow coastal areas with slow water circulation and increased levels of chronic oil pollution. These levels also occur in areas of oil discharges and accidental spills.

Using the data on oil content in bottom sediments (see Table 26) and its toxicity for benthic organisms (see Table 30), a graph similar to the one for seawater can be created. Figure 43 shows the approximate limits (zones) of oil concentrations in sediments causing possible effects in benthic biota in relation to typical oil levels in bottom sediments of different areas. Zone of no observed effect concentrations is defined by upper limit of approximately 10 mg/kg of oil in sediments, while acute toxic effects are usually inevitable at the oil levels higher than 1,000 mg/kg.

Suggested ecotoxicological diagrams summarize the main bulk of available information about the typical levels of oil contamination and their effects on marine biota. In spite of the conditional nature of any environmental ranging, these diagrams can be used for solving many ecological problems of marine oil pollution (e.g., hazard and risk assessment, monitoring, and regulation).

Conclusions

1. From the chemical perspective, oil is a complex mixture of hundreds of organic substances dominated by hydrocarbons (up to 98%). When entering the marine environment, oil is quickly separated into different fractions. These fractions exist in the forms of surface slicks, dissolved and suspended substances, emulsions, solid and viscous components deposited on the bottom, and compounds accumulated in water organisms. The dominant migrational form is usually emulsified and dissolved oil.

2. Biogeochemical behavior, distribution, and migration of oil compounds in the marine environment depend upon a complex and interconnected processes. These include physical transport, dissolving and emulsification, oxidation and decomposition, sedimentation and biosedimentation, and microbial degradation and aggregation.

3. In contrast with xenobiotics, oil hydrocarbons are continuously released into the marine environment due to natural oil seepage from the seafloor. Identical or similar hydrocarbons (mainly aliphatic) are produced as a result of biosynthetic processes in living organisms. The existence of this changeable natural background, that is still poorly studied, considerably complicates interpretation of the results of assessment and monitoring of marine contamination by oil.

4. The main anthropogenic sources of oil and oil products in the marine environment include land-based discharges, shipping, offshore oil production and transportation, accidental spills, combustion of oil products, and atmospheric precipitation. The global and regional estimates of the absolute and relative contribution of these sources vary significantly because of the lack of reliable data.

5. Global distribution of oil hydrocarbons in the World Ocean is characterized by general increasing concentrations when moving from the ocean pelagic areas to enclosed seas, coastal waters, and estuaries. The increased levels of oil pollution are also typical for areas of intense shipping and tanker oil transportation. The phenomena of oil localization on the water-atmosphere (thin surface microlayer), water-bottom (bottom sediments), and water-shore (littoral zone, beaches) boundaries are found everywhere.

6. Some data indicate that the inputs of oil pollution tend to decrease over the last 10–15 years. However, the high inconsistency of available estimates and the variety of pollution situations raise some doubts about the stability of this tendency. Further research is required both on the regional and global levels.

7. Bioavailability and bioaccumulation of oil directly and strongly depend on hydrophobic and lipophilic properties of oil hydrocarbons. The concentrations of polyaromatic hydrocarbons (PAHs) in marine organisms are usually 3–4 orders of magnitude higher than their concentrations in seawater. Bivalve mollusks-filtrators have the strongest ability to accumulate PAHs without noticeable metabolic decomposition occurring in tissues.

8. From the ecotoxicological perspective, oil is a complex multi-component toxicant with nonspecific effects. Its toxic properties are defined by the groups of hydrocarbons with higher molecular weight and complex structure as well as high water solubility and bioavailability. High-molecular-weight PAHs (with five and more benzene rings) cause carcinogenic and mutagenic effects, especially in benthic organisms that are in contact with chronically polluted sediments. The primary responses of marine biota to the trace quantities of PAHs in the water and bottom environments can be reliably detected with the help of molecular and cellular biomarkers. Sensitive parameters used in these techniques include, in particular, the level of cytochrome P-450, activity of some enzyme systems, and the integrity of lysosomes in individual organs and tissues of water organisms.

9. The available results of experimental and field observations indicate the increased vulnerability of early developmental stages (eggs, larvae, and young specimens) of most marine organisms exposed to oil hydrocarbons. They also show the obvious and persistent disturbances in benthic communities under chronic oil pollution of bottom sediments.

10. Several processes can intensify hazardous consequences of oil pollution. These include photochemical degradation of oil in the marine environment and metabolic transformations of oil compounds in living organisms (mainly in the biochemical system of cytochrome P-450). These processes can lead to the formation of substances with higher toxicity than that of the original hydrocarbons.

11. The lowest levels of oil hydrocarbons that cause primary (mostly reversible) physiological and biochemical responses or no effect at all range within 10^{-3}–10^{-2} mg/l for seawater and 10–100 mg/kg for bottom sediments. These ranges can be roughly considered as the limits of maximum permissible concentration (MPC) of oil hydrocarbons dissolved in seawater and accumulated in bottom sediments, respectively.

12. Available experimental data make it possible to distinguish the following toxicological zones:

- no effect concentrations;
- primary threshold responses;
- sublethal effects; and
- acute toxicity zone.

The zoning makes it possible to compare the levels of oil hydrocarbons causing particular biological effects (acute toxicity, sublethal impairment, etc.) with actual levels of oil pollution found in different areas of the World Ocean. The results of comparison suggest the absence of any detectable biological effect when the concentrations of the sum of oil hydrocarbons are below 10^{-3} mg/l in seawater and below 10 mg/kg in bottom sediments. These concentrations are typical for marine pelagic areas remote from the shore. In relatively clean coastal areas, primary threshold effects are possible. In bays, estuaries, ports, and other shallow areas with limited water circulation and relatively high levels of chronic oil pollution (over 0.1 mg/l for seawater and over 100 mg/kg for sediments), persistent structural and functional disturbances are potentially possible and, in fact, do actually develop in marine communities and ecosystems.

References

Abdullah, A. R., Tahir, N. M., Wei, L. K. 1994. Hydrocarbons in seawater and sediments from the west coast of Peninsular Malaysia. *Bull.Environ. Contam. and Toxicol.* 53(4):618–626.

Addison, R. F. 1992. Detecting the effects of marine pollution. In *Science Review 1990–1991.*—Dartmouth, Nova Scotia, pp.9–12.

Al-Saad, H. T. 1990. Distribution and sources of aliphatic hydrocarbons in fish from the Arabian Gulf. *Mar.Pollut.Bull.* 21(3):155–157.

Ambrose, Ph. 1994. Tarred loggerhead turtles. *Mar.Pollut.Bull.* 28(5):273.

Ameijeres, A. H., Gandara, J. S., Hernandez, J. L., Lozano, J. S. 1994. Classification of the coastal waters of Galicia (NW Spain) on the basis of total aliphatic hydrocarbon concentrations in mussels (*Mytilus galloprovincialis*). *Mar.Pollut.Bull.* 28(6):396–398.

Anderson, J. W. 1985. Oil pollution: effects and retention in the coastal zone. In *Proceedings of the International Symposium on Utilization of Coastal Ecosystems: Planning, Pollution and Productivity.*—Rio Grande, Brazil, pp.197–211.

Anderson, J. W., Gosett, R. W. 1986. *Polynuclear aromatic hydrocarbons contamination in sediments from coastal waters of Southern California. Final Report to the California State Water Resources Control Board.* Sacramento, California.

Anonymous. 1993. US production water disposal problem. *Mar.Pollut.Bull.* 26(9):473.

Archakov, A. I., Davidov, D. P., Koryavkin, A. V. 1988. Cytochrome P-450: the problems of ecology and toxicology. In *Ecological chemistry of the water environment.*—Moscow: Institut khimicheskoi fisiki AN SSSR, pp.62–78. (Russian)

Armstrong, H. W., Fuick, K., Anderson, J. W., Neff, J. M. 1979. Effects of oil-field brine effluent on sediments and benthic organisms in Trinity Bay, Texas. *Mar.Envir.Res.* 2:55–69.

Arnold, I. N. 1897. *Water pollution by oil products and its impact on the fisheries.* Spb., 63 pp. (Russian)

Asuaquo, F. E. 1991. Tar balls on Ibeno-Okposo beach of Southern Nigeria. *Mar.Pollut.Bull.* 22(3):150–151.

Atlas, R. M. 1993. Bacteria and bioremidation of marine oil spills. *Oceanus* 36(2):71.

Atlas, R. M. 1994. Petroleum biodegradation and oil spill bioremediation. *Mar.Pollut.Bull.* 31(4–12):178–182.

Atwood, D. K., Burton, E. J., Harvey, G. R., Mata-Jimenez, A. J., Vasquez-Botello, A., Wade, B. A. 1987. Results of the CARIPOL petroleum pollution monitoring project in the Wider Caribbean. *Mar.Pollut.Bull.* 18(10):540–548.

Augenfeld, J. M. 1980. Effects of Prudhoe Bay crude oil contamination on sediment working rates of *Abarenicolla pacifica. Mar.Environ.Res.* 3:307–313.

Badawy, M. I., Al-Mujainy, I. S., Hernandez, M. D. 1993. Petroleum-derived hydrocarbons in water, sediment and biota from the Mina al Fahal coastal waters. *Mar.Pollut.Bull.* 26(8):457–460.

Baker, J. M., Clark, R. B., Kingston, P. F., Jenkins, R. H. 1990. *Natural recovery of cold water marine environments after an oil spill. A paper presented at the Thirteenth Annual Arctic and Marine Oil Spill Program Technical Seminar.* 111 pp.

Balci, A. 1993. Dissolved and dispersed petroleum hydrocarbons in the Eastern Aegean Sea. *Mar.Pollut.Bull.* 26(4):222–223.

Barnung, T. N., Grahl-Nielson, O. 1987. The fatty acids profile in cod (*Gadus morhua L.*) eggs and larvae developmental variations and responses to oil pollution. *Sarsia* 72(3–4):415–417.

Benzhitski, A. G. 1980. *Oil contaminants in the hyponeustal of the seas and oceans.* Kiev: Naukova dumka, 120 pp. (Russian)

Bernard, D., Jeremie, J. J., Pascaline, H. 1995. First assessment of hydrocarbon pollution in a mangrove estuary. *Mar.Pollut.Bull.* 30(2):146–150.

Bewers, J. M., Karau, J. 1993. *Sediment quality guideline development in Canada. International Council for the Exploration of the Sea.* ICES E:17, 7 pp.

Boesch, D. F., Rabalis, N. N., eds. 1988. *Long-term environmental effects of off-shore oil and gas developments.* New York: Elsevier Applied Science, 708 pp.

Bogdashkina, V. I., Petrosyan, V. S. 1988. Ecological aspects of pollution of the water environment by the oil hydrocarbons, pesticides and phenols. In *Ecological chemistry of the water environment.*—Moscow: Institut Khimicheskoi Fisiki AN SSSR, pp.62–78. (Russian)

Bohle, B. 1982. *Avoidance from petroleum hydrocarbons by cod (Gadus morhua L.). International Council for the Exploration of the Sea.* ICES C.M. E:56, 10 pp.

Bonsdorff, E., Bakke, T., Pedersen, A. 1990. Colonization of amphipods and polychaetes to sediments experimentally exposed to oil hydrocarbons. *Mar.Pollut.Bull.* 21(7):355–358.

Borisov, V. P., Osetrova, N. V., Ponomarenko, V. P., Semenov, V. N. 1994. *Impact of the offshore oil and gas developments on the bi resources of the Barents Sea.* Moscow: VNIRO, 251 pp. (Russian)

Borowski, B., Aitken-Ander, P., Tanacredi, J. T. 1993. The effect of low doses of waste crancase oil in *Melita nitida* Smith (Crustacea: Amphipoda). *J.Exp.Mar.Biol.Bull.* 166(1):39–46.

Bouloubassi, I., Saliot, A. 1991. Composition and sources of dissolved and particulate PAHs in surface waters from the Rhone Delta (NW Mediterranean). *Mar.Pollut.Bull.* 22(12):588–594.

Bouloubassi, I., Saliot, A. 1993. Dissolved, particulate and sedimentary naturally derived polycyclic aromatic hydrocarbons in a coastal environment: geochemical significance. *Mar.Chem.* 42:127–143.

Bruns, K., Dahlmann, G. D., Gunkel, W. 1993. Distribution and activity of petroleum hydrocarbon degrading bacteria in the North and Baltic Seas. *Deutsche Hydrographische Zeitschrift* H6:359–369.

Canby, T. Y. 1991. After the storm. *National Geographic* 180(2):2–35.

Capuzzo, J. M. 1987. Biological effects of petroleum hydrocarbons: assessment from experimental results. In *Long-term environmental effects of off-shore oil and gas development.*—London: Elsevier Applied Science Publishers, pp.342–410

Cariochia, A. M., Cremisini, C., Martini, F., Morabito, R. 1993. PAHs, PCBs, and DDE in the Northern Adriatic Sea. *Mar.Pollut.Bull.* 26(10):581–583.

Clamer, H. J. G., Fomsgaard, L. 1993. Geographical distribution of chlorinated biphenyls (CBs) and polycyclic aromatic hydrocarbons (PAHs) in surface sediments from the Humper Plume, North Sea. *Mar. Pollut.Bull.* 26(4):201–206.

Cocchieri, R. A., Arnese, A., Minicucci, A. M. 1990. Polycyclic aromatic hydrocarbons in marine organisms from Italian Central Mediterranean Coast. *Mar.Pollut.Bull.* 21(1):15–18.

Corbin, C. J. 1993. Petroleum contamination of the coastal environment of St.Lucia. *Mar.Pollut.Bull.* 26(10):579–580.

Cripps, G. C. 1992. Baseline levels of hydrocarbons in seawater of the Southern Ocean. Natural variability and regional patterns. *Mar.Pollut.Bull.* 24(2):133–145.

Danishevskaya, A. I., Smirnov, V. I., Petrova, V. I., Beltyaeva, A. N. 1989. Geochemistry of organic substance in bottom sediments of the Weddell Sea. *Oceanologia* 29:322–327. (Russian)

Davies, J. M., Hardy, R., McIntyre, A. D. 1981. Environmental effects of North Sea oil operations. *Mar.Pollut.Bull.* 12:412–416.

De Dominico, L., Crisafi, E., Magazzu, G., Puglisi, A., La Rosa, A. 1994. Monitoring of petroleum hydrocarbon pollution in surface waters by a direct comparison of fluorescence spectroscopy and remote sensing techniques. *Mar.Pollut.Bull.* 28(10):587–591.

Desederi, P., Lepri, L., Chechini, L. 1991. Organic compounds in seawater and pack ice in Ternova Bay (Antarctica). *Annal.Chem.* 81:395–416.

Eletski, B. D., Khoroev, V. V. 1992. Anthropogenic pollution of the coastal zone of the Black Sea in the summer of 1989. In *Ecology of the coastal zone of the Black Sea.*—Moscow: VNIRO, pp.234–249. (Russian)

Emara, H. I. 1990. Oil pollution in the Southern Arabian Gulf and the Gulf of Oman. *Mar.Pollut.Bull.* 21(8):399–401.

Emerson, R. 1994. Oil spill risk analysis. In *Environmental risk assessment for oil and gas development on the continental shelf of the Russian Far East (Seminar presentations, February 1994, Magadan).*—Magadan, pp.34–40.

Esters, J. L., Commendatore, M. J. 1993. Total aromatic hydrocarbons in waters and sediments in a coastal zone of Patagonia, Argentina. *Mar.Pollut.Bull.* 26(6):341–342.

Falk-Petersen, I. B., Kjorsvik, E. 1987. Acute toxicity tests of the effects of oil and dispersants on marine fish embryos and larvae. *Sarsia* 72(6):411–414.

Farrington, J. W., Tripp, B. W. 1993. International mussel watch. *Oceanus* 36(2):62–66.

Foeyn, L., Serigtad, B. 1989. Fish stock vulnerability and ecological evaluations in light of recent research. In *Proceedings of International Conference on Fisheries and Offshore Petroleum Exploitation, Bergen 23–25 October 1989.*—Bergen: Chamber of Commerce and Industry, p.23.

Fowler, S. W. 1993. Pollution in the Gulf: monitoring the marine environment. *IAEA Bull.* 35(2):10–13.

Galleogo, A., Cargill, L. H., Heath, M. R., Hay, S. J., Knutsen, T. 1995. An assessment of immediate effect of the *Braer* oil-spill in the growth of herring larvae using otholith microstructure analysis. *Mar.Pollut.Bull.* 30(8):536–542.

GESAMP. 1977. *Impact of oil on the marine environment. GESAMP Reports and Studies No.6.* Rome:FAO, 250 pp.

GESAMP. 1989. *Long-term consequences of low-level marine contamination. An analytical approach. GESAMP Reports and Studies No.40.* Rome:FAO, 17 pp.

GESAMP. 1990. *The state of the marine environment. GESAMP Reports and Studies No.39.* UNEP, 112 pp.

GESAMP. 1993. *Impact of oil and related chemicals and wastes on the marine environment. GESAMP Reports and Studies No.50.* London: IMO, 180 pp.

GESAMP. 1994. *Report of the twenty-fourth session of GESAMP. GESAMP Reports and Studies No.53.* New York: UN, 60 pp.

Golik, A., Weber, K., Salihoglu, A., Yilmaz, A., Loizedes, A. 1988. Pelagic tar in the Mediterranean Sea. *Mar.Pollut.Bull.* 19(11):567–572.

Grahl-Nielson, O. 1987. Hydrocarbons and phenols in discharge water from offshore operations. Fate of the hydrocarbons in the recipient. *Sarsia* 72(3–4):375–382.

Granby, K., Spliid, N. H. 1995. Hydrocarbons and organochlorines in common mussels from the Kattegat and the Belts and their relation to condition indices. *Mar.Pollut.Bull.* 30(1):74–82.

Grassle, J. F., Elmgren, R., Grassle, J. P. 1981. Response of benthic communities in MERL experimental ecosystems to low level, chronic additions of No.2 fuel oil. *Mar.Environ.Res.* 4:279–297.

Gray, J. S. 1987 Oil pollution studies of the Solbergstrand mesocosms. *Phil.Trans.R.Soc.Lond.* NB316:641–654.

Green, G., Skerratt, J. H., Leeming, Rh., Nichols, P. D. 1992. Hydrocarbon and coprostanol levels in seawater, sea-ice algae and sediments near Davis Station in the Eastern Antarctic. *Mar.Pollut.Bull.* 25(9–12): 293–302.

Grimm, O. A. 1891. On the hazardous oil effect on fish and the measures of its prevention. *Vestnik rybopromishlennosti* 12:379–387. (Russian)

Gupta, R. S. 1991. Gulf oil spill and India. *Mar.Pollut.Bull.* 22(9):423–424.

Gupta, R. S., Naik, S., Varadachari, V. V. R. 1989. Environmental pollution in coastal areas of India. In *Ecotoxicology and climate.—SCOPE 38*, Chichester, pp.235–246.

Gurvich, L. M. 1993. Is it possible to prevent the oil pollution of the hydrosphere? *Zemlya i Vselennaya* 2:39–43. (Russian)

Hellou, J., Stenson, G., Ni, I. H., Payne, J. F. 1990. Polycyclic aromatic hydrocarbons in muscle tissues of marine mammals from the Northwest Atlantic. *Mar.Pollut.Bull.* 21(10):469–473.

Hellou, J., Upshall, C., Ni, I. H., Payne, J. F., Huang, Y. S. 1991. Polycyclic aromatic hydrocarbons in harp seals (*Phoca groenlandica*) from the Northwest Atlantic. *Arch.Environ.Contam.Toxic.* 21(10):135–140.

Hellou, J., Upshall, C., Taylor, D., O'Keefe, P., O'Malle, V., Abrajano, T. 1994. Unstructured hydrocarbons in muscle and hepatopancreas of two crab species, *Chionoecetes opilio* and *Hyas coarctatus. Mar.Pollut.Bull.* 28(8):482–488.

Hinrichsen, D. 1990. *Our common seas. Coast in crisis.* London: Earthscan Publications, 184 pp.

Hinshery, A. K., Baghadi, A. L., Kumar, N. S. 1991. Petroleum hydrocarbons levels near offshore water of Libia. *Bulletin of National Institute of Oceanography and Fisheries* 16(3):255–262.

Howarth, R. W. 1991. Assessing the ecological effects of oil pollution from outer continental shelf oil development. In *Fisheries and Oil Development on the Continental Shelf. Amer.Fish.Soc.Symp.*—Bethesda, Maryland, pp.1–8.

Hyland, J. L., Schneider, E. D. 1978. *Petroleum hydrocarbons and their effects on marine organisms, populations, communities and ecosystems.* Narragansett: EPA, 41 pp.

ICES. 1994a. *International Council for the Exploration of the Sea. Report of the ICES Advisory Committee on the Marine Environment.* Copenhagen: ICES, 122 pp.

ICES. 1994b. *International Council for the Exploration of the Sea. Report of the Joint meeting of the Working Group on Marine Sediments in relation to pollution and the Working Group on Biological Effects on Contaminants (Nantes, France, 21–22 March 1994).* ICES C.M. 1994/ENV:2, 19 pp.

ICES. 1995. *International Council for the Exploration of the Sea. Report on the ICES Advisory Committee on the Marine Environment.* Copenhagen: ICES, 135 pp.

ICES. 1997. *International Council for the Exploration of the Sea. Report on the ICES Advisory Committee on the Marine Environment.* Copenhagen: ICES, 210 pp.

Ilnitski, A. P., Korolev, A. A., Khudoley, V. V. 1993. *Carcinogenic substances in the water environment.* Moscow: Nauka, 220 pp. (Russian)

Ilyin, G. V., Petrov, V. S. 1994. Studies of the sediment pollution by oil hydrocarbons. In *Ecological studies of the zone of the industrial development of Shtokmanovskoe gas condensate field on the shelf of the Barents Sea.* —Apatiti: Izd-vo KNTS RAN, pp.48–50. (Russian)

IMO. 1990. *International Maritime Organization. Petroleum in the marine environment. Document MEPC 30/ INF.13 submitted by the United States.* London: IMO.

IOC. 1986. *Intergovernmental Oceanographic Commission. Workshop on the Biological Effects of Pollutants. Workshop Report No.53.* Paris: UNESCO, 75 pp.

Izrael, J. A., Tsiban, A. V. 1988. *Anthropogenic ecology of the ocean.* Leningrad, Gidrometeoizdat, 588 pp. (Russian)

Jones, M. A. J., Bacon, P. R. 1990. Beach tar contamination in Jamaica. *Mar.Pollut.Bull.* 21(7):331–334.

Kadam, A. N., Bhangale, V. P. 1993. Petroleum hydrocarbons in northwest coastal waters of India. *Indian Journ.Mar.Sci.* 22:227–228.

Kalitovich, A. G. 1994. Chemical pollution of the coastal zone near Murmansk. In *Ecological situation and protection of the flora and fauna of the Barents Sea.*—Apatiti: Izd-vo KNTS RAN, pp.106–114. (Russian)

Karataeva, B. B. 1993. Impact of environmental factors on fish. In *Bioproductive and economical issues of the world fisheries.*—Moscow: VNIERKH, pp.6–22. (Russian)

Kennicutt, M. C., McDonald, S. J., Sweet, S. T. 1992. The *Bahio Paras* spill in Arthur Harbor, Anvers Island. *Antact. Journ.* 27(5):331–332.

Khailov, K. M. 1971. *Ecological metabolism in the sea*. Kiev, Naukova dumka, 320 pp. (Russian)

Khan, R. A., Kiceneuk, J., Dawe, M., Williams, U. 1981. *Long-term effects of crude oil on Atlantic cod. International Council for the Exploration of the Sea*. ICES C.M. 1981/E:40. 10 pp.

Kingston, P. F. 1992. Impact of offshore oil production installations on the benthos of the North Sea. *ICES J.Mar.Sci.* 49(1):45–53.

Klekowski, E. J., Corredor, J. E., Morel,l J. M., Del Castillo, C. A. 1994. Petroleum pollution and mutation in mangroves. *Mar.Pollut.Bull.* 28(3): 166–169.

Knoppe, B. A., Lacenda, L. D., Patchinealam, S. R. 1990. Nutrients, heavy metals and organic micropollutants in an eutrophic Brazilian lagoon. *Mar.Pollut.Bull.* 21(8):381–384.

Knyazeva, N. S., Zaitseva, V. A., Aleksyuk, T. G. 1991. Concentration of arenes in the organs and tissues of the fish from the Ob River under pollution. In *Theses of the Second All-Union Conference on Fisheries Toxicology. Vol.1.*—Spb., pp.268–269. (Russian)

Konovalov, S. M. 1995. Anthropogenic impact and ecosystems of the Black Sea. *Les mers tributaires de Mediterrannee. CIESM Science series. No.1. Bulletin de l'Institut Oceanographique*, Monaco. No.15 (special):53–84.

Kornberg, H. 1981. *Royal Commission on Environmental Pollution, 8th report.* London: H.M. Stationary Office.

Koronelli, T. V., Ilyinski, V. V., Semenenko, M. M. 1994. Oil pollution and stability of the marine ecosystems. *Ecologia* 4:78–81. (Russian)

Kostrov, B. P., Magomedov, A. K., Samudov, S. M. 1991. Concentrations of the oil hydrocarbons in the water and organisms of the Middle and Southern Caspian Sea. In *Theses of the Second All-Union Conference on Fisheries Toxicology. Vol.1.*—Spb., pp.291–292. (Russian)

Laane, R. W. P. M., ed. 1992. *Background concentrations of natural compounds in rivers, sea water, atmosphere and mussels. Report DGW-92.033.* The Hague, 78 pp.

Laevastu, T., Fukuhara, F. 1984. *Oil on the bottom of the sea. International Council for the Exploration of the Sea.* ICES C.M. 1984/E:6. 19 pp.

Law, A. T., Arshad, A., Yusoff, F. M. 1995. Coastal oceanographic studies off Perlis, Straits of Malacca. In *International Seminar on Marine Fisheries Environment, 9–10 March 1995, Rayong, Thailand.* Emdec and JICA, pp.263–281.

Law, R. J. 1986. *Polycyclic aromatic hydrocarbons in the marine environment: overview. International Council for the Exploration of the Sea.* ICES MCWG 1986/10.1, 12 pp.

Law, R. J., Blackman, A. A. 1981. *Hydrocarbons in water and sediments from oil-producing areas of the North Sea. International Council for the Exploration of the Sea.* ICES C.M. 1981/E:16, 20 pp.

Law, R. J., Hudson, P. M. 1986. *Preliminary studies of the dispersion of oily water discharges from the North Sea oil production platforms. International Council for the Exploration of the Sea.* ICES C.M. 1986/E:15, 12 pp.

Law, R. J., Waldock, M. J., Allchin, C. R., Laslett, R. E. 1992. *Contaminants in seawater around England and Wales: results from the monitoring surveys, 1990–91 International Council for the Exploration of the Sea.* ICES C.M. 1992/E:31, 14.

Law, R. J., Whinnett, J. A. 1992. Polycyclic aromatic hydrocarbons in muscle tissue of harbor porpoises (*Phocoena phocoena*) from UK waters. *Mar.Pollut.Bull.* 24(11):550–553.

Levy, E. M. 1984. Oil pollution in the world oceans. *Ambio* 13:226–235.

Lin, Q. J. X. 1991. Petroleum hydrocarbons in marine organisms from Hong-hai Bay, the northeastern South China Sea. *Mar.Pollut.Bull.* 10(1):33–38.

Linden, O., Laughlin, R., Sharp, j. R., Neff, J. M. 1980. The combined effects of salinity, temperature and oil on the growth pattern of embryos of the killifish—*Fundulus hetroclitus* Walbaum. *Mar.Environ.Res.* 3:129–144.

Liplatou, E., Saliot, A. 1991. Hydrocarbon contamination of the Rhone Delta and Western Mediterranean. *Mar.Pollut.Bull.* 22(6):297–304.

Lotufo, G., Fleeger, J. 1995. Acute and sublethal effects of PAHs to the estuarine harpacticoid copepod *Schizopera knabeni* Lang. In *Twenty Third Benthic Ecology meeting.*—Rutgers the State Univ., New Brunswick, NJ (USA) Inst.Marine Coastal Science.

Madany, I. M., Al-Haddad, A. Jaffar, A., Al-Shirbini, E. -S. 1994. Spatial and temporal distribution of aromatic petroleum hydrocarbons in the coastal waters of Bahrain. *Arch.Environ.Contam. and Toxicol.* 26(2):185–190.

Makarov, E. V., Semenov, A. D., Aleksandrova, Z. V., Kharkovski, V. M., Soyer, V. G. 1995. Pollution of the coastal areas of the north-eastern part of the Black Sea. In *Theses of the International Symposium on Mariculture.*—Moscow: VNIRO, pp.34–36. (Russian)

Matsueda, H., Handa, N. 1986. Source of organic matter in sinking particles from the Pacific sector of the Antarctic Ocean. *Mem.Nat.Inst.Polar.Res.* 40:364–379.

Matthiessen, P. 1994. *The use of sediment bioassays in marine environmental monitoring.* MAFF Fisheries Laboratory. Prepared for ICES WGBEC. 21 pp.

Mazmanidi, N. D. 1991. Fish as the object of monitoring of the ecological state of the Black Sea. In *Theses of the Second All-Union Conference on Fisheries Toxicology. Vol.1.*—Spb., pp.22–23. (Russian)

Mazur, I. I. 1993. *Ecology of oil and gas development.* Moscow: Nedra, 494 pp. (Russian)

McDonald, S. J., Kennicutt, M. C., Brooks, J. M. 1992. Evidence of polycyclic aromatic hydrocarbon (PAH) exposure in fish from the Antarctic Peninsula. *Mar.Pollut.Bull.* 25(9–12):313–317.

McKinnon, M., Vine, P. 1991. Tides of War. *Eco-disaster in the Gulf.* London: Boxtree Ltd, 192 pp.

Mikhailova, L. V. 1991. Impact of the water-soluble fraction of the oil from Ust-Balik on the early ontogenesis of fish *Acipenser ruthenus. Gidrobiologicheski zhurnal* 27(1):77–87. (Russian)

Minas, W., Gunkel, W. 1995. Oil pollution in the North Sea—a microbiological point of view. *Helgolander Meeresuntersuchungen* 49(1–4): 143–158.

Mironov, O. G. 1972. *Biological resources of the sea and the oil pollution.* Moscow: Pishchepromizdat, 105 pp. (Russian)

Mironov, O. G. 1989. Stability and drift of the oil slicks in the sea during small amounts of the spilled oil. In *Complex study of the World Ocean in connection with development of its mineral resources.*—Leningrad, pp.84–88. (Russian)

Mironov, O. G., Shchekaturina, T. A., Pisareva, N. A., Kopilenko, L. P. 1991. Arenes in the fish and mussels of the Black Sea. *Biologicheskie nauki* 5:75–79 (Russian)

MLA. 1993. *Marine Laboratory Aberdeen. Annual Review (1991–1992).* Aberdeen: Agriculture and Fisheries Department. The Scottish Office, 68 pp.

Moles, A., Rice, S., Nocross, B. L. 1994. Non-avoidance of hydrocarbons laden sediments by juvenile flat-fishes. *Neth.J.Sea.Res.* 32(3–4):361–367.

Monina, Y. I. 1991. Geography of the oil pollution of the ocean. *Priroda* 8:70–74. (Russian)

Moore, D. C. 1983. *Biological effects on benthos around the Beryl oil platform. International Council for the Exploration of the Sea.* ICES C.M. 1983/E:43, 25 pp.

Moore, M. N., Lowe, D. M., Bucke, D., Dixon, P. 1991. *Molecular and cellular markers of pollutant exposure and liver damage in fish.* ICES CM 1991/E:23, 18 pp.

Moyano, M., Moresco, H., Blanco, J., Rosadilla, M., Caballero, A. 1993. Baseline studies of coastal pollution by heavy metals, oil, and PAHs in Montevideo. *Mar.Pollut.Bull.* 26(8):461–464.

Murray, A. P., Richardson, B. J., Gibbs, C. F. 1991 Bioconcentration factors for petroleum hydrocarbons, PAHs, LABs and biogenic hydrocarbons in the blue mussel. *Mar.Pollut.Bull.* 12:595–603.

Neff, J. M. 1979. *Polycyclic aromatic hydrocarbons in the aquatic environment: sources, fates and biological effects.* London: Applied Science Publ. Ltd, 262 pp.

Neff, J. M. 1993. *Petroleum in the marine environment: regulatory strategies and fisheries impact.* Battelle Ocean Sciences Laboratory. Duxbury, 13 pp.

Nelson-Smith, A., trans. 1977. *Oil pollution and marine ecology.* Moscow: Progress, 301 pp.(Russian)

Nemirovskaya, I. A. 1991. Hydrocarbons in the waters of the Indian Ocean. *Oceanologia* 31(2):239–244. (Russian)

Nemirovskaya, I. A. 1994. Hydrocarbons in the Gulf of Riga and Gulf of Finland. *Oceanologia* 34(3):383–390. (Russian)

NERC. 1994. *Natural Environmental Research Council. Report of the Plymouth Marine Laboratory (1993–1994).* 70 pp.

Nesterova, M. P., Nemirovskaya, I. A., Anuphrieva, N. M., Maslov, V. U. 1979. Oil products in the surface waters of the Pacific and Indian

Oceans. In *Chemical pollution of the water environment.*—Leningrad: Gidrometeoizdat, pp.129–136. (Russian)

Nesterova, M. P., Nemirovskaya, I. A., Gurevich, L. M. 1986. Oil pollution of the Baltic Sea, monitoring and control methods. In *Methodology of prognoses of ocean and seas pollution.*—Moscow: Gidrometeoizdat, pp.35–39. (Russian)

Nicholson, G. J., Theodoropoulos, T., Fabris, G. J. 1994. Hydrocarbons, pesticides, PCB and PAH in Port Phillip Bay (Victoria) sand flathead. *Mar.Pollut.Bull.* 28(2):115–120.

Nicols, P. D., Volkman, J. K., Palmisano, A. C., Smith, G. A., White, D. C. 1988. Occurrence of an isoprenoid C-25 diunsaturated alkene and high neutral lipid content in Antarctic sea-ice diatom communities. *Journ. Phycol.* 24:90–96.

NRC. 1985. *National Research Council. Oil in the sea. Inputs, fates and effects.* Washington, DC. National Academy Press, 601 pp.

NSTF. 1994. *North Sea Task Force. The Quality Status Report on the North Sea.* Fredensborg. Olsen and Olsen, 250 pp.

Olla, B. L., Bejda, A. J., Studhiolme, A. L., Pearson, W. J. 1984. Sublethal effects of oiled sediment on the sand warm, *Nereis (Neanthes) virens:* induced changes in burrowing. *Mar.Environ.Res.* 13:121–139.

Oradovski, S. G. 1986. Results and perspective of developing the methods of chemical monitoring of marine pollution. In *Methodology of prognoses of ocean and seas pollution.*—Moscow: Gidrometeoizdat, pp.7–11. (Russian)

Palatnikov, G. M., Mamedov C. A., 1988. Neuron correlation of the olfactory behavior of the Russian sturgeon under condition of the oil pollution. In *Theses of the Fifth All-Union Conference on Water Toxicology.*—Moscow: VNIRO, pp.141–142. (Russian)

Panov, G. E., Petryashin, L. F., Lysyani, G. N. 1986. *Environmental protection in the oil and gas industry.* Moscow: Nedra. (Russian)

Patin, S. A. 1979. *Pollution impact on the biological resources and productivity of the World Ocean.* Moscow: Pishchepromizdat, 305 pp. (Russian)

Patin, S. A. 1982. *Pollution and the biological resources of the oceans.* London: Butterworth Scientific, 320 pp.

Patin, S. A. 1988. Establishing fisheries standards for the water environment quality. In *Water toxicology and optimization of bioproductive processes in aquaculture.*—Moscow: VNIRO, pp.5–18. (Russian)

Patin, S. A. 1994. *Evaluation of ecological situation, strategy and priorities in the protection of the marine environment from pollution.* Moscow: Minpriroda RF, 80 pp. (Russian)

Patin, S. A. 1994. Oil and gas production on the continental shelf: ecofisheries analysis. *Rybnoe khosyaistvo* 5:30–32. (Russian)

Patin, S. A. 1995. Global pollution and biological resources of the World Ocean. In *World Fisheries Congress Proceedings.* New Delhy: Oxford and IBH Publ.Co., pp.69–75.

Patin, S. A., Lesnikov, L. A. 1988. Main principles of establishing the fisheries standards for the water environment quality. In *Methodical guidelines on establishing the maximum permissible concentrations of pollutants in the water environment.*—Moscow: VNIRO, pp.3–10. (Russian)

Pelletier, E., Ouelett, S., Paquet, M. 1991. Long-term chemical and cytochemical assessment of oil contamination in estuarine intertidal sediments. *Mar.Pollut.Bull.* 22(6):273–281.

Pendoley, K. 1992. Hydrocarbons in the Rowley Shelf (Western Australia) oysters and sediments. *Mar.Pollut.Bull.* 24(4):210–215.

Perey, I. A., Wells, P. G. 1985. Effects of petroleum in polar marine environments. *Mar.Technol.Soc.J.* 18(3):51–61.

Peters, L. D., Coombs, S. H., McFadzen, I., Sole, M., Albaiges, J., Livingstone, D. R. 1992. *Toxicity studies, 7-ethoxyresorufin o-deethylase (EROD) and antioxidant enzymes in early life stages of turbot (Scophthalmus maximus L.) and sprat (Sprattus sprattus l.). International Council for the Exploration of the Sea.* ICES C.M.1992/E:12, 10 pp.

Peters, L. D., Porte, C., Albaiges, J., Livingstone, D. R. 1994. 7-Ethoxyresorufin O-deethylase (EROD) and antioxidant enzyme activities in larvae of sardine (*Sardina pilchardus*) from the north coast of Spain. Mar.*Pollut.Bull.* 28(5):299–304.

Petrov, A. A. 1984. *Petroleum hydrocarbons.* Springer-Verlag, 255 pp.

Phillips, D. J. H., Richardson, B. J., Murray, A,P., Fabris, J. G. 1992. Trace metals, organochlorines and hydrocarbons in Port Phillip Bay, Victoria: a historical review. *Mar.Pollut.Bull.* 25(5–8):200–217.

Polikarpov, G. G., Benzhitski, A. G. 1975. Oil aggregates—new ecological niche in the ocean. *Khimia i zhizn* 11:75–78. (Russian)

Preston, M. R., Chester, R., Bradshaw, G. F., Merrett, J. L. 1992. A possible tool for the assessment of anthropogenic influence on marine aerosols. *Mar.Pollut.Bull.* 249(3):164–166.

Quintero, S., Diaz, C. 1994. Aliphatic hydrocarbons in fish from the Canary Islands. *Mar.Pollut.Bull.* 28(1):44–49.

Quintero, S., Diaz, C. 1994. Hidrocarburos aromaticos policiclicos en peces de costers canarias. *Scientia Marina* 58(4):307–313.

Ram, A., Kadam, A. N. 1991. Petroleum hydrocarbon concentration in sediments along the north-western coast of India. *Indian J.Mar. Sci.* 20(3).

Ramade, F. 1978. *Elements d'ecologie appliquee.* McGrow Hill Publ., 497 pp.

Register of MPC. 1995. *Register of the maximum permissible concentrations (MPC) of hazardous substances for the water environment.* Moscow: Medinor, 220 pp. (Russian)

Reiersen, L.-O., Grey, J. S., Palmork, K. N., Lange, R. 1989. Monitoring in the vicinity of oil and gas platforms: results from the Norwegian sector of the North Sea and recommended methods for forthcoming surveillance. In *Drilling Wastes.*—London: Eslevier Applied Science, pp.91–117.

Rice, S. D. 1985. Chapter 5. Effects of oil on fish. In *Petroleum effects in the Arctic environment.*—London: Eslevier Applied Science, pp.157–182.

Romankevich, E. A. 1977. *Geochemistry of organic substance in the ocean.* Moscow: Nauka. (Russian)

Rovinski, F. Y., Teplitskaya, T. A., Alekseeva, T. A. 1988. *Background monitoring of the polycyclic aromatic hydrocarbons.* Leningrad: Gidrometeoizdat, 224 pp. (Russian)

Saliot, A. 1981. Natural hydrocarbons in sea water. In *Marine organic chemistry.*—Amsterdam: Elsevier, pp.327–374.

Savin, A. D., Podplyetnaya, N. F. 1991. Characteristic of the oil pollution near Odessa. In *Theses of the Second All-Union Conference on Fisheries Toxicology. Vol.2.*—Spb., pp.139–140. (Russian)

Sapozhnikov, V. V., Sokolova, S. A. 1994. Distribution of pollutants in the water and bottom sediments of the White Sea. In *Complex studies of the ecosystem of the White Sea.*—Moscow: VNIRO, pp.104–108. (Russian)

Semenov, A. D., Aleksandrova, Z. V., Kishkinova, T. S., Romova, M. G., Katalevski, N. I., Pavlenko, L. F., Morozova, G. M. 1991. Pollution level in the ecosystem of the Sea of Asov according to new assessments. In *Theses of the Second All-Union Conference on Fisheries Toxicology. Vol.2.* —Spb., pp.155–157. (Russian)

Semenov, V. N. 1991. Kinds of anthropogenic impacts on the marine ecosystems and some ways of its revealing. In *Ecological situation and protection of the flora and fauna of the Barents Sea.*—Apatiti: Izd-vo KNTS AN SSSR, pp.115–121. (Russian)

SEPA. 1990. *Swedish Environmental Protection Agency. Marine pollution. Action programme.* Swedish Environmental Protection Agency, 166 pp.

Sergeev, B. L., Bogovski, S. P. 1993. The changes in activity of the monooxygenage enzyme system of the flounder liver as a pollution indicator in the Baltic Sea. In *XV Mendeleev Congress on the General and Applied Chemistry (Minsk, May 24–29, 1993), vol.3.*—Minsk. (Russian)

Serigstad, B. 1987. Effects of oil-exposure on the oxygen uptake of cod (*Gadus morhua L.*) eggs and larvae. *Sarsia* 72(3–4):401–403.

Shchekaturina, T. L., Khesina, A. L., Mironov, O. G., Krivosheeva, L. G. 1995. Carcinogenic polycyclic aromatic hydrocarbons in mussels from the Black Sea. *Mar.Pollut.Bull.* 30(1):38–41.

Sharblom, P. M., Kelly, D., Pierce, R. H. 1995. Baseline survey of pesticide and PAH concentrations from Sarasota Bay, Florida, USA. *Mar.Pollut.Bull.* 30(8):568–573.

Shparkovski, I. A. 1993. Biotesting of the quality of the water environment using fish. In *Arctic seas: bioindication of the state of the environment, biotesting and technology of the pollution destruction.*—Apatiti: Izd-vo KNTS RAN, pp.11–30. (Russian)

Simonov, A. I., Zubakina, A. N. 1979. Scientific foundation of the problem of marine pollution. In *Chemical pollution of the marine environment.* —Leningrad: Gidrometeoizdat, pp.9–24. (Russian)

Singh, J. G., Chang-Yen, I., Stoute, V. A., Chatergoon, L. 1992. Hydrocarbon levels in edible fish, crabs and mussels from the marine environment of Trinidad. *Mar.Pollut.Bull.* 24(4):270–272.

Sinyukov, V. V. 1993. *Development of the marine hydrochemical studies (the Black Sea, the Sea of Asov, Arctic seas).* Moscow: Nauka, 223 pp. (Russian)

Smith, J. N., Levy, E. M. 1990. Geochronology for polycyclic aromatic hydrocarbon contamination in sediments of the Saguenay Fijord. *Envir.Sci.Technol.* 24:874–879.

Snedaker, S. C., Glynn, P. W., Rumbold, D. G., Corcoran, E. F. 1995. Distribution of n-alkanes in marine samples from Southern Florida. *Mar.Pollut.Bull.* 30(1):83–89.

Sokolova, S. A., Storozhuk, N. G., Goryunova, V. B., Nemirovskaya, I. A. 1995. Distribution of pollutants in the Western Bering Sea. In *Complex studies of the ecosystem of the Bering Sea.*—Moscow: VNIRO, pp.247–256. (Russian)

Solberg, T., Tilseth, S., Serigstad, B., Westrheim, K. 1982. *Effects of low levels of a heavy fraction of Ekofisk crude oil on eggs and yolk sac larvae of cod (Gadus morhua L.). International Council for the Exploration of the Sea.* ICES C.M. 1982/ E:59, 13 pp.

Somerville, H. J., Bennett, D., Davenport, J. N., Holt, M. S., Lynes, A., Mahie, A., McCourt, B., Parker, J. G., Stephenson, J. G., Watkinson, T. G. 1987. Environmental effect of produced water from North Sea oil operations. *Mar.Pollut.Bull.* 18(10):549–558.

Stebbing, A. R. D., Dethlefsen, V. 1991. *Interim Report on the ICES/IOS Bremerhaven Workshop on biological effects techniques. International Council for the Exploration of the Sea.* ICES C.M. 1991/ E:6, 36 pp.

Stegeman, J. J., Moore, M. N., Hahn, M. E., eds. 1993. *Responses of marine organisms to pollutants. Part 2. Proceedings of International Symposium on Responses of Marine Organisms to Pollutants. Woods Hole, 24 Apr. 1992. Mar.Environ.Res.* 35(1–2), 245 pp.

Stradomskaya, A. G., Semenov, A. D. 1991. Level of the oil pollution of water and bottom sediments of the shallow areas of the Caspian Sea. In *Theses of the Second All-Union Conference on Water Toxicology.Vol.2.* —Spb., pp.194–195.(Russian)

Stroemgren, T., Nielsen, M. V. 1991. Spawning frequency, growth and mortality of *Mytilus edulis* larvae, exposed to copper and diesel oil. *Aquat.Toxicol.* 21(3–4):171–180.

Swan, J. M., Neff, J. M., Young, P. C., eds. 1994. *Environmental implications of offshore oil and gas development in Australia—the findings of an independent scientific review.* Sydney: Australian Petroleum Exploration Association Limited, 696 pp.

Tavares, T. M., Rocha, V. C., Porte, C., Barcelo, D., Albaiges, J. 1988. Application of the mussel watch concept in studies of hydrocarbons, PCBs, and DDT in the Brazilian Bay of Todos os Santos (Bahia). *Mar.Pollut.Bull.* 19(11):575–578.

Teal, J. M. 1993. A local oil spill revisited. *Oceanus* 36(2):65–70.

Theobald, N., Lange, W., Jerzycki-Brandes, K. 1992. Baseline value of fossil hydrocarbons in the Northeast Atlantic Ocean. *Dtcsch.Hydrogr.Z.* 44(1):17–33.

Tilseth, S., Solberg, T. S., Westrheim, K. 1981. *Sublethal effects of the water-soluble fraction of Ecofisk crude oil on the early larval stages of cod (Gadus morhua L.). International Council for the Exploration of the Sea.* ICES C.M.1981/E:52. 21 pp.

Tinsli, I. 1982. *Behavior of the chemical pollutants in the environment.* Moscow: Mir, 280 pp. (Russian)

Tkalin, A. V. 1993. Background pollution characteristics of the N.E.Sakhalin Island Shelf. *Mar.Pollut.Bull.* 25(1–4):704–706.

Topping, G. 1992. The role and application of quality assurance in marine environmental protection. *Mar.Pollut.Bull.* 25(1–4):61–66.

Tsiban, A. V., Volodkovich, U. L., Panov, G. V., Khesina, A. A., Ermakov, E. A., Pfeifere, M. U. 1985. The distribution and microbial oxidation of the carcinogenic hydrocarbons (benzo(a)pyrene in particular) in some areas of the World Ocean. In *Ecological consequences of the ocean pollution.*—Leningrad: Gidrometeoizdat, pp.88–112. (Russian)

UNEP. 1990. *The state of the environment. UNEP Regional Seas Reports and Studies No.15.* 85 pp.

UNESCO. 1984. *Manual for monitoring oil and dissolved/dispersed petroleum hydrocarbons in marine waters and beaches. IOC Manuals and Guides No.13.* 35 pp.

Veldre, I. A. 1991. General pattern of accumulation of PAHs and organochlorines in the organisms of the Baltic Sea. In *Theses of the Second All-Union Conference on Fisheries Toxicology. Vol.1.*—Spb., pp.82–83. (Russian)

Venkatesan, M. I., Kaplan, I. R. 1987. The lipid geochemistry of Antarctic marine sediments: Bransfiekld Strait. *Marine Chemistry* 21:345–375.

Vignier, V., Vandermeulen, J. H., Fraser, A. J. 1992. Growth and food conversion by Atlantic salmon parr during 40 days exposure to crude oil. *Trans.Amer.Fish.Soc.* 121(3):322–332.

VNIIP. 1994. *The state of the environment and environmental protection activities in the former USSR.* Moscow: VNIIP, 111 pp. (Russian)

Wang, Z., Fingas, M. 1994. Study of the effects of weathering on the chemical composition of a light crude oil. In *17. Arctic and Marine Oil Spill Program Technical Seminar. Vol.1.*—Ottawa: Environment Canada, pp.133–171.

Weber, R. R., Bicego, M. C. 1990. Petroleum aromatic hydrocarbons in surface waters around Elephant Island, Antarctic Peninsula. *Mar.Pollut. Bull.* 21(9):448–449.

Weber, R. R., Bicego, M. C. 1991. Survey petroleum aromatic hydrocarbons in the San Sebastiano Channel, SP, Brazil, November 1985 to August 1986. *Bohm.Inst.Oceanogr., S.Paulo* 39(2):117–121.

Wheeler, R. B. 1978. *The fate of petroleum in the marine environment.* Exxon Production Research Company Special Report. Houston: Exxon Production Research Company.

Widbom, B., Oviatt, C. A. 1994. The "World Prodigy" oil spill in Narragansett Bay, Rhode Island: acute effects on macrobenthic crustacea populations. *Hydrobiologia* 291(2):115–124.

Widdows, J. 1992. *Role of physiological energetics in ecotoxicology and environmental pollution monitoring. Report of the Working Group on Biological Effects of Contaminants.* ICES C.M. 1992/POLL:5, pp.60–71.

Witt, G. 1995. Polycyclic aromatic hydrocarbons in water and sediment of the Baltic Sea. *Mar.Pollut.Bull.* 31(4–12):237–248.

Yankyavichus, K., Pakalnis, R., Baranauskene, A., Yuknyavichus, L., Yankyavichus, G. 1992. Impact of the oil pollution of the Baltic Sea on the plankton organisms and the role of the water organisms in the self-purification of the marine environment. *Ecologia* 4:2–6. (Russian)

Yunker, M. B., MacDonald, R. W. 1995. Composition and origins of polycyclic aromatic hydrocarbons in the Mackenzie River and on the Beaufort Sea shelf. *Arctic* 48(2):118–129.

Zheng, W., Van Fleet, E. S. 1988. Petroleum hydrocarbon contamination in the Dry Tortugas. *Mar.Pollut.Bull.* 19(3):134–136.

Chapter 5

Biogeochemical and Ecotoxicological Characteristics of Natural Gas in the Marine Environment

In contrast with oil hydrocarbons, which have been an object of wide and detailed ecotoxicological studies worldwide, natural gas and its components have been left outside the sphere of environmental analysis, control, and regulation. Scientific publications on this topic are rare. Modern knowledge about the behavior of hydrocarbon gases in the natural waters and especially their impacts on water organisms, populations, and ecosystems is very limited. This, in turn, limits the possibility of prognoses, assessments, and prevention of environmental effects of accidental gas blowouts and leakages. At the same time, the input of natural gas and products of its combustion into the biosphere is one of the typical and global factors of anthropogenic impact. This factor is present at all stages of oil and gas production, gas transportation, processing, and energetic use in different areas.

This chapter will describe the sources, scales, and biogeochemical characteristics of natural gas and its components in the biosphere. It will also discuss the ecotoxicological aspects of natural gas's impact on water organisms and give the preliminary characteristics of natural gas as an ecological factor in the water environment. Finally, the chapter will outline some methodological approaches to assessing the toxicity of gas hydrocarbons and establishing their permissible levels in the sea.

5.1. Sources, Composition, and Biogeochemistry of Natural Combustible Gas in the Biosphere

Natural gas is closely related to crude oil. Both substances are thought to have formed in the earth's crust as a result of transformation of organic matter due to the heat and pressure of the overlying rock. All oil deposits contain natural gas, although natural gas is often found without oil.

Gas hydrocarbons can also be produced as a result of microbial decomposition of organic substances and, less often, due to reduction of mineral salts. Many of these gases are released into the atmosphere or hydrosphere, or they accumulate in the upper layers of the earth's crust.

Enormous amounts of gas hydrocarbons are present in the dissolved state in underground waters. For example, the average content of methane in the underground waters of the West-Kubanski sag varies within 1–10 m^3/m^3. The total content of methane dissolved in the formation waters is many times higher than all its reserves in gas and oil fields [Sokolov, 1966].

The composition of natural gas varies. It depends on the origin, type, genesis, and location of the deposit, geological structure of the region, and other factors. As Table 32 shows, the chemical composition of natural gas chiefly consists of saturated aliphatic hydrocarbons, i.e., methane and its homologues. The deeper the location of gas deposit, the higher the number of methane homologues. In gas condensate fields, the content of methane homologues is usually considerably higher than the level of methane. In gases associated with

Table 32

Composition of Natural Combustible Gas from the Ust-Tominskoe Oil and Gas Field in the Far East Region [Obzhirov, 1993] and the Shtokmanovskoe Gas Condensate Field in the Barents Sea [Borisov et al., 1994]

	Content (%)		
	Ust-Tominskoe Field at the Well Depth of		
Component	2,786 m	328 m	Shtokmanovskoe Field
Methane	93.30	95.00	96.21
Ethane	0.76	2.07	1.33
Propane	0.07	0.62	0.37
Butane	0.04	0.04–0.13	0.19
Carbon dioxide	0.07	1.42	0.27
Hydrogen	0.19	0.04	–
Helium	0.004	0.001	0.01
Argon	0.12	0.04	–

oil, the content of methane homologues is comparable with the content of methane. Large amount of gases associated with oil is dissolved in this oil. During oil extraction, as the pressure goes down, gases come to the surface of the oil. They are released in the environment in volumes of 30–300 m^3 for every ton of extracted oil. These gases give about 30% of the gross total production of combustible gases in the world. However, over 25% of this amount are flared off because of the absence of the needed capacities and equipment for gas collection and processing.

Other components commonly found in natural gas are carbon dioxide, hydrogen sulfide, nitrogen, and helium. Usually, they constitute an insignificant proportion of natural gas composition. However, in some areas, their concentrations can be considerably higher. For example in the Orenburgskoe and Astrakhanskoe gas condensate fields, a high reservoir drive and large amounts of condensate (up to 6%) lead to a drastic increase in the amount of hydrogen sulfide (up to 30%), carbon dioxide (up to 14%), and mercaptan in the gas's composition [Karamova, 1989]. At the same time, such anomalies occur rather rarely and have an exclusively local nature.

Besides the previously mentioned sources of natural gas (transformations of organic matter in the earth's crust, microbial decomposition of organic substances, and reduction of mineral salts), gas hydrates are another extremely promising source of gas hydrocarbons on the sea bottom. According to some estimates [Zubova et al., 1990; Kellard, 1994], the reserves of gas hydrates are an order of magnitude higher than potential recoverable gas resources of all conventional fields in the world.

From the physicochemical point of view, gas hydrates can be considered as a modification of ice that has a high content of gas. They are solid crystallized substances that look like compressed snow. Hydrates form during the interaction of many components of natural gas (methane, ethane, propane, isobutane, carbon dioxide, and hydrogen sulfide) with water under certain combinations of high pressure and relatively low temperature. Figures 44–46 give the phase diagrams and equilibrium curves for pure methane and natural gas. They also show the depth of hydrate formation in the bottom sediments when considering water temperature.

Hydrate formation usually accompanies and complicates gas and oil extraction and transportation because hydrates can accumulate on the sides of wells and pipelines and thus plug them. The methods used to overcome these difficulties include pumping different inhibitors (methanol, glycol, and solutions of potassium chloride)

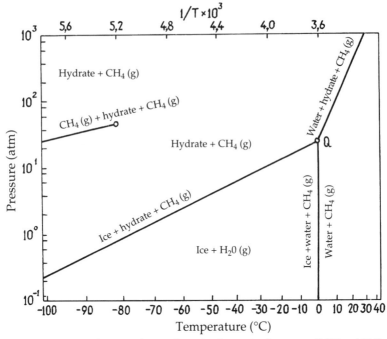

Figure 44 Phase diagram for methane hydrates in the ocean [Miller, 1974].

into the wells and pipelines, dehydrating the gas, and heating it up to temperatures higher than the temperature of hydrate formation.

Table 33 gives a general idea about the sources of gas hydrocarbons in the atmosphere and hydrosphere. Similar to oil, gas enters the environment due to both natural and anthropogenic processes. However, the lack of necessary data complicates the quantitative assessment of the contribution of these sources into the total input of methane. Tables 34 and 35 present the results of some earlier attempts of such calculations. Later studies give somewhat different estimates of the relative contribution of methane sources. For example, one of them [Zavarzin, 1995] suggests that the largest global sources of atmospheric methane include marshes (21.3%), rice fields (20.4%), and ruminant animals (14.8%) followed by the products of biomass burning (10.2%), natural gas seeping (8.3%), garbage dumps (7.4%), and coal production (6.5%). The total amount of gas annually entering the atmosphere is presently estimated at 310–990 million tons [Novozhevnikova, 1995].

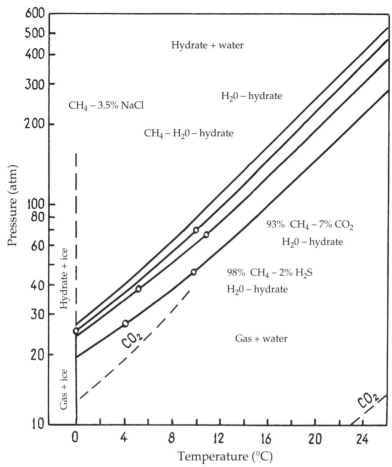

Figure 45 Phase equilibrium for the system water–gas [Claypool, Kaplan, 1974].

Among the major mechanisms of methane natural production in the biosphere, the decomposition of organic matter by methane-producing bacteria (e.g., Methanococcus, Methanosarica) deserves a special mention. These bacteria are able to get the energy by reducing carbon dioxide in accordance with $CO_2 + 4H_2 = CH_4 + 2H_2O$ reaction. Figure 47 gives a general scheme and sequence of these processes that continuously take place under anaerobic conditions in the soil on land and in the bottoms sediments in natural waters. These processes are typical for the silt deposits of lakes and marshes

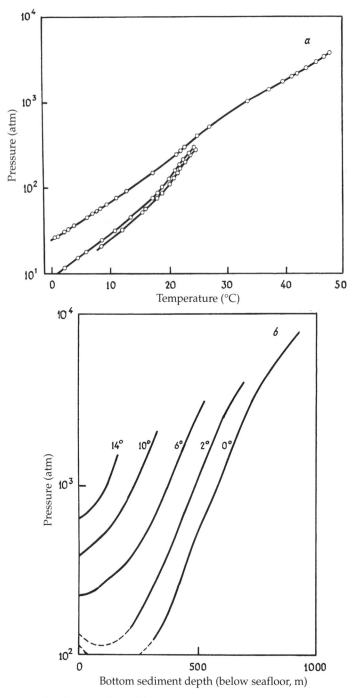

Figure 46 Conditions of equilibrium for pure methane and natural gas (a) and potential depth of gas hydrates formation in bottom sediments of the ocean (for natural gas) with consideration of bottom water temperature (b) [Stole, 1974].

216

Table 33

Sources and Scale of Input of Saturated Aliphatic Hydrocarbons (Methane and Its Homologues) into the Environment

Type and Source of Input	Environment		Scale of Distribution and Impact		
	Hydrosphere	Atmosphere	Local	Regional	Global
Natural:					
Decomposition of organic matter:					
On land	–	+	+	+	+
In the water column	+	–	+	+	?
In the bottom sediments	+	–	+	+	+
Decomposition of gas hydrates	+	–	+	+	?
Seeping from natural formations:					
On land	–	+	+	–	–
On the bottom of water bodies	+	–	+	+	?
Volcanic activity	+	+	+	?	–

Table 33 (continued)

Type and Source of Input	Environment		Scale of Distribution and Impact		
	Hydrosphere	Atmosphere	Local	Regional	Global
Anthropogenic:					
Technological gas emission and leakage:					
Into the atmosphere	–	+	+	+	+
Into the water environment	+	–	+	–	–
Accidents with gas emissions during:					
Drilling	+	+	+	–	–
Gas production and transportation	+	+	+	–	–
Water pollution by oil and oil products	+	–	+	+	?
Decomposition of organic matter:					
On land	–	+	+	+	+
On the bottom of water bodies	+	–	+	–	–
Production and burning of fossil fuel	–	+	+	+	+

Note: +, –, and ? mean, respectively, presence, absence, or uncertainty of corresponding situations.

Table 34
Annual Methane Production on Land [Freyer, 1979]

Source of Input	Annual Production (10^{12} g/year)
Coal burning	6.3–22
Lignite burning	1.6–5.7
Industrial losses	7–21
Automobile exhaust	0.5
Volcanic gases	0.2
Total input	15.6–49.4

Table 35
Annual Methane Balance on the Earth [Freyer, 1979]

Sources	Methane Input and Consumption (10^{15} g/year)
Methane input on land:	
Animal enzyme processes	0.10–0.22
Rice fields	0.28
Swamps and marshes	0.19–0.3
Freshwater lakes	0.001–0.025
Forests and tundra	0.011–0.013
C^{14}-free methane	0.016–0.21
Methane input in ocean:	
All oceans	0.001–0.018
Total annual production	0.6–1.07
Methane decomposition:	
Oxidation of OH	0.86–2.86
Decomposition in stratosphere	0.1–0.21
Total consumption	0.96–3.07

and for marine sediments that are lacking in oxygen and rich in organic matter.

Microbial methane formation in the oceans is usually accompanied by sulfur reduction and the release of hydrogen sulfide. These take place inside the upper part of sediments from the seafloor surface to tens and even hundreds of meters deep. The production of biogenic methane in the upper 2 m of bottom sediments alone totals about 325 million tons a year [Ivanov, 1988]. In regions with a cold

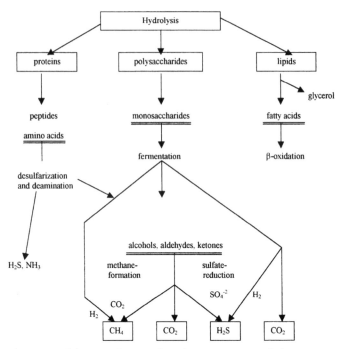

Figure 47 Scheme of anaerobic decomposition of the organic matter [Levin, Ivanov, 1981].

and moderate climate at depths of over 500 m, methane can accumulate in a form of crystal gas hydrates. In areas with a warmer climate, some methane from shallow formations is often released from the sediments into the water column and then into the atmosphere.

Methane can appear in the marine environment not only due to microbial and biochemical decomposition of the organic substance in bottom sediments. It can also occur as a result of the natural bottom seepage of combustible gases from shallow oil- and gas-bearing structures. Such seeping has been found in the Gulf of Mexico, North Sea, Black Sea, Sea of Okhotsk, and other marine areas. This process can lead to intensive vertical flows of hydrocarbon gases from the bottom to the sea surface. Sometimes it is accompanied by gas hydrate decomposition. Some researchers suggest that in the North Sea, natural seepage of methane into the water corresponds to the input of 2.6 million tons of oil and gas hydrocarbons a year [Davis, 1988; Gribbin, 1989]. The authors of this assessment admit that their result may be an overestimation. However, these figures clearly indicate that underwater methane seepage plays a much more significant role in the balance of natural gas than previously considered.

Underwater natural gas seepage takes place on the bottom of the freshwater bodies as well. Sometimes it can have a lethal impact on water organisms. Such situations were observed, for example, on the aquaculture farms in the Rostov region located in the zone of Obukhskoe gas field with shallow gas-bearing formations. In 1984, gas released from the bottom of the ponds contained over 95% methane and caused mass mortality of the pond fish [AzNIIRKH, 1986].

Gas seepage from natural formations into the atmosphere takes place everywhere. This phenomenon has been used by people for ages (for example, "eternal lights" in Dagestan, Azerbaidjan, Iran, and other places). Seepage combined with all other sources of hydrocarbon gases, especially with decomposition of organic matter (see Tables 33–35), form the global reserve of these gases in the atmosphere. The average life-time of methane in the atmosphere is about 10 years. Removal of hydrocarbon gases from the atmosphere takes place mainly as a result of their interaction with the hydroxy (OH-) radical [Koropalov et al., 1988].

Over the last 100 years, the natural processes of biogeochemical production and distribution of methane in the biosphere are under large-scale anthropogenic impact. According to some estimates, anthropogenic sources contribute as much as 40–70% of methane into the global atmospheric flow of this gas [Novozhevnikova, 1995]. Large quantities of hydrocarbon gases are released during many kinds of anthropogenic activity. These include oil, gas, and coal production and transportation, burning of fossil fuels, intensive rice cultivation, animal farming, and garbage dumping. Lately, the increased levels of methane have been found even in areas of intensive aquaculture in the coastal waters. In these areas, methane could be formed as a result of decomposition of food residuals and metabolites of cultivated water organisms.

The global consequence of all these anthropogenic impacts is the gradual increase of methane concentration in the atmosphere over the last 100 years—from $0.7 \times 10^{-4}\%$ to $1.7 \times 10^{-4}\%$ (in volume). This is almost twice as high as the rate of increase in carbon dioxide concentration [Velichko et al., 1997]. The data in Figure 48 show increased methane concentration in the lower troposphere above the Pacific Ocean in the Northern Hemisphere as compared with the Southern Hemisphere. A similar pattern, as Chapter 2 noted, is typical for practically all components of global pollution. This occurs

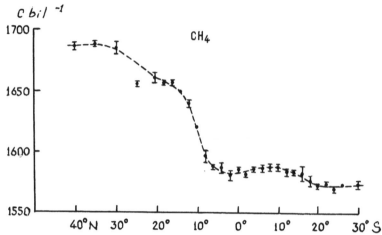

Figure 48 Distribution of methane concentration in the lower part of the troposphere above the Pacific Ocean in 1983 [Koropalov et al., 1988].

because the main anthropogenic sources of their input into the environment are located in the Northern Hemisphere. Many scientists believe that gases released due to human activities have already begun to affect the earth's overall temperature and the methane anthropogenic emission is responsible for about 30% of the total warming effect. If the concentrations of methane and other greenhouse gases in the atmosphere keep increasing, global changes in climatic conditions on the earth will be noticeable in the near future [Masood, 1995; Patin, 1997; Velichko et al., 1997].

The oil and gas activity is one of major anthropogenic sources of gas hydrocarbons. In the United States alone, fugitive emission of methane into the atmosphere from the activity of the gas industry is estimated at 4.4×10^{12} g/year [Beck, 1993] and in Russia—over 14 billion cubic meters a year [VNIIP, 1994]. The volumes and causes of these emissions as well as their distribution can differ in different situations. However, in all cases, the result is mostly air pollution. Hydrocarbons of methane series are not removed from the air masses by precipitation and their water solubility is relatively low.

Another component of natural gas—hydrogen sulfide—is water soluble in contrast with methane. It can cause hazardous pollution situations in both the atmosphere and the water environment. Its proportion in the composition of natural gas and gas condensate, as previously mentioned, sometimes reaches more than 20%. Pollution by hydrogen sulfide can lead to disturbances in the chemical composition of surface waters. This gas belongs to the group of poisons with acute effects. Its appearance in the atmosphere and hydrosphere can cause serious economic damage and medical problems among local population. Unfortunately, in Russia, air, soil, and water pollution by hydrogen sulfide and sulfur dioxide has been reported in a number of regions. Especially severe consequences for human health and biota have been observed in the basin of the low Volga River in the zone of development of the Astrakhanskoe gas condensate field [Ecology and impact of natural gas on organisms, 1989].

The sources of atmospheric pollution also include flaring of natural gas on the offshore platforms and onland terminals. Some estimates [Cairns, 1992] show that about 10% of total gas production and up to 30% of associated gases are burned here. The behavior and distribution of the products of natural gas flaring in the atmosphere, their removal by precipitation, and the impact on the water environment have not been studied. The same situation is true regarding gas emissions at different stages of its production, transportation, and processing. Some studies indicate that due to the activity of British oil companies on the shelf of the North Sea alone, about 75,000 tons of methane enter the atmosphere every year [Somerville, Shirley, 1992]. In Norway, special studies revealed that fugitive hydrocarbon emissions are equivalent to approximately 0.02% of the total gas produced in 1992 on the Norwegian continental shelf [Christensen, 1994]. However, any estimates about the possible influence of natural gas emission from the offshore oil and gas activity on atmospheric processes and global warming are absent thus far.

An important anthropogenic source of gas hydrocarbons in the water environment is the offshore drilling accidents. Their environmental consequences can be very hazardous. Especially dramatic situations developed in the Sea of Asov as a result of two large accidents on drilling rigs in the summer-autumn of 1982 and 1985. These accidents caused long-term releases of large amounts of natural gas into the water accompanied by self-inflaming of the gas. During these events, the levels of methane in surface waters exceeded the background concentrations up to 10–100 times. The air samples also showed very high concentrations of methane. These accidents drastically disturbed the composition and biomass of the water fauna and caused mass mortality of many organisms, including fish and benthic mollusks [AzNIIRKH, 1986; GLAVRYBVOD, 1983; 1986]. Similar incidents probably took place in other regions of the world as well. However, there are no publications on this topic available.

Another potential source of gas in the hydrosphere is damaged gas pipelines, both on the seafloor and on land where they cross over rivers and other water bodies. The causes of such damage can vary from corrosion processes to natural disasters (severe ice conditions, seismic activity, and earthquakes). It should be noted that hydrocarbon gases are piped over great distances totaling many thousands of kilometers. These pipelines cross hundreds of water bodies. Possible pipeline damages can lead to hazardous impacts on water ecosystems. The negative fisheries consequences in such cases may go beyond the limits of local scale. Regional problems can emerge if, for example, an accidental gas blowout or leakage blocks the spawning migration of anadromous fish.

A number of other human activities, besides the oil and gas industry, lead to methane input into the environment. In particular, coal mining contributes about 16×10^{12} g/year [Khalil et al., 1991]. According to different estimates, municipal garbage and waste dumps emit from 9×10^{12} to 70×10^{12} g/year of methane into the atmosphere from degrading of organic residuals [Bogner, Spokas, 1993]. Estimates of global methane emission from rice fields vary from 20×10^{12} g/year to 100×10^{12} g/year. This corresponds to 6–29% of the total annual anthropogenic emission of this gas on the earth [Neue, 1993].

All these anthropogenic inputs of methane predetermine the heterogeneity of its distribution in the biosphere (partially illustrated in Figure 48). They contribute to a noticeable (about 1% a year) increase in methane concentration in the atmosphere. This considerably intensifies the greenhouse effect and increases the likelihood of global climatic changes in the near future. These tendencies, being the subject of worldwide concern, justify the need to enforce regulative measures regarding gas emissions at the national and international levels [Khalil, 1993].

5.2. Concentrations and Distribution of Methane and Its Homologues in the World Ocean

Biogeochemical behavior, concentrations, and distribution of methane and its derivatives in the water environment depend on many factors. First, these gases have low solubility in the water. For example, the solubility of methane in distilled water under normal pressure and temperature of 18–20°C equals 90 ml/l. In seawater with salinity of 36 g/l, the solubility is considerably lower. It ranges from

22–44 ml/l and declines when the temperature increases from 0°C to 30°C [Atkinson, Richards, 1973]. The concentration of methane in seawater under a partial pressure of 2×10^{-6} atmospheres and a temperature of 5°C reaches its equilibrium in 10–20 minutes and equals 8×10^{-5} ml/l [Williams, Bainbrige, 1973].

Table 36 summarizes data on the concentration of methane in the surface waters of different regions. These data suggest that in the open waters of the seas and oceans, the methane content varies from 4.0×10^{-5} ml/l to 8.6×10^{-5} ml/l, 30–80% more than the corresponding equilibrium concentrations. Lowest values (4.0×10^{-5}–4.5×10^{-5} ml/l)

Table 36
Summarized Data on Methane Concentrations in the Surface Waters of Different Regions [Lamontagne et al., 1973; Geodekyan et al., 1994]

Area	Average Concentration ($\times 10^{-5}$ ml/l) Measured	Calculated Equilibrium	Measured Concentration/ Calculated Equilibrium Concentration
Norwegian-Greenland basin	6.4	4.6	1.4
Greenland pack ice	8.8	6.4	1.3
Tropical NW Atlantic	4.0	3.1	1.3
Sargasso Sea	4.3	3.4	1.3
Gulf of Mexico	5.3	3.2	1.6
Mississippi river	24.3	4.3	5.7
Caribbean Sea	4.3	3.1	1.4
Kariako trench	4.9	3.5	1.5
Chesapeake Bay	50.4	3.5	14.0
York river	84.3	4.0	21.0
Potomac river	3,800.0	6.6	575.0
Tropical NE Pacific	4.7	3.5	1.3
Lake-fiord Nittinut	86.0	4.6	19.0
Arabian Sea	5.2	3.1	1.7
Red Sea	4.8	2.8	1.7
Mediterranean Sea	5.3	4.0	1.3
Eastern Mediterranean Sea	5.0	4.0	1.3
Black Sea	7.9	4.5	1.8
Baltic Sea			
Gulf of Danzig	27.4	–	–
Gulf of Riga	52.8	–	–
Sea of Okhotsk	17–40	–	–
Korea Bay	4.0	–	–

are typical for tropical zones. At the low temperatures near the Arctic ice edge, the methane level increases up to 8.6×10^{-5} ml/l. In coastal areas, especially in bays and estuaries, the methane concentration in the surface waters can be 1–3 orders of magnitude higher. This tendency, supported by later studies [Lisitsin, 1983; Geodekyan et al., 1994], is explained not only by pure physicochemical phenomena (dependency of gas solubility upon temperature, pressure, etc.). It is also controlled by the general pattern of distribution of living matter and bioproduction processes in the hydrosphere. These ultimately define the genesis of the combustible gases and their biogeochemical cycle. Other hydrocarbons of the methane series usually accompany methane in natural concentrations and rarely exceed 10^{-6} ml/l.

The summarized data in Table 37 give some idea about the concentrations and distribution of methane and its homologues in the marine environment. These data, as well as other published materials, make it possible to distinguish three principal types of vertical distribution of these gases in the seas and oceans [Geodekyan et al., 1979; Lisitsin, 1983].

The first type can be found in the open ocean. It is characterized by the gradual decreasing of methane concentration as depth increases. The second type shows reverse correlation (i.e., the methane concentration increases with the depth). This is seen especially clear in regions with high levels of hydrogen sulfide, for example in the Black Sea. Here, methane concentrations at 1,000–1,500 m deep are more than an order higher than its levels in the surface waters [Atkinson, Richards, 1973]. The third type of vertical distribution is characterized by increased concentrations of methane in the layers both near the water surface and near the bottom, and by minimum levels in the intermediate horizons.

The remarkable feature of all types of vertical distribution of methane in the marine environment is the presence of its elevated levels in the upper (photic) zone of the water mass. These levels are usually higher (in many cases tens of times) than the equilibrium levels (under the conditions of atmospheric equilibrium). Different hypotheses explain this phenomenon. The most convincing one refers to vertical influx of methane toward the water surface from the sea bottom where large-scale processes of microbial methane formation and direct seeping from oil- and gas-bearing deposits are possible. Probably, these processes can also explain the very high concentrations of methane in sediments and subsurface waters of the Sea of

Table 37
Summarized Data on the Concentrations of Gas Hydrocarbons
($\times 10^{-5}$ ml/l) in the Water Column [Lamontagne et al., 1973;
Swinnerton, Linnenbom, 1976; Geodokyan et al., 1979]

Area	Depth (m)	Methane	Ethane
North Atlantic	500	4.9	–
Norwegian-Greenland basin	2,400	1.9	–
Western Norwegian Sea	3,196	97.6	0.227
Greenland pack ice	1,000	5.6	–
Tropical NW Atlantic	5,000	0.6	–
Gulf of Mexico	3,550	2.0	–
Caribbean Sea (Kariako Trench)	1,300	16,000	–
Central Pacific	4,200	0.3	–
Sea of Japan	5,210	3.0	0.083
Sea of Okhotsk:			
Central areas	1,180	14	–
Deryugin trench	1,480	122	0.203
South-Okhotsk Trench	100	31	Traces
Lake-fiord Nittinat	200	16,000	–
NW Indian Ocean	500	4.3	–
Continental slope of Africa	3,930	20.8	11.3
Gulf of Aden	2,170	41.8	–
Persian Gulf	153	27.4	–
Ridges	2,750	–	–
Abyssal bed	4,030	25.1	0.6
Red Sea	2,050	5,800	–
Atlantis trench	2,190	20,000	408
Discovery trench	2,170	2,190	21.1
Eastern Mediterranean Sea	2,000	6.2	–
Black Sea	2,000	17,000	–
Caspian Sea (near Apsheron)	150	10–539	0.27–1.25

Okhotsk, as shown in Figure 49 and supported by further studies on the Sakhalin shelf [Geodekyan et al., 1994].

The same studies found the effect of accumulating high amounts of gas (up to a 40–50-fold increase for methane levels) in the water column under the ice. They proved the critical role of primary biochemical processes taking place in the ice and near the water-ice boundary for self-purification of the water environment in the Far North areas [Geodekyan et al., 1994]. The rate of primary biochemical processes (based on a number of microbiological and biochemical parameters) as well as the concentrations and distribution of methane and its homologues in the water and bottom sediments

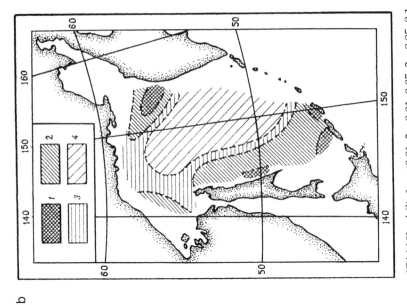

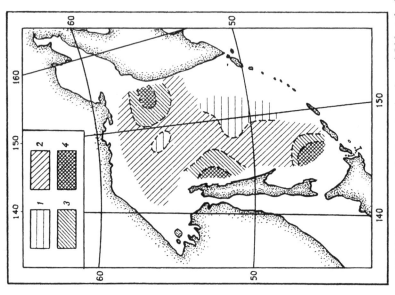

Figure 49 Methane concentrations in the Sea of Okhotsk: (a) surface waters ($C \times 10^2$ ml/l): 1—<0.01; 2—0.01-0.05; 3—0.05-0.1; 4—<0.1; (b)—bottom sediments (200±15 cm under the seafloor surface; $C \times 10^3$ ml/kg): 1—>100; 2—100-10; 3—10-5; 4—<5 [Geodokyan et al., 1979].

can be used as reliable integrative indicators. They reflect the state of the marine ecosystems and their disturbances under the impact of natural and anthropogenic impacts, including oil pollution.

American authors [Brooks, Sackett, 1973; Sackett, Brooks, 1975] revealed a dramatic (up to 5 orders of values) increase in the concentration of dissolved low-molecular-weight hydrocarbons (mainly methane and ethane) in the areas of oil production in the Gulf of Mexico as compared with background levels. Figure 50 presents some of these data. The area around one oil platform had especially high concentrations of ethane in seawater (about 0.3 ml/l, which was 85,000 times higher than the background level). This influenced a zone of over 25 square miles. The authors suggested that even such high concentrations of dissolved gases did not have a considerable impact on the life in the marine environment. However, their presence can serve as an indicator of the possible presence of other, more toxic, components of oil and gas pollution. At the same time, other studies did not reveal increased concentrations of methane and ethane in the water

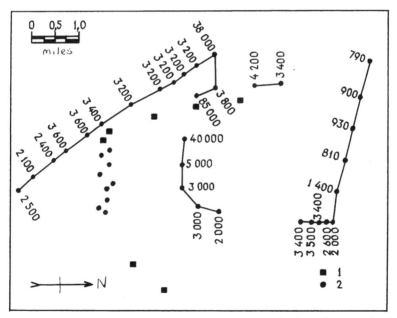

Figure 50 Ethane concentrations in seawater around oil production platforms in the Gulf of Mexico (28°58'N, 92°00'W) in relation to background level (3×10^{-6} ml/l at S.T.P.). Solid squares are locations of platforms; solid circles are exposed well heads [Sackett, Brooks, 1975].

column in the vicinity of oil production platforms [Koons, Monaghan, 1973]. Therefore, everything depends on the actual situation under specific conditions.

The geochemical studies of hydrocarbon gases (mainly methane), including the ones previously discussed, have both scientific and practical significance. They set a base for assessing the potential of oil and gas fields in marine areas. The correlation between the natural gas content/distribution in the water environment and the intensity of bottom gas seepage as well as the depth of oil and gas deposits in different areas of the World Ocean is now beyond any doubt.

A detailed study in the Far East gives convincing evidence of such correlation [Obzhirov, 1993]. The materials of this work, given partially in Table 38 and Figures 51 and 52, show that in the near-bottom waters above the shelf oil and gas fields, zones of extremely high concentrations of hydrocarbon gases exist. The content of methane and some other hydrocarbons in such zones is 10–100 times higher than the background levels. These patterns suggest that the

Table 38

Concentrations of Hydrocarbon Gases in the Bottom Waters of the Far-eastern Seas (Summary Based on [Obzhirov, 1993])

Area	Depth (m)	Concentration ($\times 10^{-5}$ ml/l)		
		Methane	Ethane	Ethylene
Sea of Okhotsk:				
Shelf of Sakhalin (1984–1985)	30–1,400	3–155	0–2	0–0.3
Shelf of Sakhalin (1988–1989)	30–1,210	7–1,355	0.03–0.1	0–0.2
Shelf of Kamchatka	75–550	7–23	0.03–0.3	0.07–0.3
Northern shelf of the Sea of Okhotsk	130–180	5–20	<0.5	<0.5
Shelf of Kuril Islands	50–800	1–20	–	–
Sea of Japan:				
Tatar Strait	–	9–18	0.2–2.0	0.2–2.0
Tatar Strait (coastal waters)	10–50	2–142	0.1–1.8	0–0.2
Korea Strait	25–2,500	2–130	<11	<11
South China Sea	30–2,500	2–154	0 0.3	–

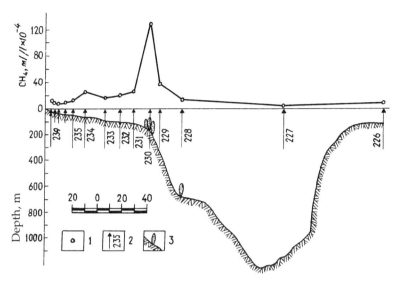

Figure 51 Geochemical profile of natural gas concentrations in the northern part of Deryugin trench (shelf of the Sea of Okhotsk near Sakhalin): 1—concentration of methane in the water near bottom (ml/1 $\times$ 10^{-4}); 2—station number; 3—bottom gas seepage [Obzhirov, 1993].

information about the zones of very high accumulations of hydrocarbons in the water can be used to locate potential oil- and gas-bearing formations. At the same time, no quantitative description of this general pattern is available. Much factual data are lacking, mainly due to analytical difficulties in determining the components of natural gases in the water environment.

These difficulties also complicate assessing the environmental situations in areas of offshore oil and gas field developments. Today, the information about the content and distribution of methane and its homologues in these areas as well as the environmental consequences of gas emissions and accidental blowouts on drilling platforms and gas pipelines is very limited. This gap in knowledge cannot be explained by the lack of such incidents (they happened in the past and can happen again). Rather, this is a result of insufficient attention to this impact factor and the absence of appropriate methodology.

Hope comes from the last analytical developments based on the use of infrared and laser absorption equipment [Wernecke et al.,

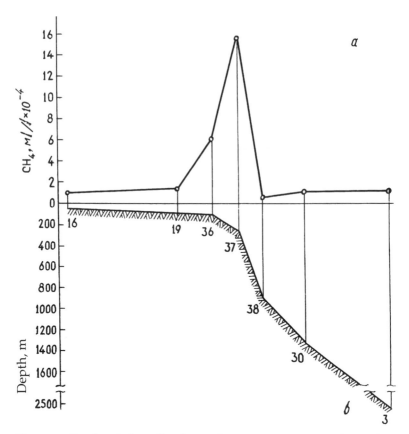

Figure 52 Geochemical profile of natural gas concentrations in the northern part of South-Konshon trench in the Korea Strait: a—concentration of methane in the water near bottom; b—seafloor profile [Obzhirov, 1993].

1991; Anonymous, 1994]. Such devices allow researchers to conduct fast (within 1.5 minute) analyses of water pumped from up to 300 m deep. The water continuously runs through an analytical system, which makes possible to measure the concentration of dissolved methane ranging from 15×10^{-6} mg/l to 5×10^4 mg/l.

In conclusion, the data on the marine biogeochemistry of natural gas, including the previously discussed materials, can and must be used to assess the environmental situation in areas of offshore oil and gas production. Measurements of hydrocarbon gases in seawater and bottom sediments should become the necessary element of monitoring in these areas.

5.3. Methane Impact on Water Organisms and Communities

The regrets, expressed in the previous sections, about the insufficient knowledge of the biogeochemistry of hydrocarbon gases in the water environment are even more justified when considering the biological effects of these gases on water organisms and communities. Few data on this issue can be found among the water toxicological studies published in Russia and abroad over the last 10 years. This suggests that water toxicology of saturated aliphatic hydrocarbons of the methane series has not been developed thus far. This gap cannot be filled by available materials on the toxicity of other gaseous poisons (e.g., carbon oxide, hydrogen sulfide, and ammonia) for fish [Metelev et al., 1971; Lukyanenko, 1983]. Clear behavioral specifics of each of these gases in the water environment do not allow us to extrapolate these data to predict the biological effects of methane and other saturated hydrocarbons.

5.3.1. General Outline and Mechanisms of Biological Response

As indicated above, available data on water toxicology of gaseous poisons other than methane and its homologues cannot be used to predict the specifics of biological responses to the presence of natural gas. However, they can help to reveal some general features of interaction between gaseous traces and marine organisms [Patin, 1993].

The first important feature is the quick fish response to a toxic gas as compared with fish response to other dissolved or suspended toxicants. Gas rapidly penetrates into the organism (especially through the gills) and disturbs the main functional systems (respiration, nervous system, blood formation, enzyme activity, and others). External evidence of these disturbances includes a number of common symptoms mainly of behavioral nature (e.g., fish excitement, increased activity, scattering in the water). The interval between the moment of fish contact with the gas and the first symptoms of poisoning (latent period) is relatively short.

Further exposure leads to chronic poisoning. At this stage, cumulative effects at the biochemical and physiological levels occur. These effects depend on the nature of the toxicant, exposure time, and environmental conditions. A general effect typical for all fish is gas emboli. These emerge when different gases (including the inert ones) oversaturate water. The symptoms of gas emboli include the rupture of tissues (especially in fins and eyes), enlarging of swim bladder, disturbances of circulatory system, and a number of other pathological changes.

These general features of fish response observed in the presence of any gas in the water environment are likely to be found for saturated gas hydrocarbons as well. Available materials derived from the medical toxicology of methane and its homologues support this suggestion. Some studies [Izomerov, 1984] note that methane from water solutions easily penetrates through the skin. Depending on its concentration, this can cause narcotic or toxic effects in mammals. These effects are based on the hypoxia that radically increases in the presence of ethane, propane, butane, and other homologues of this series. Medical toxicology distinguishes between three main types of intoxication by methane:

- light, results in reversible, quickly disappearing effects on the functions of central nervous and cardiovascular systems;
- medium, manifests itself in deeper functional changes in the central nervous and cardiovascular systems and increase in the number of leukocytes in the peripheral blood; and
- heavy, results in irreversible disturbances of the cerebrum, heart tissues, and alimentary canal as well as acute form of leukocytosis.

These types most likely adequately describe the general patterns of methane effects in vertebrates. However, its features in respect to ichthyofauna remain to be studied. Fish resistance to the presence of gas at different life stages is of special interest. With most toxicants, the most vulnerable periods are the early life stages [Patin, 1979]. Some ichthyotoxicological studies showed this for the poisonous gases, in particular ammonia [Solbe, Shurber, 1989]. The question of whether this general pattern is typical for saturated hydrocarbons still remains open. The importance of this issue in assessing biological effects of natural gas in the water environment is quite obvious.

During toxicological studies of different gases, including methane and its derivatives, one must take into consideration the influence of other factors (especially temperature and oxygen regime) that can radically change the direction and symptoms of the effect. In particular, increasing temperature usually intensifies the toxic effect of practically all substances on fish because of the direct correlation between the level of fish metabolism and water temperature. Experiments showed this for some poisonous gases, for example, for hydrogen sulfide [Metelev et al., 1971]. The increased temperature considerably shortened the time when the first symptoms of fish poisoning and the lethal effects developed. From the physiological perspective, this can be explained not only by the general intensification of fish metabolism but also by the increased permeability of the tissues for the poisons and increased oxygen consumption under high temperatures. Thus, toxicant concentrations that do not cause any effect under low temperatures can become lethal with increasing water temperature. This circumstance should be taken into consideration during ecotoxicological assessment of the potential impact of natural gas and other toxicants, especially when studies are conducted in high latitudes. In such regions, methane hydrates may be accumulated during the winter and dissociate during the increased temperatures in the summer. This may be followed by the releasing of free methane with corresponding environmental consequences.

Another critical environmental factor that directly influences the gas impact on water organisms is the concentration of dissolved oxygen. Numerous studies show that the oxygen deficit directly controls the rate of fish metabolism and decreases their resistance to many organic and inorganic poisons. This decrease sometimes depends more on the species characteristics and the rate of their gas metabolism rather than on the nature of the poison. From the physiological perspective, such a phenomenon is explained by the fact that the level of hemoglobin in fish blood and the rate of blood circulation through the gills increase under oxygen deficit. Clearly, such effects are of special interest when interpreting the data on fish response to natural gas in situations of significant change in the oxygen regime (e.g., during eutrophication of water bodies or seasonal and weather variations of the oxygen content).

5.3.2. Field and Laboratory Studies

Field and experimental studies support the previously described general pattern of fish response to the presence of methane and its homologues in the environment. In the Sea of Asov, researchers conducted detailed observations after accidental gas blowouts on drilling platforms during summer–autumn of 1982 and 1985 [GLABRYBVOD, 1983; AzNIIRKH, 1986]. The results of these observations indicate the existence of a cause-effect relationship between mass fish mortality and large amounts of natural gas input into the water after the accidents. Fish in the zones of the accidents developed significant pathological changes. In particular, they displayed impaired movement coordination, weakened muscle tone, pathologies of organs and tissues, damaged cell membranes, disturbed blood formation, modifications of protein synthesis, radically increased total peroxidase activity, and some other anomalies typical for acute poisoning of fish. These pathological changes were found even in the fish collected at a considerable distance from the place of accident. Similar anomalies were observed in fish (flounder, sturgeon) kept for 4–5 days in the net cages in the direct vicinity of the mouth of the accidental gas well. Fish caught on the control stations and fish kept in the control cages did not show any deviations from the norm.

Significantly, some fish showed species-specific features of response to natural gas exposure. For example, flounder was more sensitive to the effect of natural gas than sturgeon. In 1982 and 1985 respectively, 69% and 28% of the flounders kept in the experimental net cages died. However, no sturgeon mortality was observed for the time of the experiments.

Besides the ichthyotoxicological data, studies on gas accidents in the Sea of Asov give some idea about the methane pollution of the water environment and its possible impact on the benthic and pelagic communities. Table 39 shows the levels and scale of methane distribution in the water environment in the impact zone of the gushing gas during September 1985 [AzNIIRKH, 1986]. Methane represented over 95% of the released gas. It was present in water in concentrations of 4–6 mg/l directly near the accidental well and in concentrations of 0.07–1.4 mg/l at a distance of 200 m from the platform. The increased

Table 39

Concentration of Dissolved Methane in the Sea of Asov During the Accidental Gas Blowout in September 1985 [AzNIIRKH, 1986]

Horizon	Distance from the Platform (m)	Direction from the Platform	Concentration of Methane %	mg/l	Wind Direction
Near bottom	0	Eastward	28	6	South-west
Surface	0	–"–	20	4	–"–
Near bottom	200	–"–	6.5	1.4	–"–
Surface	200	–"–	5.9	1.3	–"–
Near bottom	200	Southward	3.3	0.7	–"–
–"–	200	Westward	0.3	0.07	–"–
–"–	200	Northward	3.3	0.7	–"–
–"–	500	North-eastward	1.6	0.35	–"–

content of this gas (0.35 mg/l) was also found 500 m from the well in the windward direction. These results suggest that methane and its homologues can stay in the water environment for a rather long period and spread over considerable distances. Similar conclusions were made based on observations in the Gulf of Mexico, where the areas around offshore drilling rigs had extremely high concentrations of methane and ethane in the water (see Figure 50).

Information about the effects of methane and its homologues on water communities is very limited. Data indicate that benthic ecosystems have been disturbed and their trophic structure has changed in areas of methane seepage on the shelf of the North Sea and near the shore of California. Dense populations of *Beggiatoa* sp. were found in bottom sediments of these areas. These microorganisms use oil and gas hydrocarbons as a food source. In turn, they can become the base of the food chain for other benthic organisms [Davis, 1988; Howard, Thomsen, 1989].

Such symbiotic communities and ecosystems dependent on methane oxidation by microorganisms (mainly Methylococcaceae) appear to be typical for areas with high levels of methane in the bottom environment. In particular, they were recently found in areas of gas hydrate formation and gas seepage in the Black Sea and the Sea of Okhotsk [Galchenko, 1995]. The enzyme systems of bivalve mussels that were part of these ecosystems acquired some specific features due to the close symbiosis with methane-oxidizing bacteria.

Table 40 gives some information about the impact of natural gas on marine communities. It presents the results of field studies around the accidental gas well in the Sea of Asov [AzNIIRKH, 1986]. These results suggest that gas affects zoobenthic organisms more than the bacterioplankton and phytoplankton. In areas with a high concentration of methane, the biomass of benthos declined, in particular, because of the mollusk mortality. Some declining of the zooplankton biomass also occurred in the vicinity of the accidental well. However, the high variability of the zooplankton parameters and insufficient amount of available data do not allow us to make any reliable conclusion.

Experimental toxicological studies of the effects of methane and its homologues on water organisms are very limited. Some of them describe the responses of fish and zooplankton to bottled gas (mainly propane) exposure [Sokolov, Vinogradov, 1991; Umorin et al., 1991; Patin, 1993]. One of the studies suggests that under experimental conditions, low-molecular-weight hydrocarbons (methane and others) do not cause harmful effects on marine phytoplankton even at high water concentrations [Sackett, Brooks, 1975].

Table 40
Ecological Parameters in Area of Gas Blowout in the Sea of Asov in September of 1985 [AzNIIRKH, 1986]

Parameters	Near the Accidental Well	200 m Eastward	200 m Northward	200 m Westward	200 m Southward	500 m from the Well
Bacterioplankton (10^6 cells/ml)	0.25	0.81	0.14	0.64	0.35	0.35
Phytoplankton (mg/m³)	65.40	78.30	52.30	31.20	17.10	88.70
Zooplankton (mg/m³)	9.10	–	–	25.60	–	16.80
Zoobenthos (g/m²)	161.40	203.00	436.00	558.70	406.0	453.80

Laboratory experiments conducted at the Russian Federal Research Institute of Fisheries and Oceanography [Patin, 1993] imitated the conditions of gash (accidental) releases of bottled gas into the water environment. They revealed that immediately after beginning the gas input into the water, fish (young specimen of carp) showed obvious signs of excitement and increased motor activity. They scattered along the experimental vessels. The fish also stopped swallowing atmospheric air, probably because the air bladder was filling with the gas released into the water. Under the impact of subsequent gas releases, fish motor activity slowed, most specimens went down to the bottom, their movements became sluggish, and any responses on physical stimulation (knocking, touching) disappeared. By the end of the experiments, which lasted 60–120 minutes, the fish school behavior was totally disturbed. Some specimens sluggishly and chaotically moved toward the surface. Some settled on the bottom. Most fish showed signs of a balance disturbance and turned on their side.

Studies of behavioral responses to the presence of gas showed a rather high olfactory sensitivity of the fry of bream, perch, and other fish [Sokolov, Vinogradov, 1991]. For example, avoidance effects were clearly seen when concentrations of dissolved gas ranged from 0.1–0.5 mg/l. The threshold concentrations were lower (and hence the sensitivity of behavioral response was higher) for the fry of bream than for the fry of perch. After repeated exposure of fish fry to the short-term impacts of the threshold gas levels, the sensitivity of all fish increased. Avoidance effects were observed in the presence of 0.02–0.05 mg/l of gas. When gas levels rapidly increased, avoidance responses were suppressed. This led to the quick death of the fish.

The concentrations of bottled gas that caused the death of 50% of the fish during 48 hours (LC_{50}) equaled 1–3 mg/l [Umorin et al., 1991]. For zooplankton, this concentration during a 96-hour exposition was 5.5 mg/l without air pumping and 1.75 mg/l with it. These results suggest that fish are more vulnerable to the effects of methane homologues than zooplankton. They also indicate that acute toxic gas effects in fish start under minimum concentration of about 1 mg/l, which approximately match the results from field observations as previously described. Some other studies give similar values of LC_{50} (96 hour) of natural gas for zooplankton, zoobenthos, and fry of marine fish (0.6–1.8 mg/l) [Borisov et al., 1994; Kosheleva et al., 1997].

The picture of fish response to the exposure of methane and its homologues in the water agrees with the general pattern of organismal response to any toxic or stress impact. This pattern involves consequent stages of indifference, stimulation (excitement), depression, and death of the organism [Metelev, 1971; Patin, 1979; Lukyanenko, 1983]. The previously described experiments suggest that along with the general pattern, some specifics of fish response to the acute impact of natural gas can be distinguished.

In particular, the primary fish response to the gas presence develops much faster than fish response to most other toxicants in the water. Clear signs of such response—the radically increased motor activity of the fish—are observed within the first seconds after gas goes into the water. Fast penetration of methane homologues into the living cells results in the instant impact of gas on fish gills, fish skin, and some other chemoreceptors. The high speed of behavioral response is most likely associated with the similarly rapid impact on the central nervous system.

Another feature of fish response to the gas exposure is a relatively short period between the first contact with gas and persistent signs of their poisoning (latent phase). The duration of this phase in acute experiments is 15–20 minutes. After this time, clear symptoms of acute poisoning indicate the beginning of the lethal phase. This includes the loss of movement coordination, disturbances of breathing, and others [Patin, 1993]. In gas concentrations of 1 mg/l and higher, lethal effects are clearly seen after 1–2 days of exposure.

Thus, in spite on the lack of research, especially under chronic exposure, the observations of both fish behavioral responses and fish mortality suggest a relatively low resistance of ichthyofauna to the presence of natural gas in the water environment. The high speed of primary responses, their clear manifestation, and their relatively short latent phase indicate a possibly damaging impact on the central nervous system of fish. Some data show the likelihood of higher resistance of zooplankton and benthos to the impacts of methane and its homologues. However, their responses still must be studied in the future.

5.4. Toxicological Characteristics of Gas Condensate and Gas Hydrates

Gas condensate is a liquid mixture of hydrocarbons (with five and more carbon atoms) of different structure and dissolved gases of methane-butane fraction. Data, given in Table 41 for the Shtokmanovskoe gas condensate field, show that the chemical composition of gas condensate is based on the fractions of paraffins and naphthenes with traces of aromatic and heteroatomic compounds.

Table 41
Fractional Composition of Gas Condensate from the
Shtokmanovskoe Gas Condensate Field [Borisov et al., 1994]

Fractions	Concentration (%)
Low-molecular-weight paraffins and naphthenes (high volatile)	32.4
High-molecular-weight paraffins and naphthenes	54.3
Aromatic compounds (monocyclic and polycyclic)	6.5
Hydrocarbon derivatives (heteroatomic compounds)	6.8

During contact with the atmosphere or water environment, volatile fractions of gas condensate rapidly (within several hours and days) evaporate. This leads to a corresponding decrease in its toxicity. The rest of the paraffins, naphthenes, and aromatic compounds partially dissolve in the water. They then undergo physicochemical and microbial degradations similar to the ones described in the previous chapter for the crude oil. The solubility of gas condensate in seawater at 5–6°C and atmospheric pressure equals approximately 1 g/l.

Table 42 gives some idea about the toxicity of gas condensate for marine organisms. It presents the results of one of the very few publications on this topic [Kosheleva et al., 1994]. This study suggested that planktonic crustaceans were the most sensitive to the gas condensate exposure. Gastropod mollusks and fish showed more resistance. The authors stressed the high speed of most responses as well as their similarity to the responses of some organisms (especially fish) to light oils and oil products. This is quite understandable when considering that the composition of gas condensate is similar to the benzene or kerosene fraction of oil or their mixture. The lowest observed effect concentrations (LOEC) of gas condensate for most organisms were 0.05–0.5 mg/l. At the same time, the safe levels (no observed effect concentration—NOEC) for phytoplankton and the most sensitive larval stages of zooplankton were estimated at 0.01 mg/l.

The 50% mortality of adult littoral *Gammarius* was observed at the gas condensate concentration of 0.2 ml/l and the 100% mortality—at 6.4 ml/l during the 96-hour exposure. The values of LC_{50} for the adult and fry salmon equaled 0.1 mg/l and 0.05 mg/l, respectively [Borisov et al., 1994].

As previously mentioned, the toxicity and physical properties of gas condensate are similar to the characteristics of light oils. Therefore, the appearance of gas condensate in the marine environment,

Table 42

Results of Toxicological Studies of Gas Condensate from the Shtokmanovskoe Field (summary based on [Kosheleva et al., 1994; 1997])

Groups and Species of Marine Organisms	Effect Criteria Test Duration	Concentration (mg/l)		
		Lethal	LOEC*	NOEC**
Phytoplankton (*Halospaera viridis, Rizosolenia sp.*)	Photosynthesis intensity 20 days	–	0.1	0.01
	Increase in abundance 20 days	100	10	1
Crustaceans (*Acartia longiremis, Gammarus fimarchicus*)	Survival 15 days	–	0.1	–
Larvae	–"–	1–100	0.05	0.01
Adults	–"–	5–50	–	–
Mollusks (*Littorina obtusata*)	Survival	–	2.5	
Fish, fry (*Gadus morhua*)	Survival 12 hours	2,500	–	–
	Survival 30 days	–	–	0.1–500
	Blood composition 30 days	–	2.5	–
	Breathing rate 30 days	–	10.0	0.1–0.5
	Histological disturbances in gills	–	2.5–5.0	–
	Organoleptic parameters	–	500	–

Notes: *—Lowest observed effect concentration. **—No observed effect concentration.

for example during accidental spills, can cause more serious consequences than emissions of natural gas [Borisov et al., 1994].

Ecotoxicological experimental studies or field observations of gas hydrates are practically absent. We might suggest that under certain conditions, the fine fractions of gas hydrates could spread over large distances. They can then become the source of secondary water pollution by methane and its homologues, posing certain danger to marine organisms. Evidence indicates the possibility of suspended gas hydrates accumulating in zooplankton organisms. This can result in their poisoning (including impaired ability to migrate and feed actively) as well as in the death of fish feeding on such plankton [Novoselov et al., 1992]. Some observations also indicate the existing of trophic and biochemical transformations of benthic ecosystems in areas of gas hydrate accumulations and dissolution on the bottom [Galchenko, 1995].

One of the widespread methods used by the oil and gas industry to prevent hydrate formation in the systems of circulation and transportation of natural gas includes the usage of methanol (methyl alcohol). Methanol is one of the products of gas refining. It can become a component of possible pollution in areas of offshore oil and gas production. During the development of the Shtokmanovskoe gas condensate field, the amount of methanol stored in underwater tanks and transported by tankers will be thousands of tons. The toxic properties of this highly soluble compound and its impacts on marine organisms have not been studied thus far. At the same time, it is well known that it has narcotic and nervous-paralytic effects. The sanitary-toxicological maximum permissible concentration (MPC) of methanol in the natural waters, adopted in Russia, equals 0.1 mg/l. At the same time, data [Shparkovski, 1993] indicate that 0.05 mg/l of methanol in the water causes the death of salmon fry, and the threshold of physiological sensitivity to this toxicant is less than 0.005 mg/l. Such a discrepancy clearly indicates the necessity of additional studies and the corresponding correction of currently adopted MPC.

5.5. Levels, Thresholds, and Zones of Biological Effects

Logically, this chapter should conclude by summarizing data on the water levels of gas hydrocarbons in different areas of the World Ocean. It should also examine the thresholds that cause different biological effects. The previous chapter used such an approach while discussing oil hydrocarbons. Doing so has several advantages.

First, this approach allows us to present a general picture of the natural gas distribution in the hydrosphere under the influence of natural and anthropogenic factors. The graph summarizes the results of biogeochemical studies of gas hydrocarbons and their background characteristics. This representation can be very useful for monitoring the ecological situation in areas where methane and its homologues are possibly released into the water environment.

Second, this method illustrates the concept of natural variability of environmental factors (including the natural levels of hydrocarbon gases) influencing water biota. This concept is based on the notion of ecological spectrums developed in hydrobiology by Zhadin [1956]. Distinguishing ecotoxicological zones and threshold levels of methane in the marine environment helps to predict the most typical biological effects in different areas.

Third, the combined analysis of biogeochemical and ecotoxicological materials helps to develop more adequate standards of permissible water levels for a number of contaminants, including hydrocarbons of the methane series. Current methodology uses the results of only experimental studies. As noted in a number of publications [Patin, 1979; 1988; Semenov et al., 1991; Volkov et al., 1993], this ignores the natural variability in levels and distribution of the studied components in the hydrosphere. The proposed method gives the opportunity to overcome this limitation providing a sounder foundation for determining permissible levels of water pollution.

The scheme given in Figure 53 takes into account these considerations. This figure graphically interprets the previously discussed biogeochemical and toxicological data. It shows the range of methane concentrations typical for different ecological zones of the World Ocean on a background of the typical toxicological gradation of biological responses (effects) depending on different levels of toxic impact. The threshold levels and concentration ranges given in Figure 53 are approximate. They should be revised as new data are received.

The graph suggests that the zone of acute toxicity, where the death of a certain number of specimens within 2–4 days is inevitable, begins from methane levels of about 1 ml/l and higher. Below this level, within the range of approximately 0.1–1.0 ml/l, the zone of sublethal responses occurs. These concentrations don't lead to the death of water organisms but can cause clear physiological, biochemical, behavioral, and other anomalies, including the disturbance of reproductive functions. Even lower on the scale are the zone of primary responses (usually reversible after the removal of the impact) and the zone of ecological tolerance.

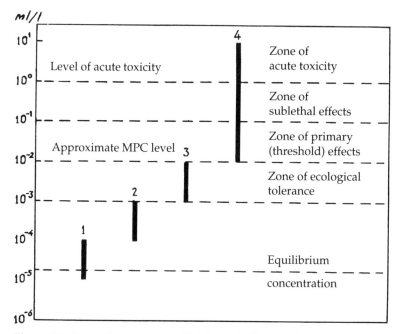

Figure 53 Approximate zones of biological effects and ranges of typical levels of methane in seawater (concentrations of 1 ml/l corresponds to the weight concentrations of 0.72 mg/l under normal conditions): 1—pelagic areas (open waters) of the seas and oceans; 2—coastal areas; 3—bays and estuaries; 4—areas of bottom seepage and accidental blowouts and emissions of gas. MPC—maximum permissible concentration.

The first three zones mentioned above correspond to very high concentrations of methane in the marine environment. These typically occur in areas of intense flows of microbial methane from bottom sediments and seepage of natural gas from gas-bearing bottom structures. Such areas can be found in the Gulf of Mexico and the Persian Gulf as well as in the North, Black, and Caspian Seas, and marine areas of the Far East. Evidence suggests that specific benthic communities, including methane-oxidizing microorganisms, develop in the environment with increased content of methane. Such methane-dependent communities serve as the base of the food chain for a host of fauna. Their species and trophic structure must be studied in the future. However, even at present, there is no doubt about the wide distribution and large scale of these phenomena and their important role in the marine ecosystems [Ivanov, 1988; Geodekyan et al., 1994; Galchenko, 1995].

The activity of the offshore industry accompanied by chronic routine or abrupt accidental releases of gas hydrocarbons can cause corresponding environmental and biological effects. These include the fast development of methane-dependent microorganisms, changes in benthic and plankton communities, disturbances of fish migrations, and others. In all probability, such phenomena have short-term and local nature.

The zone of ecological tolerance, shown in Figure 53, corresponds to natural methane levels in the bays and estuaries. As mentioned in Section 5.2, the concentrations of methane here are considerably higher than the average levels in the open waters of seas and oceans. This zone is of special interest. Its upper limit is essentially a natural, evolutionary-determined limit of the acceptable content of a natural microcomponent (in this case—methane) for marine biota. It can be called the biogeochemical threshold of ecological tolerance [Patin, 1979].

Undoubtedly, some water organisms can survive and reproduce at the levels exceeding this threshold. However, this environmental limit is distinguished not for individual species and communities. Rather, it refers to all marine biota assuming stability of the main structural and functional parameters of marine ecosystems.

With methane, the biogeochemical threshold of ecological tolerance can be roughly estimated at 0.01 ml/l. This figure can be used as an approximate level of the maximum permissible concentration (MPC) for dissolved methane in the marine environment.

It is also possible to estimate the MPC for methane using the previously discussed toxicological data. Figure 53 shows that the acute toxicity of dissolved methane begins to manifest itself at about 1 ml/l. In Russia, establishing permissible levels of water pollution requires both acute (short-term) tests and long-term (chronic) experiments with the organisms of different trophic levels [Patin, Lesnikov, 1988]. Unfortunately, the results of chronic experiments with methane are not available. To overcome this limitation, the data on lethal concentration and application factor are often used to assess the maximum acceptable (noneffective) levels [Bittel, Lacourley, 1972; Lloyd, 1979; GESAMP, 1986]. The values of application factor usually range from 0.1–0.01, depending on the type of curve that best describes the results of the short-term toxicological tests (see Figures 54 and 55). The experimental data suggest that fish responses to methane exposure most likely develop in accordance with curve *b* in Figure 54 (tendency to a threshold) and curve *a* in Figure 55 (rapid response at physiological and behavioral levels). In any case, using an

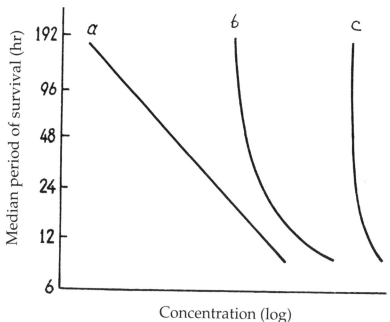

Figure 54 Different types of relationship between the toxicant concentration and the time of fish survival: a—the absence of threshold effects; b—tendency to a threshold; c—clear threshold effects [Lloyd, 1979].

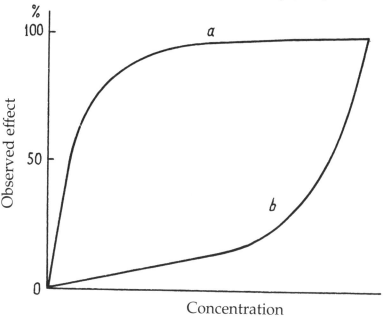

Figure 55 The response to increasing concentration of a contaminant: a—organisms with increased sensitivity of response; b—organisms with increased resistance [GESAMP, 1986].

application factor of 0.01 gives an approximate value of MPC for dissolved methane of about 0.01 ml/l. The same figure was found previously based on biogeochemical data and considerations.

The same result of two independent estimates certainly increases the reliability of the final assessment. However, it does not exclude the need for its correction with new data. In any case, the concentrations of methane in surface waters of the World Ocean are significantly (2–3 orders of magnitude) lower than the acceptable levels that are established using toxicological and ecological criteria (see Figure 53).

The combined biogeochemical and ecotoxicological materials given above may be used for environmental impact assessments during the development of the offshore oil and gas fields. They could also be helpful for environmental monitoring and predicting the consequences in case of accidental and routine gas emissions in areas of offshore platforms and underwater pipelines. The general methodology of such monitoring and prognoses is based on the notion of ecological assimilation capacity of water bodies and the acceptance of a controlled impact on them [GESAMP, 1986; Lavrick et al., 1991; Zaidiner, 1993].

Conclusions

1. The chemical composition of natural gas is rather changeable. However, in all cases, it chiefly consists of methane (up to 97%) and methane homologues, as well as inorganic gases such as carbon oxide, hydrogen sulfide, nitrogen, and others.

2. Over the last 100 years, the natural process of methane production and distribution in the biosphere has become the subject of large-scale anthropogenic impact. Already, this has noticeably increased methane concentration in the atmosphere (more than a twofold increase since the beginning of the century), especially in the Northern Hemisphere.

3. The main anthropogenic sources of methane include the production and burning of fossil fuel, decomposition of organic matter (rice fields, garbage dumps, etc.), technological and accidental gas emissions (drilling, gas production, transportation, etc.).

4. The main natural sources of direct methane input into the water environment include microbial and biochemical degradation of organic substances in bottom sediments, decomposition of marine gas hydrates, and bottom gas seepage from shallow oil- and gas-bearing formations. Structural and functional changes of benthic communities are possible in such areas.

5. A potentially hazardous source of methane series hydrocarbons in the water environment can be their release during underwater pipeline damages as well as accidental blowouts during drilling activities on the continental shelf. The latter occurred in 1982 and 1985 in the Sea of Asov, where serious ecological and fisheries consequences were observed.

6. The natural content of methane in seawater usually varies from 10^{-5} ml/l (close to the equilibrium concentration) to 10^{-3} ml/l. It is higher in the coastal areas, bays, and estuaries than in open waters of the seas and oceans. Vertical distribution of methane in the water column is characterized by its elevated concentration in the upper water layers and sometimes in the layers near the bottom. In areas of accidental blowouts and releases of natural gas, methane concentrations in seawater can reach 1–10 ml/l.

7. The toxic properties of methane and its homologues in the water environment have been very poorly studied. Available materials suggest that these hydrocarbons belong to the group of poisonous gases that have narcotic effects and damage the nervous system. Acute fish poisoning and lethal damages occur at the concentrations of gas hydrocarbons over 1 mg/l. Primary behavioral responses are observed at levels as low as 0.02–0.1 mg/l.

8. The combined analysis of available biogeochemical and ecotoxicological data suggests that 0.01 ml/l is the approximate maximum permissible concentration (MPC) for methane in the marine environment. Levels over this limit can be found both in areas of technological and accidental gas emissions and in the impact zones of natural gas flow from the sea bottom.

9. The minimum effective (lowest observed effect) concentrations of gas condensate in the water for marine organisms are 0.05–0.5 mg/l, and the safe level is about 0.01 mg/l.

10. The given materials and estimates may be used to solve the practical tasks of environmental assessments, monitoring, and prediction of consequences during oil and gas field development on the continental shelf.

References

Anonymous. 1994. Measuring of methane dissolved in sea water. *Mar.Pollut.Bull.* 28(2):128.

Atkinson, L. P., Richards, F. A. 1973. The occurrence and distribution of methane in the marine environment. *Deep-Sea Res.* 14(6).

AzNIIRKH. 1986. *Report on the results of chemico-toxicological studies regarding accidental gas blowout in the Sea of Asov.* Rostov-na-Donu, 40pp. (Russian)

Beck, L. L. 1993. A global methane emissions programme for landfields, coal mines, and natural gas systems. *Chemosphere* 26(1–4):447–452.

Beck, R. J. 1990. Natural gas to start long period of growth during next 3 years. *Oil and Gas Journ.* 91:53–63.

Bittel, R., Lacourley, G. 1972. Method of approach for the evaluation of chemical pollution levels in marine environment and marine food chains. In *Marine pollution and sea life.*—London: Fishing News (Books) Ltd, pp.349–342.

Bogner, J., Spokas, K. 1993. Landfill CH-4: Rates, fates and role in global carbon cycle. *Chemoshere* 26(1–4):369–386.

Borisov, V. P., Osetrova, N. V., Ponomarenko, V. P., Semenov, V. N. 1994. *Impact of the offshore oil and gas developments on the bioresources of the Barents Sea.* Moscow: VNIRO, 251 pp. (Russian)

Brooks, J. M., Sackett, W. M. 1973. Sources, sinks and concentrations of light hydrocarbons in the Gulf of Mexico. *Journ.Geophys.Res.* 78(24).

Cairns, W. J., ed. 1992. *North Sea oil and environment. Developing oil and gas resources, environmental impacts and responses.* London and New York: Elsevier Applied Science, 722 pp.

Christensen, R. R. 1994. Fugitive emissions from Norwegian oil and gas production. *Mar.Pollut.Bull.* 29(6–12):300–303.

Claypool, G. E., Kaplan, J. R. 1974. The origin and distribution of methane in marine sediments. In *Natural gases in marine sediments.*—New York: Plenum press.

Davis, G. 1988. North Sea hydrocarbon seeps. *Mar.Pollut.Bull.* 7(6):387.

Ecology and impact of natural gas on organisms. 1989. *Theses of the All-Union Conference on Ecology and Impact of Natural Gas on Organisms.* Astrakhan, 180 pp. (Russian)

Freyer, H. D. 1979. Atmospheric cycles of trace gases containing carbon. In *The global carbon cycle.*—Chichester, pp.101–128.

Galchenko, V. F. 1995. Bacterial cycle of methane in the marine ecosystems. *Priroda* 6:35–48. (Russian)

Geodokyan, A. A., Avilov, V. I., Avilova, S. D. 1994. Geo-ecological monitoring of the water basin on the Sakhalin shelf. *Doklady AN SSSR* 334(1):63–86. (Russian)

Geodokyan, A. A., Trotsyuk, V. A., Verkhoyanskaya, Z. I. 1979. Hydrocarbon gases. In *Chemistry of the ocean. Vol.1.*—Moscow: Nauka, pp.164–175. (Russian)

GESAMP. 1986. *Environmental capacity. An approach to marine pollution prevention. GESAMP Reports and Studies No.30.* Rome: FAO, 50pp.

GLAVRYBVOD. 1983. *Report on protection and reproduction of the fishery stock.* Moscow: Glavrybvod. (Russian)

GLAVRYBVOD. 1986. *Report on protection and reproduction of the fishery stock.* Moscow: Glavrybvod. (Russian)

Gribbin, J. 1989. The methane mill. *BBC Wildlife* 7 (6):387.

Howard, M., Thomsen, E. 1989. Hydrocarbon-based communities in the North Sea? *Sarsia* 74:29–42.

ICES. 1993. *International Council for the Exploration of the Sea. Report of the ICES Advisory Committee on Marine Environment.* Copenhagen: ICES, 84 pp.

Ivanov, M. V. 1988. Microbiological components of the cycle of sulfur and carbon in the ocean. In *Complex analyses of the environment: Proceedings of the Fifth Soviet-American Symposium.*—Leningrad: Gidrometeoizdat, pp.112–116. (Russian)

Izomerov, N. F. 1984. *Preventive toxicology. Vol.1.* Moscow: Tsentr mezhdunarodnikh proektov GKNT. (Russian)

Karamova, L. M. 1989. Natural gas and the problems of ecology. In *Theses of the All-Union Conference on Ecology and Impact of Natural Gas on Organisms.* Astrakhan. (Russian)

Kellard, M. 1994. Natural gas hydrates: energy for the future. *Mar.Pollut.Bull.* 29(6–12):307–311.

Khalil, M. A. K. 1993. Methane from coal burning. *Chemosphere* 26(1–4): 473–477.

Khalil, M. A. K., Rassmussen, R. A., Sharer, M. J. 1991. The global methane cycle: trends of sources, sinks and concentrations. *World Resour.Rev.* 3:391–405.

Koons, C. B., Monoghan, P. H. 1973. Petroleum derived hydrocarbons in Gulf of Mexico waters. *Trans. Gulf Coast Assoc.Geol.Soc.* 23:170–181.

Koropalov, V. M., Nazarov, I. M., Kazakov, U. E., Gammon, R. H., Kron, D. R. 1988. Complex Soviet-American studies of the global atmospheric distribution of the gaseous traces that are able to bring the climatic change. In *Complex analysis of the environment: Proceedings of the Fifth Soviet-American Symposium.*—Leningrad: Gidrometeoizdat, pp.117–128. (Russian)

Kosheleva, V. V., Novikov, N. A., Migalovski, I. P., Gorbacheva, E. A., Lapteva, A. N. 1994. Evaluation of toxicity of gas condensate from the Shtokmanovskoe field for the organisms of the Barents Sea. In *Report of NIR PINRO in 1993.*—Murmansk: PINRO, pp.267–275. (Russian)

Kosheleva, V. V., Novikov, N. A., Migalovski, I. P., Gorbacheva, E. A., Lapteva, A. N. 1997. *Responses of marine organisms to the environmental pollution during oil and gas development on the shelf of the Barents Sea.* Murmansk: PINRO, 95 pp. (Russian)

Lamantagne, R. L., Swinnerton, I. W., Linnenbom, V. I., Smith, W. 1973. Methane concentrations in various marine environment. *Journ.Geophys.Res.* 78(24).

Lavrik, V. I., Merezhko, A. I., Serenko, L. A., Timchenko, V. M. 1991. Ecological capacity and its quantitative assessment. *Gidrobiologicheski zhurnal* 27(3):13–22. (Russian)

Lein, A. U., Ivanov, M. V. 1981. *Dynamic biogeochemistry of anaerobic diagenesis of sediments. Lithology at the new stage of developing the geological knowledge.* Moscow: Nauka. (Russian)

Lisitsin, A. P. 1983. Biogeochemistry of gases in the ocean. In *Biogeochemistry of the ocean.*—Moscow: Nauka, pp.274–301. (Russian)

Lloyd, R. 1979. The use of the concentration-response relationship in assessing acute fish toxicity data. In *Analyzing the hazard evaluation process.*—Washington: American Fisheries Society, pp.59–61.

Lukyanenko, V. I. 1983. *General ichthyotoxicology.* Moscow: Legkaya i pishchevaya promishlennost, 320 pp. (Russian)

Masood, E. 1995. Climate panel confirms human role in warming, fights off oil states. *Nature* (Gr.Brit) 378(6557):524–525.

Metelev, V. V., Kanaev, A. I., Dzasokhova, N. G. 1971. *Water toxicology.* Moscow: Kolos, 150 pp. (Russian)

Miller, S. L. 1974. The nature and occurrence of clarate hydrates. In *Natural gases in marine sediment. Vol.3.*—New York: Plenum press, pp.151–177.

Neue, H.-U. 1993. Methane emission from rice fields. *Bioscience* 43(7):466–474.

Novoselov, S. U., Bondarenko, I. V., Kuzmin, A. U. 1992. Ecological problems of the oil and gas development on the shelves of the Barents and Kara Seas. In *Report of NIR PINRO in 1991.*—Murmansk: PINRO, pp.237–248. (Russian)

Novozhevnikova, A. N. 1995. Garbage dumps—"methane bombs" of the planet. *Priroda* 6:25–34. (Russian)

Obzhirov, A. I. 1993. *Gas-chemical fields of the bottom horizon of the seas and oceans.* Moscow: Nauka, 138 pp. (Russian)

Patin, S. A. 1979. *Pollution impact on the biological resources and productivity of the World Ocean.* Moscow: Pishchepromizdat, 305 pp. (Russian)

Patin, S. A. 1988. Establishing fisheries standards for the water environment quality. In *Water toxicology and optimization of bioproductive processes in aquaculture.*—Moscow: VNIRO, pp.5–18. (Russian)

Patin, S. A. 1993. *Ecotoxicological characteristic of natural gas as ecological factor of the water environment.* Moscow: VNIRO, 40 pp. (Russian)

Patin, S. A. 1997. Marine ecosystems, bioresources and global climate in the twenty-first century. *Rybnoe khozaistvo* 2:32–40. (Russian)

Patin, S. A., Lesnikov, L. A. 1988. Main principles of establishing the fisheries standards for the water environment quality. In *Methodological guidance on establishing the maximum permissible concentrations of pollutants in the water environment.*—Moscow: VNIRO, pp.3–10. (Russian)

Sackett, W. M., Brooks, J. M. 1975. Use of low molecular-weight hydrocarbons as indicators of marine pollution. In *NBS Spec.Publ.409. Marine pollution monitoring (Petroleum). Proceedings of Symposium and Workshop held at NBS, Gaithersburg, Maryland, May 13–17, 1974.* NBS, pp.172–173.

Semenov, A. D., Romova, M. G., Soier, V. G., Kisikinova, T. S. 1991. Fisheries MPC and control of the pollution of the water bodies. In *Theses of the Second All-Union Conference on Fisheries Toxicology. Vol.2.*—Spb., pp.183–184. (Russian)

Shparkovski, I. A. 1993. Biotesting of the quality of the water environment with the use of fish. In *Arctic seas: Bioindication of the state of the envi-*

ronment, biotesting and technology of destruction of pollution.—Apatiti: Izd-vo KNTS RAN, pp.11–30. (Russian)

Sokolov, V. A. 1966. *Geochemistry of the gases in the lithosphere and atmosphere.* Moscow: Nauka.

Sokolov, V. A., Vinogradov G. A. 1991. Effect of the bottled gas on the behavior responses of the juvenile fish. In *Theses of the Second All-Union Conference on Fisheries Toxicology. Vol.2.*—Spb., pp.183–184.

Solbe, J. F., Shurben, D. G. 1989. Toxicity of ammonia at early stage of rainbow trout. *Water Res.* 23(1):121–123.

Somerville, H. J., Shirley, D. 1992. Managing chronic environmental risks. In *North Sea oil and environment. Developing oil and gas resources, environmental impacts and responses.*—London and New York: Elsevier Applied Science, pp.643–661.

Stole, R. D. 1974. Effects of gas hydrates in sediments. In *Natural gas in marine sediments. Vol.3.*—New York: Plenum press, pp.235–248.1

Swinnerton, I. W., Linnenbom, V. I. 1976. Determination of the Cl-C4 hydrocarbons in sea water by gas chromatography. *Journ.Gas.Chromatogr.* 5(6).

Umorin, P. P., Vinogradov, G. A., Mavrin, A. S., Verbitski, V. B., Bruznitski, A. A. 1991. Impact of the bottled gas on ichthyofauna and zooplankton organisms. In *Theses of the Second All-Union Conference on Fisheries Toxicology. Vol.2.*—Spb., pp.222–224. (Russian)

Velichko, A. A., Borisova, O. K., Zelikson, E. M., Kremenitzki, K. V., Nechaev, V. P. 1997. Methane emission in global warming. *Doclady RAN* 365(3):387–389. (Russian)

VNIIP. 1994. *The state of the environment and environmental protection activities in the former USSR.* Moscow: VNIIP, 111 pp. (Russian)

Volkov, I. V., Zalicheva, I. N., Ganina, V. S., Ilmast, T. B., Kaimina, N. V., Movchan, G. V., Shustova, N. K. 1993. On principles of regulation of anthropogenic impact on water ecosystems. *Vodnye resursy* 20(6):707–713. (Russian)

Williams, R. T., Bainbridge, A. E. 1973. Dissolved carbon oxide, methane and hydrogen in the Southern Ocean. *Journ.Geophys.Res.* 78(15).

Wernecke, G., Floeser, G., Korn, S., Michaelis, W. 1991. *Device for the in-situ determination of methane in seawater.* Communaute Urbaine de Brest (France). NB19, 10 pp.

Zavarzin, G. A. 1995. World cycle of methane under cold conditions. *Priroda* 6(3–14). (Russian)

Zaidiner, U. I. 1993. Methods of assessment of anthropogenic hazard to the water bioresources. *Rybnoe khozaistvo* 3:18–20. (Russian)

Zhadin, V. I. 1956. *Life of the fresh waters of the USSR.* Moscow: Izd-vo AN SSSR, 320 pp. (Russian)

Zubuva, M., Ginsburg, G., Soloviev, V., Shadrina, T. 1990. Gas hydrates. In *Geology and mineral resources of the World Ocean.*—Warsaw: INTERMORGEO, pp.461–468. (Russian)

Chapter 6

Ecotoxicological Characteristics of Related Chemicals and Wastes from the Offshore Oil Industry

As Chapter 3 showed, the assortment of chemicals and wastes from the offshore oil and gas complex is extremely wide and variable. In spite of the technological progress and stricter environmental requirements, a considerable part (up to 80%) of these substances enter the marine environment one way or another. Produced, ballast, and injection waters as well as drilling cuttings and fluids polluted by hundreds of different chemicals often go directly overboard at the production site. The majority of available studies of liquid and solid wastes of the offshore oil and gas industry give very different estimates of their toxicities. They are difficult to compare due to the variability of the chemical composition of these discharges. At the same time, studying the toxic properties and behavior of all these wastes in the marine environment is extremely important. This information can be used to select environmentally safe technological formulations and their components, to regulate their use, and to establish environmental standards.

6.1. Drilling Discharges

Since the beginning of large-scale offshore activities, drilling wastes and their components have been the object of especially detailed and extensive ecotoxicological studies both in the laboratories and on the shelves of many countries. Special books [Ray, 1985; Boesch, Rabalais, 1987; Lichtenberg et al., 1988], international conferences and symposiums [Proceedings, 1980; Engelhardt et al., 1989], and hundreds of publications have been devoted to this topic. However, similar to the situation with oil hydrocarbons, most of these publications deal with individual features and details of toxicological testing of different mixtures and their components. Summarizing studies occur rather rarely [NRC, 1983; Thomas et al., 1984; EPA, 1985; Duke, 1985; Neff, 1987; GESAMP, 1993; Swan et al., 1994].

6.1.1. Drilling Fluids (Muds)

The overwhelming majority of studies on the toxicology of drilling fluids are based on acute experiments (48–96 hour exposure). The most common species of fish and invertebrates are usually used as test objects. The results of such tests are often represented in a form of LC_{50} (lethal concentrations causing the death of 50% of test organisms during a certain exposure time). The data of acute experiments make it possible to compare toxic properties, reveal the species responses, and assess possible biological effects of different fluids. Sometimes other information supplements these results. This includes data from experiments using model ecosystems, mesocosms, and microcosms as well as materials from field observations in areas of drilling discharges. Some countries have developed and adopted uniform standard procedures for ecotoxicological testing of drilling fluids and their components [Duke, 1985; GESAMP, 1993; Swan et al., 1994]. Besides, substances and compounds are classified according to their toxic properties to compare toxicity of drilling fluids and their components. Table 43 shows one of the classifications recently adopted by a number of international organizations.

Table 44 and Figures 56 and 57 show the acute toxicity of widespread drilling muds for major groups of marine fauna. These data indicate that most drilling formulations have 96-hour LC_{50} values ranging within $10^4–10^6$ g/kg (1–10%). The exceptions are the drilling fluids based on diesel fuel. They have considerably lower values of LC_{50}. This was the reason for limiting (in some countries—even prohibiting) the use of diesel-based drilling fluids in the offshore oil and gas production practice. Diesel base has been replaced by other formulations that do not include highly toxic products in their composition (see Chapter 3).

Ecological consequences of the discharges of oil-based drilling muds can also be hazardous due to their persistence in the marine environment [Østgaard, Jensen, 1985]. Experiments showed that 180 days after the discharge of oil-based drilling wastes, they had biodegraded less than 5%. At the same time, drilling wastes based on esters of fatty acids had almost completely (99%) lost their organic

Table 43

Classification of Different Agents and Substances According to Their Toxic Properties [GESAMP, 1997]

	Acute Toxicity	Chronic Toxicity	
Rating	48–96-hr LC_{50}/EC_{50} (mg/l)	Rating	No Observed Effect Concentration (mg/l)
(0) Non-toxic	>1,000	–	–
(1) Practically non-toxic	100–1,000	–	–
(2) Slightly toxic	10–100	(2) Low chronic toxicity	>1
(3) Moderately toxic	1–10	(3) Moderate chronic toxicity	0.1–1.0
(4) Highly toxic	0.1–1.0	(4) High chronic toxicity	0.01–0.1
(5) Very highly toxic	0.01–0.1	(5) Very high chronic toxicity	0.001–0.01
(6) Extremely toxic	<0.01	(6) Extremely high chronic toxicity	<0.001

Table 44
Summarized Results of Toxicological Studies of Some
Drilling Fluids

Type of Drilling Fluids	Concentrations, Test Duration, Test Organisms	Effects	Reference
Water-based clay fluids	2–15 mg/l 1-2 min. Salmon fry	Changes in the rates of respiration and heart contraction	Kozak, Shparkovski, 1991
	5 mg/l 10–30 days Cod, flounder	Reduced survival	
Water-based clay-bentonite fluids	15–40 mg/l 2–5 min. Cod, salmon, haddock, ray	Threshold changes in respiratory and cardiac activities	Shparkovski et al., 1989
Lygnosulfonate drilling fluids	30–100 mg/l Chronic tests Cod fry, flounder, salmon	Persistent disturbances of physiological functions and behavioral responses	Kozak, 1991
Water-based (lygnosulfonate and ammonium) drilling fluids	5–22 g/l 48–96 hours Salmon fry, amphipod	50% mortality	Borisov et al., 1994
Polymer muds	45–93 g/l 48–96 hours Salmon fry, amphipod	–"–	–"–
Polymer-clay muds	10–25 g/l 48–96 hours Salmon fry, amphipod	–"–	–"–
Drilling fluids based on emultane and hydrolyzed polyacrylonitrile and polyacrylamide	5 mg/l	MPC* according to sanitary-toxicological criteria of hazard	Register of MPC, 1995

Table 44 (*continued*)

Type of Drilling Fluids	Concentrations, Test Duration, Test Organisms	Effects	Reference
Water-based muds	$10–10^3$ mg/l Chronic tests Copepod, amphipod, bivalve, cod fry	No observed effect	Kosheleva et al., 1997
Drilling fluids of different types	100–200 g/l Acute tests Fish, invertebrates	100% mortality	Vekilov et al., 1990
Drilling fluids of different types	>100 mg/l Acute tests	Reduced survival	Gusseynov, Alekperov, 1989

Note: *—MPC—maximum permissible concentration of a pollutant in natural waters (Russian standard).

fraction due to microbial and physicochemical decomposition [Bakke, Laake, 1991].

Although new drilling fluids include a number of less toxic compounds (e.g., paraffin, fatty acids amide), this still does not ensure complete and rapid degradation of the oil always associated with these muds. For example, when less-toxic, paraffin-based drilling fluids were mixed with bottom sediments, the level of the oil fraction in the upper 1-cm layer of sediments declined only twice (from 200 mg/kg to 100 mg/kg) after 70 days of exposure. In the lower layers of sediments, the concentration of oil hydrocarbons remained the same and even slightly increased [Peterson et al., 1991].

Using water-based formulations does not fully eliminate the environmental hazards. Some comparative studies showed that water-based fluids do not always meet strict ecological requirements [Sorbye, 1989]. In particular, they can include some toxic biocides and heavy metals in their composition. Besides, in contrast with oil-based muds, these fluids display a higher capacity for dilution in the marine environment. Finally, large volumes of water-based muds are usually disposed of overboard, while muds on the oil base are largely recycled. At the same time, the experimental and field studies show that acute toxic effects of water-based drilling muds can be manifested only at high concentrations [Patin, 1998]. Such concentrations

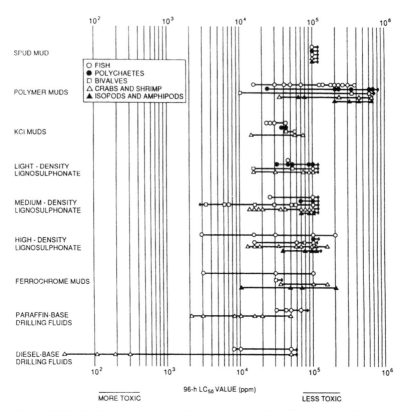

Figure 56 Lethal toxicity ranges of generic types of drilling mud with major groups of marine fauna [Thomas et al., 1984].

can be found only in direct vicinity to the discharge point (within a radius of several meters).

Comparative studies [NRC, 1983; Neff, 1987; Swan et al., 1994] showed that marine organisms were more sensitive to the suspended particulate phase of drilling muds than to the liquid phase. This indicates that suspended particles in drilling fluids may contribute substantially to their toxicity. Toxicity testing of drilling fluids, conducted both in Russia and other countries [NRC, 1983; Serigstad et al., 1988; Gusseinov, Alekperov, 1989; Khomichuk, Vidre, 1989; Koefoed, 1989; Kozak, Shparkovski, 1991; Gudimov, 1993; Kozak, 1993; Moseichenko, Abramov, 1994; Kosheleva et al., 1997], give values of LC_{50} close to the ones shown in Figures 56 and 57. These results of acute (short-term) toxicological tests based on measuring organisms' mortality provide valuable preliminary information.

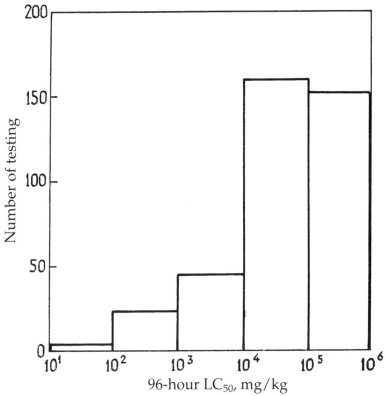

Figure 57 Distribution of LC_{50} values (96-hour) of drilling muds for major groups of marine organisms (plotted based on summarized data [Ray, 1985] on 398 test with 61 species of water organisms and 71 types of drilling fluids).

However, these data are not sufficient for comprehensive evaluation of the environmental hazards of drilling fluids. Such an evaluation is possible when using the results of long-term experiments according to the methodology adopted for establishing the maximum permissible concentrations (MPC) of pollutants [Patin, Lesnikov, 1988]. This methodology is widely applied in Russia to evaluate the toxicity of individual components of drilling muds. At the same time, with respect to a drilling fluid as a whole it is used rather rarely.

When an exposure duration increases from 48–96 hours to 10–30 days, the levels of drilling fluids that cause lethal and sublethal effects decrease considerably (up to 2–3 orders of magnitude) (see, for example, data in Table 44). Applying eco-physiological methods to assess the organism responses also reveals considerably lower observed effect concentrations than LC_{50} values determined by standard acute tests. Some of these studies found the following responses to the impact of drilling muds:

- biochemical and hematological changes in fish fry exposed to low levels of some drilling fluids [Akhmedova et al., 1989];
- significant (3–20-fold) increase in sensitivity of some marine crustaceans (shrimps, mysids, lobsters) at the early (larval) stages of their development [Duke, 1985];
- accumulation of oil hydrocarbons in the organs and tissues of fish and invertebrates in areas of oil-based mud discharges [Vogt et al., 1988];
- changes in the settling rate of plankton larvae of benthic invertebrates onto solid substrates; alternations in community structure; and development of anaerobic conditions in bottom sediments chronically polluted by oil-based drilling muds [Blackman et al., 1986].

These results suggest that despite the relatively high values of LC_{50} (usually 10^4–10^6 g/l) obtained in acute experiments for most drilling muds (see Figures 56 and 57), they cannot be considered completely nontoxic wastes. Under natural conditions and chronic impacts, these muds can cause effects that short-term toxicological testing is unable to predict.

Certain hope regarding environmental safety of drilling formulations has recently emerged due to the development of new generations of drilling fluids based on new synthetic products (see Chapter 3, Section 3.3.3). Laboratory and field studies of 60 samples with such formulations [Burke, Veil, 1995] showed that their toxicity was lower than the standard toxicity limit adopted in the US (96-hour LC_{50} must exceed 30 g/kg). Almost 80% of these samples were considered practically nontoxic because their LC_{50} exceeded 1,000 g/kg.

During the discharge of spent drilling muds into the marine environment, other processes besides pure physical dilution in the water occur. These include dissolving, sedimentation, adsorption, chemical and bacterial degradation, and interaction of numerous components of these fluids. These processes are so complex and diverse that any reliable modeling and prediction of the ecological consequences of such discharges appear to be a very difficult task. One should take into account the high variability in the chemical composition of these fluids and the simultaneous presence of other wastes with varying composition (drilling cuttings, produced waters, produced sand, and others). Thus, the methodology based on observing the dilution of fluorescent dyes (e.g., rhodamine) in areas of drilling discharges used to assess the environmental hazard of these wastes [Gordon, 1989] seems not to be quite adequate. Probably, the approach to overcome methodological difficulties during environmental assessment of drilling discharges should involve a combination of ecotoxicological methods, modeling data, and field study results. An example of such integrated approach is presented in Figure 58. This plot shows the curve of possible distribution of drilling muds in waters near Sakhalin [Smith et al., 1997] in combination with the results of toxicity testing of these muds [Patin, 1998].

6.1.2. Components of Drilling Fluids

The formulations of drilling fluids vary widely and include hundreds of different components. Numerous studies of these substances usually apply two main approaches to their toxicological assessment. One includes short-term (acute) tests on marine organisms by

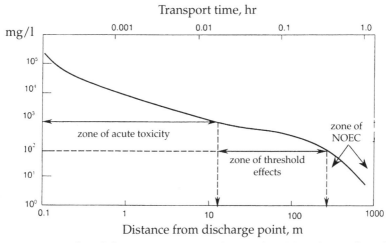

Figure 58 Predicted decrease in concentrations and toxicity of water-based drilling muds in seawater at the end of 160 m³/hour discharge under hydrological conditions of the north-eastern Sakhalin shelf. The zones of possible effects of the mud were distinguished based on the results of acute (96-hour) toxicity testing. Ecological risk (probability of the organisms' death) is significantly less than 1% within 10 m from point of discharge [Patin, 1998].

measuring the LC_{50} under relatively standardized conditions. The technique is similar to the one described above for drilling fluids. The other approach, widely used in Russia, is based on rather complex and time-consuming procedures. It involves testing discharged water pollutants according to the protocol adopted for determining maximum permissible concentrations (MPC) [Patin, Lesnikov, 1988].

Tables 45 and 46 summarize toxicities of drilling fluid components to water organisms in acute and chronic experiments conducted in Russia and Australia. These data, as well as materials on the maximum permissible concentrations (MPC) adopted in Russia for chemicals and agents, used during the oil and gas production (see Appendix), allow us to make some general conclusions regarding the ecotoxicological properties of drilling components.

The first conclusion based on all of the available information is an extremely wide range of concentrations that cause different toxic effects. The levels of acute (lethal) and threshold (reversible) effects as well as the MPC for different components vary from 10^{-4}–10^{-3} mg/l to 10^3–10^4 mg/l (i.e., within 7–8 orders of magnitude). Even taking into account the inevitable variability due to the diversity in experimental conditions and methods, the components of drilling fluids cause an extremely wide spectrum of responses—from practical absence of toxic effects to lethal toxicity. The interpretation of this extremely variable and multifactorial picture is even more complicated by the presence of other constituents and additives (with changing and often unknown composition) in the formulations of most components.

Broadly speaking, the variety of substances shown in Table 45 can be divided into three main groups depending on their relative content in drilling formulations and the degree of toxicity.

The first group includes major components that quantitatively dominate and define the generic type of a drilling fluid. These are drilling clay mixtures (bentonite and other organophilic clays), organic polymers (e.g., carboxy-methyl-cellulose, starch, gum polymers, tannins, lignins), weighting agents (e.g., barite, calcite, carbonates, soluble salts). These components regulate the main chemical and physicochemical parameters of drilling fluids and form their basic composition. Each of them constitutes 1% to 10% of the drilling fluid volume. One of the typical characteristics of the majority of these basic components is their low toxicity. Their LC_{50} values in acute experiments usually exceed 100–1,000 mg/l, and their MPC range within 1–100 mg/l.

The other group, radically contrasting to the first one, includes minor additives that constitute a relatively low portion of the total

Table 45

Summarized Results of Toxicological Studies of the Drilling Fluid Components Used in Russia

Component and Its Primary Function	Concentration (mg/l), Test Duration, Test Organisms	Effects	Reference
Barite (barium sulfate)—major weighting component	0.1–10 Standard chronic tests	No observed effects—at <2 mg/l	Register of MPC, 1995
Bentonite—major viscosifier in water-based clay fluids	100–100,000 2–4 days Zoobenthos, fish	50% mortality—at 10–100 g/l, primary responses—at 100–200 mg/l	Kozak, 1993; Shparkovski, 1993
FCLS (ferro-chrome lignosulfonates)—deflocculant and fluid loss control additive	10–1,000 Several days Flounder (eggs, larvae)	Reduced survival, morphological anomalies	Ghuravleva, Savinova, 1988
	0.02–0.1 3 days Cod	Changes of gematological parameters	Lega, 1988
	100–500 4 days Pink salmon (fry)	50% mortality—at 275 mg/l	Moseychenko, Abramov, 1994
	15–40 2–5 min. Marine fish	Primary physiological responses	Shparkovski et al., 1989
	2,500–4,800 Several days Fish, invertebrates	50–100% mortality	Vekilov et al., 1990; Kozak, 1993
	0.1–10 Standard chronic tests	No observed effects—at <1 mg/l	Register of MPC, 1995

Table 45 (continued)

Component and Its Primary Function	Concentration (mg/l), Test Duration, Test Organisms	Effects	Reference
CMC (carboxy-methyl-cellulose)—filtration control additive	1–20 10 days White sturgeon (fry)	Reduced oxygen consumption, biochemical changes	Dzhabarov, 1988
	10–2,000 1–75 days Caspian fish fry (salmon, goby, sturgeon, Caspian roach)	Death—at 1,000–2,000 mg/l in acute tests; physiological changes—at 12–50 mg/l	Dokholyan et al., 1988
	30–1,000 Chronic tests Trout (juvenile)	Death—at 125–1,000 mg/l; histological and biochemical disturbances, pathological changes in organs and tissues—at 30–125 mg/l	Chinareva, 1989
	15–40 2–5 min Marine fish	Primary behavioral and physiological responses	Shparkovski et al., 1989
	5–1,200 2–4 days Salmon fry	50% mortality—at 800–1,200 mg/l; threshold effects—at 5–20 mg/l	Kozak, 1993
	1–20 Standard chronic tests	No observed effects	Register of MPC, 1995

Table 45 (*continued*)

Component and Its Primary Function	Concentration (mg/l), Test Duration, Test Organisms	Effects	Reference
CSMT (condensed sulfite-methylated tannins)—thinner and viscosifier	12–1,500 Up to 75 days Caspian fish fry	Death and reduced survival—at 100–1,500 mg/l, physiological disturbances—at >12 mg/l	Dokholyan et al., 1988
	15–40 2–5 min. Marine fish	Primary physiological responses	Shparkovski et al., 1989
	1–100 Standard chronic tests	No observed effects—at <12 mg/l	Register of MPC, 1995
CSMT-1, CSMT-2	10–10,000 2–4 days Salmon and cod fry	Death—at >1,000 mg/l, threshold effects—at 10–100 mg/l	Kozak, 1993
STPF—stabilizer of technological parameters	8–75 ml/l Several days Fish, invertebrates	100% mortality—at 70–75 ml/l, no effects—at 8.0–8.4 ml/l	Vekilov et al., 1990
SNPH-7214R (oxy-alkylated alkylphenols and oil sulfonates in aromatic solvent)—descaler for the underground well equipment	0.01–10 Up to 120 days Daphnia	Death and reduced survival—at 1–10 mg/l, reduced fertility and growth—at 0.1–1.0 mg/l	Kabanova, Kuznetsova, 1988
	0.01–10 Standard chronic tests	No observed effects—at <0.01 mg/l	Register of MPC, 1995

267

Table 45 (*continued*)

Component and Its Primary Function	Concentration (mg/l), Test Duration, Test Organisms	Effects	Reference
SNPH 7225 M—descaler for the underground well equipment	0.01–5 Up to 120 days Organisms of the main trophic levels	Death and reduced survival— at 0.5–5.0 mg/l; disturbances of reproduction, feeding, and growth—at 0.1–0.5 mg/l	Artemyeva et al., 1988
SDEB-2 (composition based on fatty acids)— lubricant	0.1–10 2–4 days Fry of salmon and cod	50% mortality—at 1.6 mg/l, threshold effects—at <1mg/l	Kozak, 1993
OKZIL (lignosulfonate derivative)— deflocculant/thinner additive	0.1–4.2 ml/l Several days Fish, invertebrates	100% mortality—at 2.0–4.2 ml/l, no effects— at 0.1 ml/l	Vekilov et al., 1990
	5–1,000 1–75 days Caspian fish fry	Death and reduced survival— at 600–1,000 mg/l, biochemical changes—at >6 mg/l	Doholyan et al., 1988
Alkylsulfates— detergent/emulsifier additives	0.05–10 Several days Fish eggs	Reduced fertility, embryonic anomalies—at 0.1–1.0 mg/l	Kuzmina, Shcherbakov, 1988

Table 45 (continued)

Component and Its Primary Function	Concentration (mg/l), Test Duration, Test Organisms	Effects	Reference
GKG-11 (solution of monosodium salt of methyl-silane-triol)—stabilizer of structural-rheological parameters	1–500 Up to 120 days Organisms of the main trophic levels	Death—at 100–500 mg/l, sublethal effects—at 10–100 mg/l, threshold reversible disturbances—at 1–10 mg/l	Buslenko et al., 1991
	0.1–1.0 ml/l 4 days Fry of salmon and pink salmon	50% mortality—at 0.36–0.53 ml/l depending on salinity	Moseychenko, Abramov, 1994
Phosphoxit-7, polyphos 108N, foamformer POA, syntanol DS-6, copolymer GDPE-106, neonols AF9-6 and AF9-10—surfactants for corrosion inhibition, descaling, foamforming regulation, etc.	0.005–500 Up to 120 days Organisms of the main trophic levels of the Barents Sea	Reduced fish eggs fertility—at 5 mg/l; 50% mortality of embryos and larvae—at 0.25–10 mg/l of non-ionactive surfactants and at 0.1–0.5 mg/l of anionactive surfactants; acute lethal effect on juveniles—at 1–5 mg/l; mutagenic effects—at 0.25–1.0 mg/l of phosphoxit-7, DS-6, AF9-6 and AF9-10	Migalovski et al., 1991

269

Table 45 (*continued*)

Component and Its Primary Function	Concentration (mg/l), Test Duration, Test Organisms	Effects	Reference
Dissolvan HOEF 1877-4, Progalit DEM 15/100—all-purpose surfactants	0.1–50 Several hours Embryos, larvae, and juvenile fish	LC_{50}—within 0.3–13.5 mg/l for Dissolvan and within 7.9–16.8 mg/l for Progalit	Lesyuk et al, 1989
	0.05–0.5 Standard chronic tests	No observed effects	Register of MPC, 1995
PAF-32 and PAF-41—polyelectrolytes based on phosphonic acids	1–340 30 days Zooplankton	Death—at 34–70 mg/l, impaired reproduction—at lower levels	Igumentsova et al, 1988
	0.03–0.2 Standard chronic tests	No observed effects	Register of MPC, 1995
Neonols (1020-3, AN-1214-5, A-1620-4)—complex surfactants on aminophenol and alcohol derivatives	10^{-4}–1 Up to 120 days Organisms of the main trophic levels of the Caspian Sea	Acute toxic effects—at 0.1–1.0 mg/l, safe levels—within 10^{-4}–10^{-2} mg/l	Akmedova et al, 1988; Register of MPC, 1995

Table 45 (continued)

Component and Its Primary Function	Concentration (mg/l), Test Duration, Test Organisms	Effects	Reference
Sulfanol—all-purpose additive	1–5 4 days Salmon fry	50% mortality—at 2 mg/l	Moseychenko, Abramov, 1994
	0.1–0.5 Standard chronic tests	No observed effects	Register of MPC, 1995
RPK-1, Polygum-K—compounds with multiple application (stabilizers, inhibitors, etc.)	10–30 2–4 days Crustaceans and salmon fry	50% mortality—at 22–31 mg/l, threshold effects—at $<$10 mg/l	Kozak, 1993
Trixan (defoamer), T-66 and T-80 (floatation agents), Sprint (all-purpose agent)	0.01–1 2–4 days Gammarus and salmon fry	Toxic and threshold effects—at all exposure concentrations	Kozak, 1993
ICB-2-2, ICB-6-2—corrosion inhibitors	10^{-4}–10^{-1} Acute and chronic tests Organisms of different trophical levels	Maximum no observed effect concentrations—within 10^{-4}–10^{-3} mg/l, mutagenic and teratogenic effects—at higher levels	Petukova et al., 1991

Table 46

Acute Toxicities of Used Drilling Fluids and Drilling Fluid Components to Marine Organisms Found in Australian Waters [Swan et al., 1994]

Species	Fluid Description	Concentration (mg/l)	Test Type	Toxicity Rating*
Skeletonema costatum (planctonic marine alga)	Imco LDLS/SW	1,325–4,700	96-hr EC_{50}	4
	Imco Lime/SW	1,375	–"–	4
	Imco nondispersed/SW	5,700	–"–	4
	Lightly treated LS/SW-FM	3,700	–"–	4
	Barite	385–1,650	96-hr LC_{50}	3–4
	Aquagel	9,600	–"–	4
Carcinus maenas (marine shore crab)	LDLS	89,100	96-hr LC_{50}	5
	LDLS (suspended WM)	15,000	–"–	5
	LDLS (MAF)	100,000	–"–	6
	MDLS	68,000–100,000	–"–	5–6
	MDLS (suspended WM)	15,000	–"–	5
	MDLS (MAF)	100,000	–"–	6
	HDLS (MAF)	100,000	–"–	6
Mytilus edulis (marine mussel)	Spud mud (MAF)	100,000	96-hr LC_{50}	6
	Seawater LS (MAF)	100,000	–"–	
	MDLS (suspended WM)	15,000	–"–	5
	HDLS (MAF)	100,000	–"–	6
	HDLS (suspended WM)	15,000	–"–	5

272

Table 46 (*continued*)

Species	Fluid Description	Concentration (mg/l)	Test Type	Toxicity Rating*
Salmo gairdneri (juvenile freshwater rainbow trout)	Kipnik-KCl polymer	24,000–42,000	95-hr LC$_{50}$	5
	Seawater polymer	130,000	-"-	6
	KCl-XC polymer	34,000	-"-	5
	Weighted shell polymer	16,000	-"-	5
	Pelly gel chemical-XC	42,000	-"-	5
	Weighted gel XC-polymer	18,000–48,000	-"-	5
	Imnac gel XC-polymer	42,000	-"-	5
Crassostrea gigas (pacific oyster) 3–10 mm	MWL	53,000	96-hr LC$_{50}$	—
	HWL	74,000	-"-	—
	Spud	**	-"-	—
-"- 10–25 mm	MWL	50,000	-"-	—
	HWL	73,000	-"-	—
	Spud	**	-"-	—
Panaeus monodon (tiger prawn)	Bentonite Polymer Seawater	1,000,000	96-hr LC$_{50}$	6
	8% KCL 0.7 μg/1 PHPA powder	76,000	96-hr LC$_{50}$	5

Notes: *—Toxicity rating: 1—very toxic (1 mg/l); 2—toxic (1–100 mg/l); 3—moderate toxic (100–1,000 mg/l); 4—slightly toxic (1,000–10,000 mg/l); 5—almost non-toxic (10,000–100,000 mg/l); 6—non-toxic (100,000 mg/l). **—Insufficient mortality to compute 96-hr LC$_{50}$ value.

weight (volume) of drilling fluids (usually less than 0.1%). However, they can make a significant contribution in the fluids' total toxicity. This group includes the most toxic biocides (e.g., carbamate, sodium sulfide, chlorinated phenols, aldehydes), corrosion and scale inhibitors (e.g., IKB-2-2, IKB-6-2, phosphoxit-7), some of the defoamers and scavengers (e.g., PO-A, KE-10-12, trixan), and other minor agents with MPC within 10^{-4}–10^{-2} mg/l and LC_{50} within or below 0.1–1.0 mg/l. This group also includes most heavy metals (mercury, lead, cadmium, chromium, zinc) that are part of some components of drilling fluids (e.g., barite and other weighting agents) and/or can get into drilling muds from drilling cuttings.

The last group unites the largest assortment of compounds with medium toxicity and very variable composition and function. These include lubricants, emulsifiers, dispersants, viscosity control agents, polyelectrolytes, solvents, stabilizers, detergents, neonols, sulfonols, as well as oil, oil products, and their derivatives. The MPC values for most of these ingredients vary within the range of 10^{-2}–1 mg/l. These values are often close to or the same as the MPC value for dissolved and emulsified fractions of oil, which equals 0.05 mg/l.

Figure 59 gives a more detailed reflection of this toxicological classification of drilling fluid components. Instead of using MPC values, this figure uses 1/MPC values, which some authors [Khomichuk, Vidre, 1989] call the factor of ecological hazard (FEH).

The suggested grouping of components of drilling fluids, as well as other known classifications of substances according to the degree of their toxicity [Khomichuk, Vidre, 1989; Register of MPC, 1995; GESAMP, 1997], is relative and conditional. However, it still gives a certain reference point for the assessment of hazard of such complex, multicomponent systems as drilling fluids.

The groups, shown in Figure 59, unite the components of three levels of toxicity—low/moderate, intermediate, and high. Within each group and even within a set of agents with a similar composition and function, a more detailed toxicological rating is possible. For example, the components of the first group can be presented in the following sequence of decreasing toxicity: barite—Okzil—carboxy-methyl-cellulose (CMC)—ferro-chrome lignosulfonates (FCLS)—bentonite. Even substances with a similar purpose can differ considerably in their toxicity. For example, about 20 corrosion inhibitors and descalers have been tested according to the full standard protocol of ecofisheries assessment. Their MPC vary within several orders—from 10^{-4} mg/l to 0.3 mg/l [Register of MPC, 1995]. Some inhibitors, for example, phosphoxit-7, EKB-2-2, and EKB-6-2, are characterized not only by high toxicity (MPC values of 10^{-3} to 10^{-4} mg/l) but

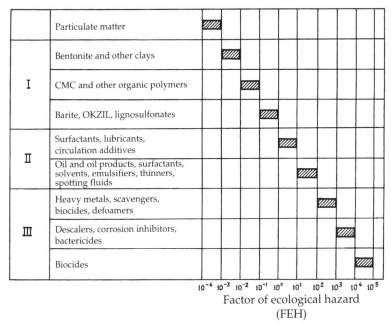

Figure 59 Classification of the drilling fluids components and discharges depending on their ecotoxicological hazard: I—low and moderate toxicity; II—intermediate toxicity; III—high toxicity (shaded areas show the ranges of FEH=1/MPC for corresponding groups of substances and agents).

by their ability to cause genetic and teratogenic damages as well [Petukhova et al., 1991]. Such properties were also revealed for a number of surface-active substances (surfactants) from a group of neonols— AF9-6, AF9-10, and others [Migalovski et al., 1991]. Clearly, this information should be used to develop the biologically safest formulations of drilling fluids and other technological agents.

The results of toxicological studies of drilling fluids and their components can be also used to regulate and control the limits of permissible discharges of hazardous substances during drilling and other oil production operations in the sea. In particular, the current practice of regulating and limiting the pollution of natural waters in Russia uses a methodology based on calculations of the maximum permissible discharges (MPD) and of individual components and their bioassays (i.e., toxicity testing with different water organisms) [Patin, 1991; Filenko, 1991]. However, the application of these approaches and methods for solving the environmental problems in relation to the offshore industry is still extremely limited and requires further improvement.

6.1.3. Drilling Cuttings and Suspended Substances

Drilling cuttings.   From the toxicological point of view, cuttings by themselves (i.e., pieces of rock crushed by the drill and brought up to the surface) do not pose any special threat. The environmental consequences of their discharge include increasing turbidity and smothering the benthic organisms. These effects are often similar to the situations emerging during bottom dredging and dumping operations or during some natural events (e.g., resuspension of bottom sediments as a result of storm and wave activities in shallow coastal waters). However, cutting discharges from the offshore oil activities can cause effects that are more hazardous due to cutting contamination by oil and toxic components of drilling muds. Even after separation and cleaning in special units, drilling cuttings still contain a wide array of organic and inorganic traces, especially when oil-based fluids are used. Drilling cuttings usually go overboard the offshore oil platforms in thousands and tens of thousands of tons (see Chapter 3). Hundreds of tons of oil and dozens of tons of chemicals for each drilled well can enter the marine environment with these discharges. This raises serious concerns about the possible ecotoxicological disturbances in areas of offshore production. Chapter 7 will discuss this in more detail.

The toxic properties of drilling cuttings are first defined by their chemical composition. This, in turn, depends on the particle size, sorption capacity of the crushed rock, and on a number of technological factors. These factors include type and formulation of drilling fluids, physicochemical parameters in the drilling zone, conditions of the mud and cuttings contact with extracted hydrocarbons, methods of cuttings separation and treatment, and others. Obviously, the composition of cuttings will vary a lot even while drilling a single well, which might explain the absence of any generalized information about this issue. As an example, cutting discharges from the oil wells in the Caspian Sea included 60–80% of drilled rock solids that contained 8% organic matter; 6% mineral salts, clay, and weighting agents; oil products; stabilizers; viscosity regulators; and other components of drilling fluids (carboxy-methyl-cellulose, lignosulfonates, and other ingredients) [Gusseinov, Alekperov, 1989].

Most researchers believe that the main toxic agents in drilling cuttings are oil and oil products. These accumulate in the solid phase of drilling cuttings when crude oil and drilling fluids contact cuttings during oil extraction. According to some national and international standards [GESAMP, 1993], the permissible content of oil in discharged drilling cuttings should not exceed 100 g/kg. Even if this requirement were observed during actual industrial operations, this concentration is much higher (100–1,000 times) than the thresholds of acute and sublethal toxic effects of oil-polluted bottom sediments as discussed before (see Chapter 4, Table 30, and Figure 43). Probably, this circumstance was one of the reasons to enforce stricter environmental requirements for discharges of oil-containing solid wastes of drilling activity. The enforcement has been introduced both in a number of countries and at the international level. These discharges have been either completely prohibited or limited to 10 g/kg of oil [Melberg, 1991]. Replacing oil-based drilling fluids by safer formulations that do not include toxic oil products has been in progress since the middle 1980s. Hopefully, this will considerably decrease the oil input into the marine environment with drilling muds and cuttings.

Toxicological data on the produced drilling cuttings (before their discharge) are not available except some mention of low toxicity of the particles of these cuttings, suspended in the water, in concentrations of about 500 mg/l [Gusseinov, Alekperov, 1989]. The attention of researches has been concentrated on assessing the ecological effects of oil-containing drilling cuttings after they are discharged and distributed in bottom sediments around the offshore oil platforms. Here, the levels of oil pollution are hundreds and thousands of times higher than any background characteristics. These levels can cause obvious disturbances in the structure and functions of benthic communities up to 10 km away from the place of discharge [McIntosh et al., 1990; GESAMP, 1993].

Lately, via national and international programs that monitor the environmental situation in areas of offshore oil and gas production, bioassays of bottom sediments polluted by discharges of drilling cuttings from the offshore platforms have become more common. One of such programs was conducted recently on the U.S. shelf. Testing with sea urchin embryos revealed significant toxicity of bottom sediments around the platform that directly correlated with the levels of pollution caused by production operations on the platforms [ICES, 1994]. Significantly, the increased toxicity of sediments containing drilling cuttings was connected with the presence of not only oil residuals but heavy metals as well.

Suspended materials. When drilling wastes are discharged into the marine environment, the dispersed solid phase mainly contains particles of clay minerals, barite, and crushed rock. This solid phase differentiates, and large and heavy particles are rapidly sedimented

(Figure 60). Simultaneously, small (pellet) fractions gradually spread over large distances. Particles less than 0.01 mm in size can glide in the water column for weeks and months. As a result, large zones of increased turbidity are created around drilling platforms. These phenomena, on even a larger scale, happen during the laying of underwater pipelines, construction of artificial islands, bottom dredging, and some other activities that accompany offshore oil production operations. At the same time, the increased turbidity can pose a certain ecological risk. In Russia, "Rules for water protection" [1991] do not allow increasing the suspended matter in fisheries water bodies over 0.25–0.75 mg/l higher than the background natural levels.

As direct observations in areas of exploratory drilling on the eastern shelf of Sakhalin showed [Sapozhnikov, 1995], the persistent plumes of increased turbidity disturb the balance of production-destruction processes in the surface (photic) layer of seawater. It can also cause disturbances at the ecosystem level. Experimental evidence shows the negative effects of pellet suspension (particles with

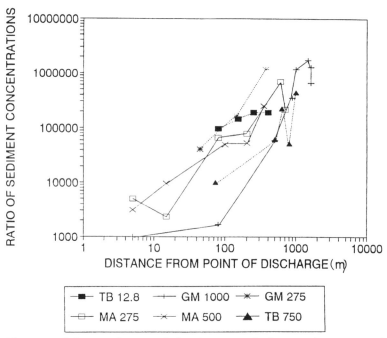

Figure 60 Dilution of suspended sediment with distance from point of discharge [EPA, 1985]: TB—Tanner Bank; GM—Gulf of Mexico; MA—Mid-Atlantic.

a size of 0.005–0.01 mm) on marine organisms. A short-term increase in concentration of such suspension above the levels of 2–4 g/l caused quick adverse effects and death of fry of salmon, cod, and littoral amphipod [Shparkovski, 1993; Matishov et al., 1995]. In 500 mg/l of this suspension, the exposure time leading to 50% mortality was 24 hours for fry of cod, 12 hours for fry of salmon, and 8 hours for amphipod. Note that in this study, the suspended materials were not polluted. Their hazard to marine organisms was connected with physical damage to filtration and respiratory organs. The impact of solid drilling wastes is usually more damaging due to their contamination by drilling muds and oil.

Zones of increased turbidity often form in shallow areas of the shelf where drilling is normally conducted without return of the drilling fluids or even without drilling fluids at all (an open drilling system, as illustrated in Figure 61). For such cases, the levels of maximum permissible discharges of suspended solids (in a form of clay and rock cuttings) into water bodies can be calculated with the help of biogeochemical approaches to assessing the state of the marine environment [Khomichuk, Vidre, 1989]. These approaches use the notion of threshold of ecological tolerance [Patin, 1979]. Results of such calculations, shown in Table 47, give a certain reference point for substantiation of environmental requirements in relation to discharges of nontoxic suspended material during drilling.

6.2. Produced Waters and Their Components

In contrast with drilling fluids, the ecotoxicological studies of produced waters are rather limited. At the same time, the amounts of produced water discharges can be enormous. Everyday volumes of produced waters discharged from a single platform can reach 2,000–7,000 m^3 and total hundreds of thousands of tons a year, as Chapter 3 showed. The oil content in these discharges usually varies within the range of 23–37 mg/l [Law, Hudson, 1986].

In broad terms, produced waters are formation waters produced along with oil during petroleum extraction. They include solutions of mineral salts along with oil, gas, low-molecular-weight hydrocarbons, organic acids, heavy metals, suspended particles, and numerous technological compounds (including biocides and corrosion inhibitors) of changeable and often unknown composition used for well development and production. Prior to discharge, produced waters can be combined with injection water, deck drainage, and

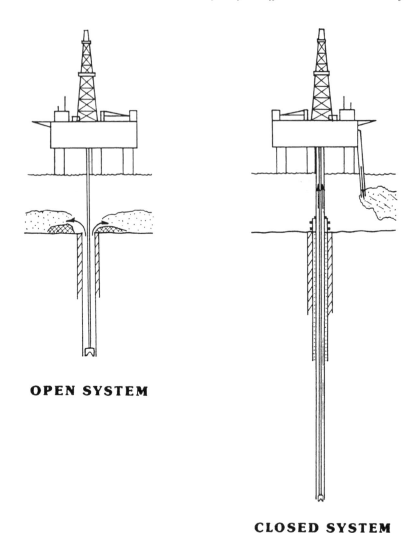

OPEN SYSTEM

CLOSED SYSTEM

Figure 61 Open and closed drilling fluid systems [Swan et al., 1994].

ballast and displacement waters that complicates the chemical composition of these discharges even more (see Chapter 3).

    Table 48 summarizes the available, rather limited, experimen-
tal studies on the toxic properties of produced waters from offshore
activities in different areas of the World Ocean, while Table 49 gives
information about produced waters discharged in the Gulf of Mexico. Since the chemical composition of produced waters varies

Table 47

Permissible Increase in Concentration of Suspended Solids Above the Background Level (t/km^2) Depending on the Type and Depth of Water Body [Homichuk, Vidre, 1989]

Type of Water Body*	Depth of Water Body (m)										
	1	5	10	15	20	25	30	35	40	45	50
A**	0.25	1.25	2.5	3.75	5.0	6.25	7.5	8.75	10.0	11.25	12.5
B**	0.75	3.75	7.5	11.25	15.0	18.75	22.5	26.25	30.0	33.75	7.5
C**	0.05	0.25	0.5	0.75	1.0	1.25	1.5	1.75	2.0	2.25	2.5

Notes: *—Classification according to "Rules for the surface water protection" [Rules for water protection, 1991]. A**—fishery water bodies of the highest and I category. B**—fishery water bodies of the II category. C**—water bodies containing over 30 mg/l of natural suspended solids during the period of the lowest water level.

Table 48
Summarized Results of Toxicological Studies of Produced Waters
Discharged in the Sea

Area of Discharges	Concentration, Test Duration, Test Organisms	Effects	Reference
North Sea	Dilution to 25 μg/l of oil Hours and days Larvae of Atlantic cod	Reduced survival	Davies et al., 1981
	0.15% with 5–15 μg/l of oil 100 days Natural zooplankton	Reduced survival at populational level	–"–
	5% 48 hours Larvae and embryos of oyster *Crassostrea gigas*	50% mortality	Somerville et al., 1987
	10% 24 hours Copepod *Calanus finmarchicus*	50% mortality	–"–
	5% 10 days Hydroid *Campularia flexuosa*	50% decrease in the population growth	–"–
	5–6% 15 minutes Luminescent bacteria	50% decrease in photoluminescence intensity	–"–
	5–6% 48 hours Daphnia *Daphnia magna*	Immobilization of 50% of test organisms	–"–
	10% 24 hours Trout *Salmo gaigdneri*	50% mortality	Somerville et al., 1987
	0.1–0.3% 60–130 days Natural communities of plankton in mesocosms	Reduced abundance of copepod larvae, disturbances of trophic structure, biochemical changes in fish larvae	Gamble et al., 1987

Table 48 (*continued*)

Area of Discharges	Concentration, Test Duration, Test Organisms	Effects	Reference
Gulf of Mexico	3–30% 96 hours Invertebrates and fish (*Panaeus aztecus, P. sectiferus, Balanus tintinnabalulum, Hypopleuorochilus germinatus*)	50% mortality	Middleditch, 1981
Caspian Sea	20–70 mg/l 1–2 days Fish (eggs, larvae, fry)	Death of juveniles— at 60–70 mg/l, larvae—at 35–40 mg/l, eggs— at 20–25 mg/l	Gusseynov, Alekperov, 1989
	2–30 mg/l Several days Invertebrates	Death of crustaceans—at 30 mg/l, reduced survival—at 2 mg/l	–"–
California shelf, USA	1–10 % Chronic tests Macrophytes *Macrocystis pyrifera*	Reduced ability of zoospores to settle on the bottom	Lewis, Reed, 1994
Different regions	0.1–1.0% 96 hours Marine organisms of different groups	50% mortality	GESAMP, 1993 (summarized data)
	10–20% 7 days *Menidia beryllina* (embryos)	Teratogenic effects	Middaugh et al., 1988
	Chronic exposure Urchin *Strongylocentrotus purpuratus* (embryos)	Biochemical, cytological, and other changes in embryonic development	Baldwin et al., 1991

Table 49

Summarized Data for Toxicity of Produced Waters to Mysid and Sheepshead Minnows [Dorn, Compernole, 1995]

Overall Mean*	Produced Water Study		Existing Louisiana Database	
	Mysid	Sheepshead	Mysid	Sheepshead
96-hr LC_{50}	7.08	21.55	10.36	14.82
7-d LC_{50}	5.77	19.72	–	–
NOEC** survival	3.14	11.70	3.39	6.28
NOEC** growth	1.60	2.75	2.44	5.23
NOEC** fecundity	2.20	–	2.67	–
Number of tests	24	24	>400	>400

Notes: *—In % produced water. **—No observed effect concentrations.

widely, it is not surprising that the toxicity of these discharges also varies within very wide limits. The values of LC_{50} in acute experiments range from 10^{-2}% to 30%. The high toxicity of some produced waters is probably explained by the presence of the most toxic substances in their composition, for example heavy metals, biocides, and other similar compounds. In particular, it was shown that toxicity of produced water samples treated with biocide was roughly 100–fold higher than the toxicity of untreated effluents [Neff, 1987].

A number of studies [Gamble et al., 1987] revealed an increased sensitivity of zooplankton organisms (copepod and others) exposed to produced waters. They are especially vulnerable at the embryonic and larval stages of their development. According to some authors [Davies, Kingston, 1992], this could be a result of accumulation of lipophilic hydrocarbons in the lipid fractions of developing embryos' organs and tissues. The level of these hydrocarbons radically increases in larvae when lipid reserves are being exhausted at the stage of larval transfer to active feeding. This causes impairing effects in zooplankton organisms. Similar processes probably occur at the embryonic and postembryonic stages of fish development predetermining their low resistance to the impacts of most toxicants, including the organic ones [Patin, 1979].

Field observations of the distribution of produced waters discharged from offshore platforms [Law, Hudson, 1986; Somerville et al., 1987; Davies, Kingston, 1992] reveal the rather rapid dilution of the discharges in the water. This occurs due to advective transport and turbulent mixing. As Figure 62 shows, 100 m from the point of discharge, produced waters can be diluted 1,000 or more times.

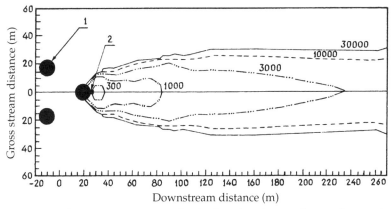

Figure 62 Dilution contours at 5 m depth of produced water dispersal in a scale model: current velocity 0.3 m/s: 1—rig support structure; 2—discharge opening [Somerville et al., 1987].

The rapid dilution of produced waters is usually used as evidence to prove the limited and insignificant environmental impact of these discharges. However, hydrological conditions in different areas and even at the same place at different periods could be extremely diverse and must be taken into account. For example, the rate of dilution in shallow waters or in areas with limited water circulation can be lower than the rate shown in Figure 62 for the open sea. Besides, and this is critically important, the long-term biological effects of low concentrations of produced waters have not been studied yet. Research in this direction may radically change the presently dominating concept about the insignificance of ecological disturbances in the marine environment caused by produced waters.

6.3. Oil Spill-Control Agents

Chapter 3 already stressed the inevitability of accidents occurring during oil extraction and transportation that often result in oil spills, sometimes catastrophic in nature. This situation explains the critical importance of uniting national, regional, and international efforts and creating special services to respond rapidly in case of oil spills.

The array of modern methods and equipment used to control oil spills is very diverse. It includes mechanical, chemical, and biological means that usually supplement each other, depending on specific conditions and circumstances. From the ecological safety perspective, the most preferable methods are mechanical. The oil slick spreading is prevented with the help of floating booms. Oil is then collected in the open waters from special ships (oil collectors) by floating separating units and other similar equipment. However, these methods are seldom effective, especially under unfavorable weather conditions. Usually mechanical means must be supplemented by chemical spill-control methods.

6.3.1. Composition, Properties, and Conditions of Applying Spill-Control Agents

The list of spill-control agents officially approved in different countries presently includes over 200 compounds. According to their chemical composition, properties, and function, they can be divided into several groups:

- dispersants: for reducing oil/water tension and hence stabilizing oil droplets dispersed in the water (oil-in-water emulsion) and removing oil from the water surface;
- de-emulsifiers: for breaking water-in-oil emulsions (mousses) and preventing their formation;
- recovery enhancers: for enhancing the viscoelastic properties of oil, reducing the spreading of oil slicks, and improving the performance of oil skimmers;
- herders: for decreasing the spreading pressure of oil and collecting it into a smaller area before mechanical removal;
- gelling agents: for giving oil a jelly-like texture for its mechanical removal and for repressing volatilization of light fuel oils that can cause fire;
- pyrotechnic materials: for igniting an oil slick on the sea surface, improving flame propagation, and increasing combustion efficiency;
- biological and microbiological agents: for promoting the process of oil biodegradation in the water environment; and
- cleaning agents: for washing oil from onshore surfaces and structures and for removing it from birds.

The leading position in this array of spill-control means belongs to chemical dispersants. These consist of mixtures of solvents and surface-active substances (surfactants). Due to their ability to reduce the oil/water interfacial tension, surfactants stabilize oil droplets in the water column. This way, they disperse the spilled oil. Dispersion prevents the formation of oil slicks on the sea surface and radically intensifies the processes of chemical and microbial degradation of oil. Some surfactant-based agents, including DN-75 (developed in the Institute of Oceanography of Russian Academy of Science) [Mochalova et al., 1987], have multifunctional properties and can be used both for oil dispersion and for its localization and removal.

The molecule of a surfactant usually consists of a long lipophilic chain with hydrophilic group at the end. Figures 63 and 64 schematically show the molecular structures of some typical surfactants and mechanisms of oil dispersion in the water.

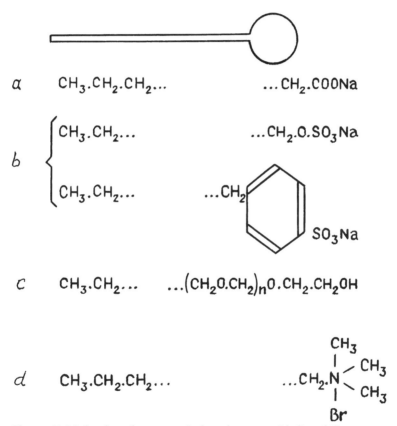

Figure 63 Molecules of some typical surfactants with lipophilic groups (left) and hydrophilic "heads" (right): a—salt of a fatty acid with long chain (detergent); b—alkylsulfate, alkylbenzene sulfonate (anionic); c—product of condensation of alkylphenol with ethylene oxide (non-ionic); d—salt of trimethyl-alkyl-ammonium (cationic).

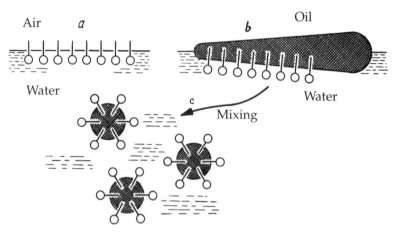

Figure 64 Surfactant behavior in the process of oil dispersion in the water. The molecules of surfactant are located on the water surface with hydrophilic "heads" down into the water and hydrophobic (lipophilic) groups up (a). Lipophilic groups project toward the floating oil and promote oil spreading by decreasing the surface tension (b). When surfactant mixes with oil (c), the oil is broken into tiny droplets and oil is held in suspension by the external layer of hydrophilic "heads" [Nelson-Smith, 1977].

Dozens of dispersant formulations have been developed and tested. Some dispersants are produced commercially and stored for application in case of accidental situations. At the same time, the practical experience of the last 30 years shows both serious difficulties in applying chemical means in such cases and the absence of any standard procedures and methods applicable to different accidental situations. These difficulties and limitations are easy to understand if one remembers the extreme complexity, dynamism, and multifactorial nature of oil's behavior after it spills in the marine environment. Each such episode develops according to its own unique scenario (see Chapter 3). The diversity of such scenarios is especially high when the spilled oil washes ashore. When this happens, the environmental effects primarily depend on the nature of the shoreline and the choice of adequate response should take it into account (see Chapter 7).

The main reasons and factors diminishing the efficiency of chemical means and methods to control oil spills include:

- impediments in prompt response in most situations because of weather conditions, remoteness of the spill location, and others;

- formation of large and thick slicks or stable water-in-oil emulsions (mousses) with increased viscosity;
- absence or insignificance of dispersing effect in case of viscous oils, including weathered-oil residuals (changed under the influence of natural factors), emulsified crude oil, and heavy oils;
- difficulties of applying chemical agents (from ships, airplanes, and helicopters) on the surface of oil spills in optimal volumes and proportions, which is needed for effective oil dispersion; and
- complex dependency of dispersant action on the composition and properties of the spilled oil and the environmental parameters (e.g., temperature, turbulence).

Numerous publications, including a review of data on using dispersants in over 100 experimental oil spills [Fingas et al., 1991], describe the difficulties and limitations of applying oil spill-control agents. Although some skepticism regarding the use of chemical methods is understandable, they have certain advantages. Chemical means help remove oil from the water surface, dilute it to nontoxic concentrations, and thus reduce the environmental impact. At present, application of chemical agents is much more promising than any other response to oil spills. Studies in progress in this area are aimed toward the search for new dispersants that effectively eliminate heavy oil pollution not only in the water environment but on the shore as well.

Environmental impact of spilled, untreated oil—especially when it reaches a shallow coastal zone and contacts the shore—usually exceeds the possible harmful effects of applying chemical agents to disperse oil in open waters. Chemical means can be especially useful and effective for controlling small oil spills in areas where quick localization of spills is possible and adequate chemical agents and equipment are available. These include ports, bays, harbors, and areas of offshore oil and gas production

Lately, interest has been growing in eliminating oil spills by burning them, especially in the open sea. Long-term experimental research conducted by Canadian specialists near the shore of Newfoundland indicates that the products of oil combustion do not pose a risk to people at a distance of 150 m. In addition, the content of surfactants in the soot, settling on the sea surface, is considerably lower than in the original oil [Lazes, 1994]. These and some other similar studies showed that the positive effects clearly overweigh the negative consequences of oil spill liquidation by its combustion. These studies confirmed the advantages of this method as compared with other possible responses in some situations. Special compounds and methods to ignite oil and maintain its burning on the sea surface have been developed and tested under different natural conditions [Putorti et al., 1994]. These compounds and methods can be rather effective during the liquidation of accidental oil spills in the open waters, in icy conditions in the Arctic seas, and in other remote regions.

For over 30 years, scientists from many countries have searched for microbiological methods to enhance the decomposition of oil spilled in the marine environment. Such methods are based on the ability of some species of marine microorganisms to use oil hydrocarbons as a food substrate and actively decompose them under certain conditions. This process constantly takes place in all water bodies, maintaining to a large extent the natural self-purification processes (see Chapter 4). In fact, numerous studies on this topic have formed an independent branch of applied microbiology [NRC, 1989; Pritchard, Costa, 1991].

The final result of abundant research in this area is still inconclusive. On one hand, evidence indicates the obvious acceleration of microbial decomposition of oil and its fractions after introduction of hydrocarbon-degrading bacterial strains, enzymes, and nutrients (mainly compounds of nitrogen and phosphorus) into the polluted environment. Most often, this evidence is found in experimental studies [Venoza et al., 1991; Rosenberg et al., 1992; Marty, Martin, 1994]. Some positive field experience is known as well, for example in situation of heavy oil pollution after the accidental oil spill near the Alaskan shore [Pritchard et al., 1992; Tumeo et al., 1994]. On the other hand, many studies of applying microbial agents for intensifying oil degradation after spills in the open and coastal waters either did not reveal any effect at all or showed uncertain effects comparable with natural oil degradation [Atlas, 1991; Glasser et al., 1991; Stone, 1992].

The difficulties of adequately selecting microbial formulations to eliminate oil pollution effectively are quite clear. They originate from the complexity of oil composition, the feeding specifics of most microbial communities, and the high dependency of microorganisms' hydrocarbon-degrading ability on environmental factors. Nevertheless, the search in this direction continues. Its success may eventually ensure radical improvement of the whole system of oil spill control, including its ecological safety. As a promising example, microbial enhancer Putidoil, developed in Russia, has been successfully applied in eliminating oil spills in a variety of situations, including the accidental fuel oil spill in Aniva Bay near the Sakhalin shore [Ecogeos-1, 1995].

6.3.2. Ecotoxicological Aspects of Applying Dispersants

The toxic properties of dispersants take a special place among the factors that should be considered when applying chemical agents to control oil spills. The concerns regarding this issue were first expressed in the beginning of the 1960s after the first unsuccessful experience in applying dispersants (for example, during the *Torrey Canyon* spill). This experience revealed the high toxicity of dispersants available at that time. These agents had low molecular weights, were relatively water soluble, and included toxic aromatic hydrocarbons in their composition. Later, with the progress in developing synthetic surfactants, the creation of relatively safe dispersants became possible. New dispersants had lower toxicity and a higher rate of degradation in the water environment. At present, three agents are widely used in many countries: Corexit-7764, Corexit-9527, and BP-1100 X. These compounds have been the objects of especially profound toxicological testing. Table 50 gives the summarized results of these tests.

These data demonstrate that studied dispersants, especially Corexit-7764 and BP-1100 X, are considerably less toxic (100–1,000 times) than the oil that they are supposed to disperse. This conclusion is very important because it addresses the often-expressed concern about the possibility of more serious ecological consequences from dispersants than from the untreated oil itself. However, this certainly does not mean that all reasons for such concern are completely

Table 50
Acute Toxicity Thresholds for Some Common Dispersants and Compared to Petroleum [GESAMP, 1993]

		Threshold Concentrations (mg/kg)					
		$>10^4$	10^3 to 10^4	10^2 to 10^3	10 to 100	1 to 100	<1
		(less toxic)				(more toxic)	
Dispersant	Total*			Number of Studies with Threshold in Range Noted Above			
Corexit 7664	32	7	17	8	0	0	0
BP 110X	26	4	17	5	0	0	0
Corexit 9527	25	0	1	6	10	5	3
(Petroleum**)	33	0	1	0	3	15	14

Notes: *—Number of data sets reported to 1986; **—48 to 720 h LC_{50} for a variety of marine zooplankton exposed to water-soluble fraction of crude oil or oil-water dispersion.

removed. Numerous data, including the ones in Table 50, clearly indicate rather wide variability in the toxicities of different dispersants even under similar conditions of standard acute bioassays. There is also evidence of a much wider range of lethal and sublethal effects caused by both dispersants and dispersed oil [NRC, 1989], especially for the most sensitive early life stages of marine organisms [Singer et al., 1991].

Experiments with developing eggs of marine fish showed considerable increases in the number of embryonic malformations at dispersant concentrations of 1–10 mg/l. The highest toxicity levels were typical for Corexit-9527, Finasol-0SRS, and Finasol-0SR7 [Falk-Peterson, Kjoersvik, 1987]. Chronic tests with fish eggs, larvae, and fry revealed radically (50–250 times) increased toxicity of Corexit-7664 in the presence of oil as well as increased toxicity of dispersed oil [Goreva, Gorev, 1988]. It should be noted that in some studies, such effects were not found [Falk-Peterson, Kjoersvik, 1987]. At the same time, some data indicate that the toxicity of dispersed oil is explained not by the presence of dispersants but by the adverse impact of the most toxic oil fractions after the oil is dispersed. The method of oil dispersion (mechanical or chemical) does not affect its toxic properties [NRC, 1989; GESAMP, 1993].

The discrepancies in experimental data originate, in particular, from the different ways of measuring active (effective) oil concentrations that can significantly vary depending on test conditions. For example, the toxic effect on plankton organisms depends not on total concentration of the oil but rather on the amount of the nonvolatile fraction of the hydrocarbons dissolved in the water [Østgaard, 1994].

The results of comparative studies of the consequences of oil spills when dispersants were applied and when oil was decomposed by only natural factors are rather controversial [NRC, 1989; Fingas et al., 1991; GESAMP, 1993]. Similar to the situation with microbiological agents for oil elimination, positive, uncertain, and negative conclusions are made regarding the effectiveness of using dispersants and other chemical compounds for oil spill control.

The potential ecological damage from introducing these compounds into the marine environment can hardly be considered significant, at least during their application in the open waters. The locality and short-term nature of such operations, relatively small (as compared with the spilled oil) volumes of applied agents, as well as their rapid dilution in the water, relatively low toxicity, and high speed of degradation of the majority of modern dispersants suggest that their application for liquidating oil pollution is justified and, in some cases, essential. For example, when an oil slick drifts toward a large population of marine birds or mammals, mangrove ecosystem, or coastal zone, chemical means can prevent or mitigate hazardous oil impacts. At the same time, application of dispersants should be regulated according to specific natural conditions. In some countries, for example in Great Britain, the application of dispersants in coastal waters less than 20 m deep is restricted in order to ensure the dilution of introduced chemicals and dispersed oil to the safe levels and to prevent their accumulation in shallow zones with slow water circulation [Monk, Cormack, 1992].

Conclusions

1. To date, extensive ecotoxicological information is available regarding the biological effects of different wastes, agents, and compounds introduced into the sea during offshore oil and gas production. It is used to select the most ecologically safe technological formulations and to develop environmental standards and regulations. The main difficulties in solving these tasks originate from the high complexity of the chemical composition of discharged wastes and the wide variability of the results of toxicological assessments.

2. The values of LC_{50} (96-hour exposure) for the majority of drilling fluids vary within 10–100 g/kg (dilution 1–10%). Oil-based muds are characterized by much higher toxicity and have values of LC_{50} as low as 0.1 g/kg. Lethal and sublethal responses of marine organisms under chronic exposure occur under considerably lower (100–1,000 times) concentrations of drilling fluids.

3. The levels of acute (lethal) and threshold (reversible) effects of drilling fluid components on water organisms of different trophic levels vary within an extremely wide range—from 10^{-5}–10^{-4} mg/l to 10^3–10^4 mg/l. Based on toxicological characteristics (values of MPC at chronic exposure and LC_{50} in acute tests) and the relative content in drilling discharges, drilling fluid components can be divided into three main groups:

- low-toxic, basic ingredients (e.g., bentonite, barite, lignosulfonates) with MPC values within the range of 1–100 mg/l, LC_{50} values of 100–1,000 mg/l, and relative content of 1–10% of each in the formulation;
- highly-toxic compounds present in small proportions (e.g., biocides, corrosion inhibitors, descalers); MPC

values vary within the range 10^{-4}–10^{-2} mg/l, and LC_{50} values are 0.1–1.0 mg/l and less; and
- components of different composition and function with variable content and medium toxicity (e.g., lubricating oils, emulsifiers, thinners, solvents). MPC values are within the range 10^{-2}–1.0 mg/l.

This suggested classification of drilling components can be used to develop ecologically safe formulations of drilling and other technological mixtures as well as to regulate and control their application at different stages of oil production.

4. The toxic properties of drilling cuttings are defined by their chemical composition (especially by oil concentration) and cleaning procedures. The levels of oil products in drilling cuttings vary within a wide range and depend on the type of drilling fluids. Toxicological studies of solid drilling wastes and available data on the oil content in them (up to 100 g/kg) indicate the possibility of lethal and sublethal impacts on benthic fauna in the zones of discharges of such wastes. This is one of the reasons to enforce stricter environmental requirements to discharges of drilling cuttings in a number of countries or even completely prohibit them. The increased water turbidity in areas of offshore drilling can also pose a certain ecological risk.

5. The acute toxicity of produced waters is relatively low (LC_{50} values in acute tests usually vary within the range of 1–30%). They rapidly dilute in the open sea. At the same time, large volumes of such discharges (thousands of tons a day), the increased oil hydrocarbon concentration, presence of other toxicants (e.g., heavy metals, biocides) in their composition, and insufficient data on long-term biological effects of low concentrations of produced waters justify some concern about the ecological safety of their discharges, especially in shallow coastal areas with slow water circulation.

6. An array of oil spill-control agents totals to more than 200 compounds of differing composition and function. The leading place among them belongs to dispersants. These emulsify and disperse oil, accelerating the natural processes of its degradation. Practical application of such compounds has revealed serious difficulties and limitations due to the tremendous diversity of oil spill scenarios in natural conditions. From the ecotoxicological perspective, modern chemical means to control oil spills do not pose a serious risk for marine biota for three reasons. First, the toxicity of most compounds is lower than oil toxicity (LC_{50} values for main dispersants are usually 10^2–10^4 mg/l). Second, they decompose rather quickly in the marine

environment. Third, their application has a local and short-term nature. Other promising directions to control oil spills include new methods of burning oil on the sea surface (especially in the Arctic) and microbiological agents for enhancing the biodegradation of oil hydrocarbons.

References

Akhmedova, T. P., Dokholyan, V. K., Kovalenko, L. D. 1989. Adaptive changes in biochemical parameters of the juvenile sturgeon under drilling fluid pollution. In *Theses of the Seventh All-Union Conference on Ecological Physiology and Biochemistry of Fish. Vol.1.*—Jaroslavl, pp.22–24. (Russian)

Akhmedova, T. P., Dokholyan, V. K., Kovalenko, L. D., Kostrov, B. P., Shleifer, G. S. 1989. Impact of some surfactants on water organisms. In *Theses of the First All-Union Conference on Fisheries Toxicology. Vol.1.*—Riga, pp.29–31. (Russian)

Allen, T. E., ed. 1984. *Oil spill chemical dispersants: research, experience and recommendations.* Philadelphia: American Society for Testing and Materials.

Artemyeva, N. V., Ivanova, N. A., Kabanova, T. I., Kurzikina, L. G., Marchenko, L. P., Flink, L. M., Urieva, V. V. 1988. Establishing biological standard for SNPH-7215M in the water of the fisheries water bodies. In *Theses of the Fifth All-Union Conference on Water Toxicology.*—Moscow: VNIRO, pp.174. (Russian)

Atlas, R. M. 1991. Bioremediation of oil spills. In *Program and Abstracts of the Second International Marine Biotechnology Conference (IMBC 91).*—Arlington, VA (USA): Society for Industrial Microbiology, p.71.

Bakke, T., Laake, M. 1991. Test on degradation of a new drill mud type under natural conditions. In *Final report. Norsk.Inst. for Vannforskning, Oslo (Norway).*—48 pp.

Baldwin, J. D., Pillai, M. S., Cher, G. N. 1991. Response of embryos of the sea urchin *Strongylocentrotus purpuratus* to aqueous petroleum waste includes the expression of a high molecular weight glycoprotein. *Marine Biology* 114(1):21–30.

Blackman, R. A. A., Fileman, T. W., Law, R. J. 1986. *The toxicity of alternative base-oils and drill-muds on the settlement and development of biota in an improved tank test. International Council for the Exploration of the Sea.* ICES C.M. 1986/E:13, 16 pp.

Boesch, D. F., Rabalis, N. N., eds. 1987. *Long-term environmental effects of offshore oil and gas developments.* New York: Elsevier Applied Science, 708 pp.

Borisov, V. P., Osetrova, N. V., Ponomarenko, V. P., Semenov, V. N. 1994. *Impact of the offshore oil and gas developments on the bioresources of the Barents Sea.* Moscow: VNIRO, 251 pp. (Russian)

Buslenko, N. M., Barchovich, O. A., Parfenova, N. A. 1991. Establishing the standard of the maximum permissible concentration for the drilling agent GKZH-11. In *Theses of the Second All-Union Conference on Fisheries Toxicology. Vol.1.*—Spb., pp.66–67. (Russian)

Burke, C. J., Veil, J. A. 1995. Synthetic-based drilling fluids have many environmental pluses. *Oil and Gas Journ.* (Nov.) pp.59–65.

Capuzzo, J. M. 1987. Biological effects of petroleum hydrocarbons: assessment from experimental results. In *Long-term environmental effects of offshore oil and gas development.*—London: Elsevier Applied Science Publishers, pp.342–410.

Chinareva, I. D. 1989. Patohistological changes in fish under impact of some polymers. In *Theses of the First All-Union Conference on Fisheries Toxicology. Vol.2.*—Riga, pp.191–192. (Russian)

Davies, J. M., Hardy, R., McIntyre, A. D. 1981. Environmental effects of North Sea oil operations. *Mar.Pollut.Bull.* 12:412–416.

Davies, J. M., Kingston, P. F. 1992. Sources of environmental disturbances associated with offshore oil and gas developments. In *North Sea oil and the environment. Developing oil and gas resources, environmental impacts and responses.*—London–New York: Elsevier Applied Science, pp.417–440.

Dokholyan, V. K., Akhmedova, T. P., Magomedov, A. K., Shleifer, G. S. 1988. Fish response to the presence of some drilling fluid components in the environment. In *Theses of the First All-Union Conference on Fisheries Toxicology. Vol.1.*—Riga, Ch.1. pp.125–126. (Russian)

Dorn, P. B., van Compernelle, R. 1995. Effluents. In *Fundamentals of aquatic toxicology: Effects, environmental fate, and risk assessment (Second edition).*—Taylor and Francies, pp.903–938.

Duke, Th. W. 1985. Potential impact of drilling fluids on estuarine productivity. In *Proceedings of the International Symposium on Utilization of Coastal Ecosystems: Planning, Pollution and Productivity. Vol.1.*—Rio Grande (Brazil), pp.215–239.

Dzhabarov, M. I. 1988. Toxicological assessment of the impact of carboxymethylcellulose on the white sturgeon juveniles. In *Theses of the Fifth All-Union Conference on Water Toxicology.*—Moscow: VNIRO, pp.112–113. (Russian)

Ecoegeos-1. 1995. *The method of removing the oil pollution from the water and soil with the help of Putidoil.* Tumen: Zapadno-Sibirski Nauchno-Issledovatelski Neftyanoi Institut. (Russian)

Engelhardt, F. R., Ray, J. P., Gillam, A. H., eds. 1989. *Drilling wastes. Proceedings of the 1988 International Conference on Drilling Wastes, Calgary, Alberta., April 1988.* London–New York: Elsevier Applied Science, 867 pp.

EPA. 1985. *Environmental Protection Agency. Assessment of environmental fate and effects of discharges from offshore oil and gas operations. US Environmental Protection Agency,* 440/4–85/002.

Falk-Petersen, I. B., Kjorsvik, E. 1987. Acute toxicity tests of the effects of oil and dispersants on marine fish embryos and larvae. *Sarsia* 72(6):411–414.

Filenko, O. F. 1991. Practical goals of the water toxicology. *Gidrobiologicheski zhurnal* 27(3):72–74. (Russian)

Fingas, M. F., Bier, I., Bobra, M., Callaghan, S. 1991. Studies on the physical and chemical behavior of oil and dispersant mixtures. In *Proceedings of the 1991 International Oil Spill Conference, American Petroleum Institute.*— Washington, DC: A.P.I. Publ.No.4529, pp.419–426.

Gamble, J. C., Davies, J. M., Hay, S. J., Dow, F. K. 1987. Mesocosm experiments on the effect of produced waters from offshore oil platforms in the northern North Sea. *Sarsia* 72(3–4):383–387.

GESAMP. 1989. *The evaluation of hazard or harmful substances carried by ships: revision of GESAMP Reports and Studies No.17. GESAMP Reports and Studies No.35.* London: IMO.

GESAMP. 1993. *Impact of oil and related chemicals and wastes on the marine environment. GESAMP Reports and Studies No.50.* London: IMO, 180 pp.

GESAMP. 1997. *Report of the twenty-seventh session of GESAMP. GESAMP Reports and Studies No.63.* Narobi: UNEP, 46 pp.

Glasser, J. A., Venosa, A. D., Opatken, E. J. 1991. Development and evaluation of application techniques for delivery of nutrients to contaminated shorelines in Prince William Sound. In *Proceedings of the 1991 International Oil Spill Conference, American Petroleum Institute.*—Washington, DC: A.P.I. Publ.No.4529, pp.450–462.

Gordon, B. P. UK 1989. Fisheries research around offshore oil installations. In *Proceedings of the 1st International Conference on Fisheries and Offshore Petroleum Exploitation.*—Bergen, pp.1–18.

Goreva, V. A., Gorev, A. S. 1988. Assessment of the simultaneous impact of the dispersant "Correxit-7664" and the oil on fish. In *Theses of the Second All-Union Conference on Fisheries Toxicology. Vol.1.*—Spb., pp.46–48. (Russian)

Gudimov, A. V. 1993. Eco-physiological biotesting of the drilling fluids by their impact on the bottom invertebrates. In *Arctic Seas: bioindication of the state of the environment, biotesting and technology of the destruction of the pollution.*—Apatiti: Izd-vo KNTS RAN, pp.36–44. (Russian)

Gundlach, E. R. 1987. Oil-holding capacities and removal coefficients for different shoreline types to computer simulate spills in coastal waters. In *Proceedings of the 1987 International Oil Spill Conference, American Petroleum Institute.*—Washington, DC, pp.451–457.

Gusseinov, G. I., Alekperov, R. E. 1989. *Nature protection during the offshore oil and gas developments.* Moscow: Nedra, 230 pp. (Russian)

ICES. 1994. *International Council for the Exploration of the Sea. Report of the Joint meeting of the Working Group on Marine Sediments in relation to pollution and the Working Group on Biological Effects on Contaminants (Nantes, France, 21–22 March 1994).* ICES C.M.1994/ENV:2, 19 pp.

ICES. 1995. *International Council for the Exploration of the Sea. Sources of data on chemicals transported at sea.* ICES ACME/15/1. 10 pp.

Igumentsova, N. I., Ugrin, A. A., Lesuk, I. I. 1988. Impact of PAF-32 and PAF-41 used during the oil production on *Daphnia magna.* In *Theses of the*

First All-Union Conference on Fisheries Toxicology. Vol.1.—Riga, pp.161–162. (Russian)

Kabanova, T. I., Kuznetsova, O. V. 1988. Studies of the toxicity of SNPH-7214P for daphnia. In *Theses of the Fifth All-Union Conference on Water Toxicology.*—Moscow: VNIRO, p.174. (Russian)

Khomichuk, A. S., Vidre, E. M. 1989. The choice of the environmentally safe technology of the shallow drilling on the continental shelf. In *Complex development of the World Ocean in connection with developing of its mineral resources.*—Leningrad: PGO Sevmorgeologia, pp.94–101. (Russian)

Koefoed, J. H. 1989. Oil mud discharges, effects and implications. In *Proceedings of the 1st International Conference on Fisheries and Offshore Petroleum Exploitation.*—Bergen, pp.1–18.

Kosheleva, V. V., Novikov, N. A., Migalovski, I. P., Gorbacheva, E. A., Lapteva, A. N. 1997. *Responses of marine organisms to the environmental pollution during oil and gas development on the shelf of the Barents Sea.* Murmansk: PINRO, 95 pp. (Russian)

Kozak, N. V. 1991. Ichthyological issues during the offshore oil and gas drilling. In *Ecological situation and protection of the flora and fauna of the Barents Sea.*—Apatiti: Izd-vo KNTS RAN, pp.93–96. (Russian)

Kozak, N. V. 1993. Complex of the ichthyological testing methods for the drilling cuttings, muds and their components. In *Arctic Seas: bioindication of the state of the environment, biotesting and technology of the destruction of the pollution.*—Apatiti: Izd-vo KNTS RAN, pp.30–36. (Russian)

Kozak, N. V., Shparkovski, I. A. 1991. Testing of the drilling muds and their components with the use of the fish from the Barents Sea. In *Theses of the Second All-Union Conference on Fisheries Toxicology. Vol.1.*—Spb., pp.272–273. (Russian)

Kuzmina, S. S., Shcherbakov, A. A. 1988. Toxic impact of alkylsulfate on fish gametes and process of fish fertilization. In *Theses of the First All-Union Conference on Fisheries Toxicology. Vol.1.*—Riga, pp.236–238. (Russian)

Law, R. J., Hudson, P. M. 1986. *Preliminary studies of the dispersion of oily water discharges from the North Sea oil production platforms. International Council for the Exploration of the Sea.* ICES C.M. 1986/E:15, 12 pp.

Lazes, R. 1994. A study on the effects of oil fires on fire boom employed during the in-situ burning of oil. In *17 Arctic and Marine Oil Spill Program Technical Seminar. Vol.1.*—Ottawa (Canada): Environment Canada, pp.717–724.

Lega, G. A. 1988. Impact of ferrochromlignosulfonate (FCLS) on the blood of cod from the Barents Sea. In *Theses of the First All-Union Conference on Fisheries Toxicology. Vol.2.*—Riga, pp.4–5. (Russian)

Lesuk, I. I., Reshetilo, S. G., Savin, O. B., Komarinets, O. T. 1988. Assessment of the fish sensitivity to the impact of some surfactants during different periods of fish onthogeneses. In *Theses of the First All-Union Conference on Fisheries Toxicology. Vol.2.*—Riga, pp.7–8. (Russian)

Lichtenberg, J. J., Winter, F. A., Weber, C. I., Fradkin, L., eds. 1988. *Chemical and biological characterization of sludges, sediments, dredges spoils and drill muds.* Philadelphia: American Society for Testing and Materials, 402 pp.

Marty, P., Martin, Y. 1994. Seed and feed strategy against oil spills in marine environment: laboratory and simulated outdoor experiments using natural selected bacterial strains. In *3rd International Marine Biotechnology Conference: Program, Abstracts and List of Participants.*—Thomsoe (Norway): Thomsoe University, p.132.

Matishov, G. G., Shparkovski, I. A., Nazimov, V. V. 1995. Impact of the bottom deepening activities on the Barents Sea biota during the development of Shtokmanovskoe gas condensate field. *Doklady RAN* 345(1):138–141. (Russian)

McIntosh, A. D., Massie, L. C., Davies, J. M. 1990. *Assessment of fish from the northern North Sea for oil taint. International Council for the Exploration of the Sea.* ICES CM 1990/E:24, 14 pp.

Melberg, B. 1991. Reduction of pollution from drilling operations. In *ENS 91: Environment Northern Seas. Abstracts of Conference Papers.*—Stavanger (Norway): Industritrykk.

Menzie, C. A. 1982. The environmental implications of offshore oil and gas activities. *Environ.Sci.Technol.* 16(18):454–472.

Middaugh, D. P., Hemmer, M. J., Lores, E. M. 1988. Teratological effects of 2,4-dinitrophenol, produced water and naphthalene on embryos of the inland silverside *Menidia beryllina. Dis.Aquat.Org.* 4:53–65.

Middleditch, B. S. 1981. *Environmental effects of offshore oil production. The Buccaneer gas and oil field study.* New York–London: Plenum press.

Migalovski, I. P., Migalovskya, V. N., Kosheleva, V. V., Kasatkina, S. V. 1991. Impact of surfactants on the different stages of ontogeneses of marine fish under the experimental conditions. In *Theses of the Second All-Union Conference on Fisheries Toxicology. Vol.1.*—Spb., pp.46–48. (Russian)

Mochalova, O. S., Nesterova, M. P., Antonova, N. M. 1987. Means of liquidation of the accidental oil spills. *Oceanologia* 32:1042–1044. (Russian)

Monk, D. C., Cormack, D. 1992. The management of acute risk. Oil spill contingency planning and response. In *North Sea oil and the environment. Developing oil and gas resources, environmental impacts and responses.*—London–New York: Elsevier Applied Science, pp.619–142.

Moseichenko, G. V., Abramov, V. L. 1994. Resistance of the salmon juveniles and their feeding base to the drilling component impact. In *Systematic, biology and biotechnology of cultivation of salmon: proceedings of the Fifth All-Union Conference of the State Institute of Lakes and Rivers.*—Spb., pp.132–133. (Russian)

Neff, J. M. 1981. *Fate and biological effects of oil well drilling fluids in the marine environment. A literature review. Report to US EPA. Report No. 15077.*

Neff, J. M. 1987. Biological effects of drilling fluids, drill cuttings and produced waters. In *Long-term environmental effects of offshore oil and gas development.*—New York: Elsevier Applied Science.

Neff, J. M. 1988. *Bioaccumulation and biomagnification of chemicals from oil well drilling and production wastes in marine food webs: a review. Report to American Petroleum Institute.* Battelle Ocean Sciences, Duxbury, Mass.

Neff, J. M. 1998. Fate and effects of drilling mud and produced water discharged in the marine environment. In *U.S.-Russian Government Workshop on Management of Waste from Offshore Oil and Gas Operation (April 1998, Moscow).*

Nelson-Smith, A., trans. 1977. *Oil pollution and marine ecology.* Moscow: Progress, 301 pp. (Russian)

Neumann, R. 1993. Legislation passed to bill ocean polluters. *Search* 24(10):284.

NRC. 1983. *National Research Council. Drilling discharges in the marine environment.* Washington D.C.: National Academy Press, 180 pp.

NRC. 1985. *National Research Council. Oil in the sea. Inputs, fates and effects.* Washington, DC: National Academy Press, 601 pp.

NRC. 1989. *National Research Council. Using oil spill dispersants on the sea.* Washington, DC: National Academy Press, 335 pp.

Østgaard, K. 1994. The oil, the water, and the phytoplankton. *Algae and water pollution. Stuttgart (FRG): Schweizerbartsche Verlagsbuchhandl., Engeb.Limnol.,* 42:167–193.

Østgaard, K., Jensen, A. 1985. Acute phytotoxicity of oil-based drilling muds. *Oil Petrochem.Pollut.* 2:281–291.

Patin, S. A. 1979. *Pollution impact on the biological resources and productivity of the World Ocean.* Moscow: Pishepromizdat, 305 pp. (Russian)

Patin, S. A. 1991. Eco-toxicological aspects of studying and control of the water environment quality. *Gidrobiologicheski zhurnal* 27(3):75–77. (Russian)

Patin, S. A. 1998. A new approach to assessment of environmental impact of the offshore oil and gas industry. *Water Resources* (in press). (Russian)

Patin, S. A., Lesnikov, L. A. 1988. Main principles of establishing the fisheries standards for the water environment quality. In *Methodical guidelines on establishing the maximum permissible concentrations of pollutants in the water environment.*—Mocsow: VNIRO, pp.3–10. (Russian)

Petersen, S. P., Kruse, B., Jensen, K. 1991. Degradation of low toxicity drilling mud base-oil in sediment cores. *Mar.Pollut.Bull.* 22(9):452–455.

Petukhova, G. A., Tupitsina, L. S., Bulovatskaya, S. E., Gerasimova, E. L. 1991. The assessment of the gen toxicity of the corrosion inhibitors. In *Theses of the Second All-Union Conference on Fisheries Toxicology. Vol.2.*—Spb., pp.97–98. (Russian)

Pritchard, P. H., Costa, C. F. 1991. EPA Alaska oil spill bioremediation projects. *Environ.Sci.Technol.* 25(3):372–379.

Pritchard, P. H., Mueller, J. G., Rogers, J. C., Kremer, F. V., Glaser, J. A. 1992. Oil spill bioremediation: experiences, lessons and results from the *Exxon Valdez* oil spill in Alaska. *Biodegradation* 3(2–3):315–335.

Proceedings. 1980. *Proceedings of the 1980 Symposium on research on environmental fate and effects of drilling fluids and cuttings. Vol.1.* Lake Buena Vista, Fla., 690 pp.

Putorti, A. D., Evans, D. D., Tennyson, E. J. 1994. Ignition of weathered and emulsified oils. In *17. Arctic and Marine Oil Spill Program Technical Seminar. Vol.1.*—Ottawa (Canada): Environment Canada, pp.39–75.

Ray, J. P. 1985. The role of environmental science in the utilization of coastal and offshore waters for petroleum development by the US industry. In *Proceedings of the International Symposium on Utilization of Coastal Ecosystems: Planning, Pollution and Productivity. Vol.1.*—Rio Grande (Brazil), pp.127–149.

Reed, D. C., Lewis, R. J. 1994. Effects of an oil and gas production effluent on the colonization potential of giant kelp (*Macrocystis pyrifera*) zoospores. *Marine Biology* 119(2):277–283.

Register of MPC. 1995. *Register of the maximum permissible concentrations (MPC) of hazardous substances for the water environment.* Moscow: Medinor, 220 pp. (Russian)

Rosenberg, E., Legmann, R., Kushmaro, A., Taube, R., Adler, E., Ron, E. Z. 1992. Petroleum bioremediation: a multiphase problem. *Biodegradation* 3(2–3):337–350.

Rules for water protection. 1991. *Rules for the surface water protection (general principles).* Moscow: Goskompriroda SSSR, 34 pp. (Russian)

Sapozhnikov, V. V. 1995. Modern understanding of the functioning of the Bering Sea ecosystem. In *Complex studies of the ecosystem of the Bering Sea.*—Moscow: VNIRO, pp.387–392. (Russian)

Sereda, N. G., Soloviev, E. M. 1988. *Drilling of the oil and gas wells.* Moscow: Nedra, 360 pp. (Russian)

Serigstad, B., Svaeren, I., Foyn, L. 1988. *The effects of oil-based drilling and crude oil on demersal fish eggs. International Council for the Exploration of the Sea.* ICES CM 1988/E:19, 13 pp.

Shparkovski, I. A. 1993. Biotesting of the water environment quality with the use of fish. In *Arctic seas: bioindication of the state of the environment, biotesting and technology of the pollution destruction.*—Apatiti: Izd-vo KNTS RAN, pp.11–30. (Russian)

Shparkovski, I. A., Petrov, V. S., Kozak, N. V. 1989. Physiological criteria of assessment of ecological situation during drilling on the shelf. In *Theses of the First All-Union Conference on Fisheries Toxicology. Vol.2.*—Riga, pp.199–200. (Russian)

Singer, M. M., Smalheer, R. S., Tjeerdema, R. S., Martin, M. 1991. Effects of spiked exposure to an oil dispersant on the early life stages of four marine species. *Environ.Toxicol.Chem.* 10:1367–1374.

Smith, J. P., Agers, R. C., Tait, R. R. 1997. *Perspectives from research on the environmental effects of offshore discharges of drilling fluids and cuttings.* Houston: Exxon Production Research Company, 20 pp.

Somerville, H. J., Bennett, D., Davenport, J. N., Holt, M. S., Lynes, A., Mahie, A., McCourt, B., Parker, J. G., Stephenson, J. G., Watkinson, T. G. 1987. Environmental effect of produced water from North Sea oil operations. *Mar.Pollut.Bull.* 18(10):549–558.

Sorbye, E. 1989. Technical performance and ecological aspects of various drilling muds. In *Proceedings of the 1st International Conference on Fisheries and Offshore Petroleum Exploitation.*—Bergen, pp.1–18.

Stone, R. 1992. Oil cleanup method questioned. *Science* 257(5068):320–321.

Swan, J. M., Neff, J. M., Young, P. C., eds. 1994. *Environmental implications of offshore oil and gas development in Australia—the findings of an independent scientific review.* Sydney: Australian Petroleum Exploration Association Limited, 696 pp.

Thomas, D. J., Green, G. D., Duval, W. S., Milne, K. C., Hutcheson, M. S. 1984. Offshore oil and gas production waste characteristics, treatment methods, biological effects and their application to the Canadian regions. In *Final Report (IR-4). Water Pollution Control Directorate Environmental Protection Service.*—Ottawa: Environment Canada.

Tumeo, M., Braddock, J., Venator, T., Rog, S., Owens, D. 1994. Effectiveness of a biosurfactant in removing weathered crude oil from subsurface beach material. *Spill Sci.Technol.Bull.* 1(1):53–59.

Vekilov, E., Arabkina, N., Badovski, N., Guseinov, G. 1990. Studies and protection of the marine environment during the geological surveys. In *Geology and mineral resources of the World Ocean.*—Warsaw: INTERMORGEO, pp.668–680. (Russian)

Venosa, A. D., Haines, J. R., Nisamaneepong, W., Govind, R., Pradhan, S., Siddique, B. 1991. Protocol for testing bioremediation products against weathered Alaskan crude oil. In *Proceedings of the 1991 International Oil Spill Conference.*—Washington, DC: American Petroleum Institute, pp.563–570.

Vogt, N. B., Davidsen, N. B., Sjoergen, C. E. 1988. Di- and triaromatic hydrocarbons in fish liver from the North Sea: multivariate and statistical analysis. *Oil and Chemical Pollution.* 4:217–242.

Zhuravleva, N. G., Savinova, T. N. 1988. Impact of some drilling fluid components on the embryonic development of flounder from the Barents Sea. In *Theses of the Fifth All-Union Conference on Water Toxicology.*—Moscow: VNIRO, p.161. (Russian)

Chapter 7

Ecological and Fisheries Implications of Offshore Oil and Gas Development

To some extent, everything discussed so far could be considered only as a prelude to the final sections of this book. Chapters 7 and 8 will summarize the available materials both on the environmental effects in areas of offshore oil and gas production and on the criteria and methods of their assessing and monitoring.

In spite of the abundance of studies and publications on these issues (some of which have already been discussed), standard methodologies and procedures to assess environmental impact of the offshore industry vary widely in different countries and regions. This leads to inevitable discrepancies of results, conclusions, and opinions. Hence, additional efforts of many specialists in different disciplines (marine biology, ecology, toxicology, and others) are needed to find the solution to these problems.

7.1. Environmental Stress and Biological Effects

The difficulties of assessing the environmental impact of the offshore oil and gas industry arise, in particular, from the already mentioned extreme complexity of marine ecosystems. In contrast with living organisms, natural ecosystems do not have rigid mechanisms of self-regulation (e.g., central nervous system, physiological and biochemical control) that ensure the homeostatic balance of vitally important processes. Besides, environmental consequences of the offshore oil and gas activity are affected by other numerous types of anthropogenic impact on the marine environment which further complicates the interpretation of research results (see Chapter 3).

Environmental stress in the ecosystem includes chemical or physical impacts that cause different biological responses up to complex transformations at all levels of the biological hierarchy. These transformations usually start from the primary biochemical responses at the subcellular level and gradually spread to the higher levels (see Table 51 and Figure 65). At each of these levels, either partial or full compensation of stress effects happens. In case of full compensation, these effects are not transferred to the next level. When rather long-term stress impacts (such as chronic pollution) occur, hazardous effects can be transferred to the higher levels of biological hierarchy. The scenario then repeats but in different forms.

Figure 66 illustrates the previously described theoretical concepts. It gives some idea about the nature, interrelations, and sequence of the developing biological response to different impacts of the offshore oil and gas production. Practically all the responses listed in Figure 66 start at the cellular and subcellular (molecular) levels. However, direct (quick) responses at the organismal, populational, and even community levels are possible as well. For example,

Table 51

Possible Mechanisms of Toxic Effect Manifestation and Adaptive Responses at Different Levels of Biological Hierarchy in the Sea

Level of Organization	Manifestation of Hazardous Effects	Adaptive Effects and Responses at the Higher Levels
Molecular	Changes in enzyme conformation and activity; genetic mutations	DNA repair; metabolic and enzyme reactions
Cellular	Morphological anomalies in cells; inhibition of mitosis; metabolic disturbances	Detoxification; changes in cell processes; appearing of toxic metabolites
Organismal	Disturbed physiological processes; reduced rates of growth, feeding, and reproduction; low immunity, decreased survival at early stages of development	Behavioral responses (e.g., avoidance); toxicant excretion; changes in energy metabolism
Populational	Changes in the size-age structure, populational growth, abundance, and biomass	Adaptations at the populational level; changes in species inter-connections
Community	Changes in species structure, domination, trophic relations, ecological metabolism, and others	Structural and functional reconstructions; ecological modifications
Ecosystem	Changes in direction and rates of production-destruction processes and trophic status of water bodies	Alternations in succession stages

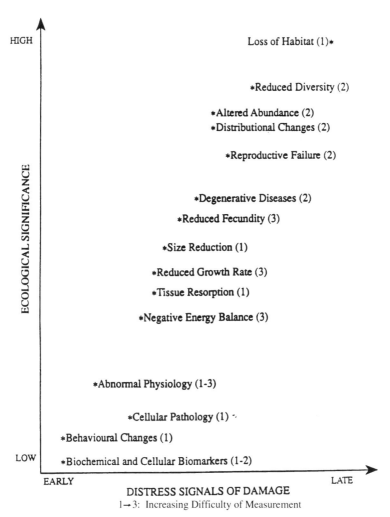

Figure 65 Levels and mechanisms of response to stress in the marine environment [GOOS, 1995].

instant lethal and sublethal damages at the organismal level or behavioral responses at the population levels (e.g., avoiding reactions of migrating fish, mammals) are possible during seismic surveys and explosion operations. Similarly, the invasion of nonindigenous organisms brought in with discharges of tanker ballast waters do not lead to cellular and subcellular responses in pristine biota. However, they can initiate direct competitive transformation of food chains and

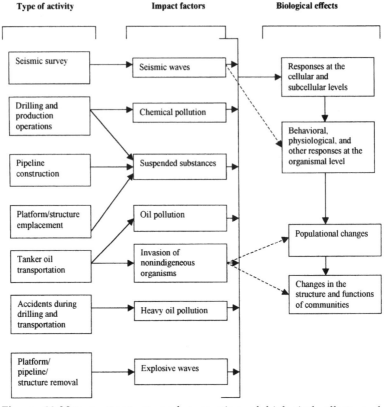

Figure 66 Nature, structure, and succession of biological effects and responses to different types of stress impacts of the offshore oil and gas industry.

subsequent deep structural and functional disturbances at the populational and even ecosystem levels. This happened, for example, in the Sea of Asov and the Black Sea as a result of jelly-fish *Mneopsis* invasion [Vinogradov et al., 1995].

The high diversity of stress impacts and corresponding biological responses in the sea explains the need for a wide range of methods to identify and assess these effects. Specific techniques vary from measuring induced responses of enzyme systems in water organisms to observations of species composition, structure, and abundance in marine communities. The detailed discussion of these methods, in particular within the framework of monitoring programs in areas of offshore oil production, is given in Chapter 8.

7.2. Ecological Situation in Areas of Offshore Oil and Gas Activity

In spite of all the attention and the large number of publications devoted to environmental implications of offshore oil and gas development, research that would use a holistic, integrated approach to these problems is extremely rare. It is difficult to find studies that would relate to a whole marine ecosystem and use an array of available ecotoxicological methods (chemical analyses, bioassays, microbiological techniques, and others). In most cases, the object is to analyze the changes only in some components of affected ecosystems (e.g., spatial-temporal changes in zoobenthos). This section attempts to summarize the available data to present an integral picture of the state of the marine environment and biota in areas affected by the offshore oil and gas activities.

7.2.1. Sources, Levels, and Consequences of Seawater Pollution

As was shown above (see Chapter 3), the volumes of discharges in areas of offshore oil and gas developments are quite impressive. They total thousands of tons of spent drilling muds, cuttings, and produced waters. These discharges contain crude oil, its fractions, and hundreds of other agents and compounds of natural and anthropogenic origin. Nevertheless, the systematic studies of water quality around oil platforms have not become a priority goal of most monitoring programs. The main argument justifying this lack of attention refers to the assumption that discharged wastes are quickly diluted to safe levels due to the hydrological processes of mixing and dispersion.

The materials discussed in Chapter 6 and the data given in Tables 52, 53 and Figures 58, 60, 62, and 67 indicate that such rapid dilution actually takes place, especially in open waters, deep locations, and under strong currents. The data in Table 52, for example, show that the original content of oil hydrocarbons of 20,000–40,000 μg/l in the waste waters drops to levels not exceeding 30 μg/l in seawater at 250–500 m from the platform. Similar results show a hundred-fold decrease in pollutant concentrations right after their discharge and a thousand-fold dilution within 50–100-m radius from the discharge point [Somerville et al., 1987; Neff et al., 1987; GESAMP, 1993]. The most rapid decreases in pollution levels typically occur for light and volatile fractions of oil [Grahl-Nielson, 1987].

In spite of this evidence, one cannot exclude the possibility that in areas of long-term and intensive oil production, rather persistent

Table 52
Levels of Oil Pollution of the Surface Seawater in Areas of Offshore Oil and Gas Production

Area and Time of Sample Collection (within 10 m from the surface)	Conditions of Sample Collection	Substance, Concentration (μg/l)	Reference
North Sea: Production platforms in the area of three fields, 1978	0.5–20 miles from platforms	TOH* 0.63–17.9	Ward et al., 1980
–"–, 1980	1–20 miles from platforms	TOH 0.3–6.6	Law, Blackman, 1981
–"–, 1980	0.5–20 miles from platforms	TOH 0.7–20.2	Massie et al., 1981
Production platforms in the area of four fields, 1981	0.5–5 miles from platforms	TOH 1.7–4.2	Law et al., 1982
Production platforms in the area of three fields, 1978–1981	0.5–10 miles from platforms	TOH 2.8–8.4	McIntosh, Ward, 1982
Production platforms, 1985	Waste waters from platforms	TOH 23,000–37,000	Law, Hudson., 1986
	250 m from the place of discharge	TOH 11–25	
	500 m from the place of discharge	TOH 4–17	
	750 m from the place of discharge	TOH 8	
	Waste waters from platforms	Benzene, toluene, xylene 25,000	Grahl-Nielson, 1987
		Non-volatile hydrocarbons (number of C-atoms >12) 26,000	
		NPD** 80	

Table 52 (*continued*)

Area and Time of Sample Collection (within 10 m from the surface)	Conditions of Sample Collection	Substance, Concentration (μg/l)	Reference
	Area of breakwater near platforms	Benzene, toluene, xylene 250	
		Non-volatile hydrocarbons (number of C-atoms >12) 400	
		NPD 0.8	
	Area beyond breakwater	Benzene, toluene, xylene 25	
		Non-volatile - hydrocarbons (number of C-atoms >12) 100	
		NPD 0.3	
Mediterranean Sea: Fixed production platforms on the shelf of Lebanon, 1986–1989	Up to 3 miles from platforms	TOH 11.2–45.6	Hinshery et al., 1991
Black Sea/Sea of Asov: Areas of drilling the exploratory wells, 1988–1990	Near drilling rigs	TOH 50–350	Semenov, Pavlenko, 1991

Notes: *—TOH—total oil hydrocarbons extracted by organic solvents, measured by the methods of UV-fluorescence and IR-spectrophotometry. **—NPD—aromatic fraction, including naphthalene, phenanthrene, dibenzothiophene, and their alkylated homologues.

Table 53
Modeled Dilution of Produced Waters Discharged to Different
Offshore Waters under Different Geographic Regimes ([Terrens,
Tait, 1993]; [Brandsma, Smith, 1996], cited from [Neff, 1998])

Location	Discharge Rate (l/day)	Current Speed (cm/sec)	Concentration at 100 m (%)
Gulf of Mexico	115,740	3.3	0.043
Gulf of Mexico	115,740	25.3	0.097
Gulf of Mexico	3,975,000	3.3	0.18
Gulf of Mexico	3,975,000	22	0.32
North Sea	9,992,000	22	1.3
Bass Strait	14,000,000	26	0.45

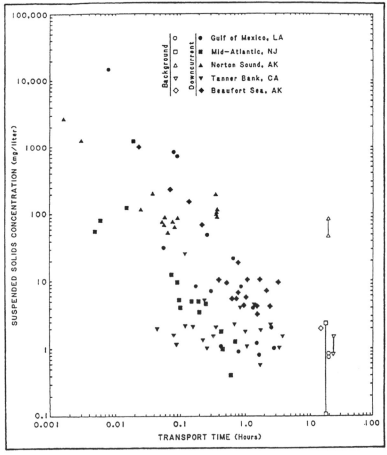

Figure 67 Suspended solids concentration versus transport time during drilling mud discharges in different regions [Neff, 1998].

zones of chemical pollution of seawater can be formed. A number of facts and circumstances support this suggestion:

- The presence of oil slicks on the water surface. These are repeatedly observed in areas of offshore oil production. These slicks form as a result of discharges of oil-containing waters from the platform, pollution by tanker transportation, and atmospheric input of the products of incomplete combustion of hydrocarbons during well testing and other operations [NSQSR, 1994; Daan, Mulder, 1994];
- The presence of large number of agents and compounds (heavy metals, biocides, corrosion inhibitors, complex products of photochemical degradation of high-molecular-weight PAHs, etc.) in liquid and solid discharges from the platforms. Many of these pollutants are persistent in the marine environment and often are not included in the sphere of regular analytical control [Semenov, Pavlenko, 1991; NSQSR, 1994];
- Highly irregular distribution of wastewaters and their individual components on the surface and in the water column in areas of oil production. These discharges can form fields with elevated levels of pollution in the coastal shallow zones or areas with slow water circulation [Somerville et al., 1987].

A number of publications describe the accumulation of PAHs and other contaminants in the tissues of organisms chronically exposed in the water column at different distances from oil platforms. This is a convincing evidence of the persistent background contamination of seawater in areas of offshore oil production. In particular, it was shown that the content of oil hydrocarbons in mussels near platforms considerably exceeded the background levels (up to 10–100 times the normal) (see Figure 68). Such changes in the chemical composition of mussel tissues were found 6–10 km (sometimes 25 km) from the place of discharge [Somerville et al., 1987]. Significantly, chemical analyses of water samples in the same locations failed to give analogous pictures of gradient changes in oil hydrocarbon concentrations depending on the distance from the discharge source. Similar results were obtained in other studies [Scholten, 1995].

Another, although indirect, proof of the possibility of persistent background contamination in areas of the offshore oil and gas activity comes from the marine microbiological research. The observations conducted in the North Sea [Bruns et al., 1993; Minas, Gunkel, 1995], Gulf of Mexico [Lizarraga-Partida et al., 1991], Oman Gulf

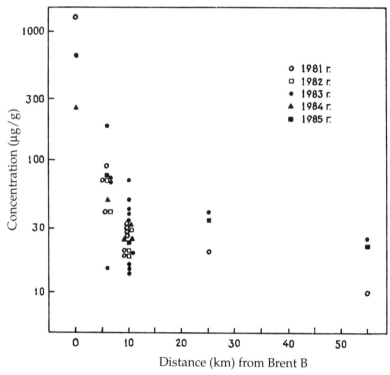

Figure 68 Concentration of oil hydrocarbons in mussels exposed at different distances "downstream" of the Brent oil field (North Sea) [Somerville et al., 1987].

[Al-Hadhrami et al., 1995], and some other regions revealed substantial restructuring of pelagic microbial communities in areas of the offshore oil and gas production. These changes happened due to the increased contribution of species and forms that use oil as a feeding substrate and decompose it. This process accelerates self-purification of seawater and helps to maintain the levels of oil pollution within the ranges shown in Table 52. The maximum values for these levels measured in areas of oil-containing discharges usually vary around several tens of micrograms per liter for the sum of oil hydrocarbons. These concentrations cannot cause any obvious and large-scale biological effects and ecological changes in the water environment detectable by modern methods. These levels fit quite well into the range of concentrations of dissolved oil hydrocarbons typical for coastal and shelf waters (see Table 25 and Figure 42). Such concentrations, as Chapter 4 showed, are either safe for many species of marine flora and fauna or can cause only reversible (threshold) effects which are compensated at the subcellular level. The question about the long-term biological consequences caused by low levels of chronic oil pollution is still open and will be discussed in Section 7.4.

Some authors [Somerville et al., 1987; Davies, Kingston, 1992] provide the evidence of disturbances in feeding, development, and behavior of zooplankton forms (especially copepod) directly affected by produced waters and other liquid wastes. Usually, these direct impacts on pelagic biota have a local nature and their consequences do not spread beyond a zone of several tens (occasionally hundreds) of meters from the place of discharge. At the same time, the high variability of composition and toxic properties of liquid wastes as well as impressive volumes of their discharges (especially in case of produced waters) explain the need for further research of the nature and scale of their environmental impact.

7.2.2. Sources, Levels, and Effects of Bottom Sediment Pollution

In contrast with seawater, bottom sediments have the ability to accumulate and localize most pollutants, especially when pollutants enter the marine environment in a rapidly sedimented solid phase. This creates favorable conditions to measure regularly both pollution levels and the biological effects in ecosystems subjected to chemical impact. That is why bottom sediments, benthic organisms, and communities are most often the focus of many programs of marine monitoring, especially in areas of oil and gas development on the shelf.

Chapter 3 discussed the leading role of drilling discharges in creating the anthropogenic press in areas of offshore oil production. The volumes of such discharges from a single platform total to thousands (sometimes tens of thousands) of tons of solid wastes often contaminated with oil, oil products, and many other pollutants. Contaminated rock cuttings and drilling muds can gradually spread up to several kilometers (sometimes over 10 km) from the place of discharge. This can considerably change the physical and chemical parameters of sediments in this area, affecting bottom biotopes and benthic populations.

The main chemical indicators of environmental stress in the areas of the offshore oil and gas production used in many countries include the sediment levels of the total oil hydrocarbons (TOH) extracted by organic solvents, sum of polyaromatic hydrocarbons (PAHs), or the concentrations of individual (usually aromatic) hydrocarbons. The methods of sampling, sample processing, and analyses (UV spectrophotometry, gas chromatography, and mass spectrometry) do not always match in different laboratories. Thus, these methods can give data that are difficult to interpret and compare. Besides, many other pollutants, including stable and toxic traces, are often not included in the sphere of analytical control. This certainly diminishes the informative value of chemical monitoring and explains the necessity of supplementing them by biological methods.

The range of oil hydrocarbon concentrations found in bottom sediments in areas of offshore oil and gas development is very wide—from less than 1 mg/kg to 10,000 mg/kg and more (see Table 54 and Figures 69 and 70). Such high variability of data is explained by differences in volumes, regime, and chemical composition of discharges (e.g., oil-based or water-based muds), type of analyzed fraction, distance from the source of discharges, hydrological conditions in the near bottom zone, analytical procedures, and others factors. The highest levels of sediment pollution as well as the most evident biological effects are usually recorded in areas where oil-based drilling fluids are used. However, according to some publications [Kingston, 1987; Reiersen et al., 1989; Daan et al., 1994], other types of drilling fluids and associated drilling cuttings improve the environmental situation on the sea bottom only to a limited extent. The maximum levels of oil hydrocarbons in sediments (up to 10,000 mg/kg and more) practically coincide with the initial oil content in drilling cuttings discharged directly near platforms. The lower concentrations occur several kilometers from the place of discharge.

Interpreting data on chemical contamination of the sediments in areas of long-term oil and gas production is a rather complicated task. The difficulties arise from the variability of volumes and composition of discharged wastes, complexity and dynamism of their behavior in the marine environment, overlapping effects caused by activities on neighboring oil fields, the impacts of onshore and ship discharges, accidental spills, and other factors and circumstances. The typical features of chemical contamination of the sediments in areas of intense oil and gas developments include:

- the localization of the maximum levels of oil hydrocarbons and other pollutants (thousands times exceeding the natural background concentrations) in discharge zones close to drilling platforms;
- distinct gradient of declining oil concentration in bottom sediments within 500 m from the discharge point [Davies, Kingston, 1992];
- the possibility of forming areas with slightly increased levels of hydrocarbons in sediments (within a radius up to 15 km), usually as a result of oil-based waste discharge, atmospheric fallout,

Table 54

Levels of Oil Pollution of Bottom Sediments in Areas of Oil and Gas Production in the North Sea

Area and Time of Sample Collection (within 10 m from the surface)	Conditions of Sample Collection	Substance, Concentration (μg/l)	Reference
Production platforms in the area of three fields, 1978	0.5–29 miles from platform	TOH* 0.6–26.2	Ward et al., 1980
Production platforms in the area of six fields, 1977–1980	Samples of cuttings before discharge	TOH 2,100–13,200	Blackman, Law, 1981
Production platforms in the area of three fields, 1980	1–20 miles from platforms, at the depths of 70–130 m	TOH 0.08–57	Law, Blackman, 1981
Production platforms in the area of three fields, 1978–1980	0.5–20 miles from platforms	TOH 0.6–257 PAH 0.1–33.5 Naphthalene 0.03–14.8	Massi et al., 1981
Production platforms in the area of three field, 1981	0.5–35 miles from platforms, at the depths of 90–183 m	TOH 8–1,100	Law et al., 1982
	0.5–5 miles from platforms	Naphthalene 0.02–74.7	
	Directly under the platform at the depth of 153 m	TOH 15,000–38,000	
Production platforms in the area of four fields, 1978–1981	0.5–10 miles from platforms	TOH 11–239 N-alkanes (C_{15}–C_{33}) 1.0–326.3	McIntosh, Ward, 1982
Production platforms in the area of two fields, 1980–1982	0.5–8 km from platforms	Naphthalene 1–10,000	Davies et al., 1984

Table 54 (*continued*)

Area and Time of Sample Collection (within 10 m from the surface)	Conditions of Sample Collection	Substance, Concentration (μg/l)	Reference
Production platform	1–8 km from the platform	TOH 1–8,000	Mair et al., 1984
Area of exploratory drilling on the shelf of Netherlands	0–1 km from the platform	TOH 20–400	Stebbing, Dethlefsen, 1991
Production platforms on the shelf of the Great Britain, 1980–1989	0.5–8 km from platforms	TOH 1–10,000	Davies, Kingston, 1992

Note: *—See *Notes* for Table 52.

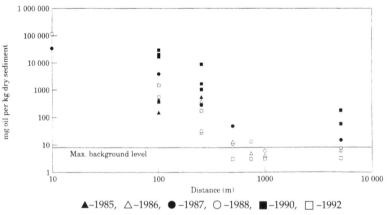

▲–1985, △–1986, ●–1987, ○–1988, ■–1990, □–1992

Figure 69 Oil concentrations in relation to distance from a production platform in the Dutch sector of the North Sea in 1985–1992 [Daan, Mulder, 1996].

and other inputs of pollutants; this happened, for example, on the shelf of the Shetland Islands [Kingston, 1991] and near the Norwegian shore [Gray, 1991];

• a considerably larger (at least 4–5 times) zone of contamination of bottom sediments along the prevailing current in multi-well

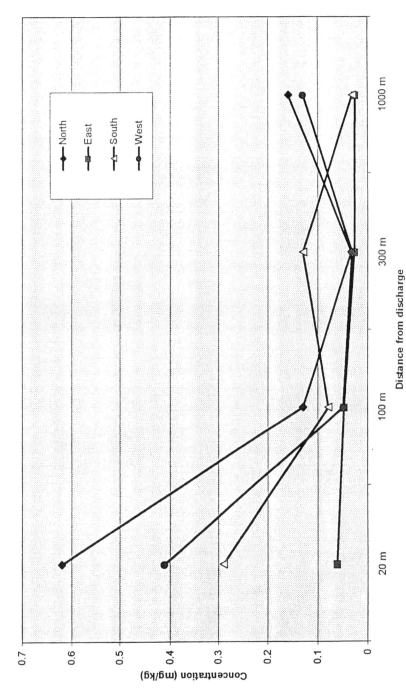

Figure 70 Concentrations of total PAHs in sediments at different distances from a production platform in 8 m of water in the Gulf of Mexico [Neff, 1998].

317

sites used for production drilling as compared with single-well sites used for exploratory drilling [GESAMP, 1993]; and

- the probable presence of a wide range of substances and agents (besides oil and oil products) in bottom sediments contaminated by drilling wastes.

The content and distribution of most trace ingredients in the bottom sediments around the platforms has not been studied in detail with the exception of some heavy metals (mercury, lead, cadmium, chromium, and zinc) and barium (rather easily determined analytically). The data in Tables 55 and 56 show that the levels of trace metals in discharges of drilling waste and produced waters vary within very wide ranges. For most elements, these ranges are close to natural variations. At the same time, the elevated levels of some heavy metals and barium can be found within a radius of up to 100 m from the platforms [Boothe, Presley, 1989]. An example of barium distribution in bottom sediments in the area of an oil field development in the North Sea is shown in Figure 71.

The levels of measured oil pollution of sediments in the zones of oil and gas activity are considerably higher than the minimum oil concentrations that cause different biological effects (see Chapter 4, Table 30, and Figure 43). This suggests the existence of a stressful

Table 55

Concentration Ranges of Metals in Water-based Drilling Fluids and in Typical Marine Sediments [Neff, 1988]

Metal	Concentration Range for Drilling Fluids (mg/kg dry weight)	Concentration Range for Marine Sediments (mg/kg dry weight)
Barium	720–49,000	60–8,100
Chromium	10.9–1,159	10–200
Cadmium	0.5–3.5	0.3–1
Copper	2.8–119	8–700
Iron	16,000–27,000	20,000–60,000
Mercury	0.015–2.8	0.05–3.0
Lead	5.0–241	6–200
Zinc	42–397	5–4,000
Nickel	3.8–19.9	2–10 (10–1,000)*
Arsenic	1.8–2.3	2–20
Vanadium	14–28	10–500
Manganese	290–400	100–10,000

Note: *—Nickel concentrations of 10–1,000 mg/kg are typical of deep-sea sediments.

Table 56

Concentrations ($\mu g/l$) of Several Metals in Produced Water Samples from the Gulf of Mexico and in Clean Coastal Seawater* [Neff, 1998]

Meal	Produced Water	Ambient Seawater
Arsenic (As)	0.05–31	0.6–3.0
Barium (Ba)	11,500–320,000**	7–88
Cadmium (Cd)	<0.1–14	0.001–0.2
Chromium (Cr)	<0.1–1.1	0.13–0.25
Copper (Cu)	<0.2–6.4	0.2–1.7
Iron (Fe)	12,000–30,000**	0.008–4.0
Mercury (Hg)	<0.01–0.08	0.0001–0.019
Manganese (Mn)	700–7,100**	0.03–1.6
Molybdenum (Mo)	0.4–1.7	3.2–13.0
Nickel (Ni)	0.4–3.1	0.2–1.2
Lead (Pb)	0.2–20	0.001–1.0
Zinc (Zn)	3.0–3,000**	0.006–4.5

Notes: *—Samples were analyzed by advanced analytical methods in 1994–1996. **—May be present in some produced waters at a concentration greater than 1,000 times higher than the concentration in clean ambient seawater.

chemical impact on benthic organisms in these areas. The manifestation of this stress depends on many factors and circumstances. These include volumes, chemical composition, and physical properties of discharged materials, water depths in the place of discharge, bottom geomorphology, type of bottom sediments, speed and direction of currents, and so on. The most typical responses of benthic biota usually involve the accumulation of oil hydrocarbons and subcellular changes in benthic organisms. Certain effects at the higher levels, including modifications of benthic communities, are also possible (see Table 51). Structural changes at the populational and community levels reflect in a cumulative form the integral response of biota on any stress. Measuring these changes is one of the main methods applied by most programs of marine monitoring [Izrael, 1984; Konstantinov, 1986; Abakumov, 1991; Warwick, Clarke, 1993; GESAMP, 1995].

Most field studies of ecological situation in areas of oil production use simple biological parameters. These include the total abundance (biomass) of benthic organisms, the number of species or larger taxons, species diversity, and some combination of these parameters based on multifactorial statistical approaches [Gray et al., 1990; Clarke, 1993]. Lately, new methods of bioassaying bottom sediments have been introduced into the practice of marine monitoring

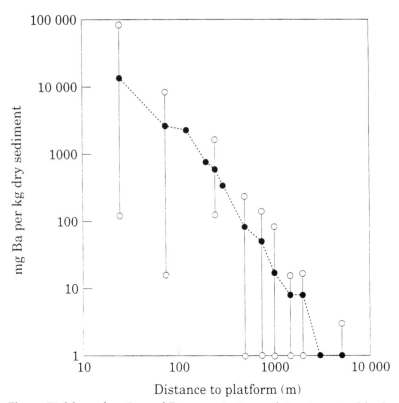

Distance to platform (m)

Figure 71 Mean elevations of Ba concentrations relative to natural background levels in the sediment at increased distances from platforms in the Dutch sector of the North Sea [Daan, Mulder, 1996].

[Stebbing, Dethlefsen, 1991; MacDonald, 1993; ICES, 1994]. Simultaneously applying a combination of chemical, toxicological, and ecological methods seems to give the most reliable results of environmental assessments [Chapmen et al., 1991].

One of the examples of applying ecological methods involves estimating the changes of index of diversity (H_S) for benthic communities in areas of offshore oil and gas production. The index of diversity is a measure of the relative proportions of species and individual specimens in a community. It is calculated using the formula:

$$H_s = -\sum_{i=1}^{s} \frac{B_i}{B} \log_2 \frac{B_i}{B},$$

where B_i is the abundance (biomass) of all specimens of species *i*, B is the abundance (biomass) of all species in the sample, and *s* is the number of species in the sample. Increasing the stress impact usually leads to reducing species diversity, which is reflected by the index of diversity. The degree of structural disturbances in benthic eco-systems can also be estimated by a simpler method. It is based on ob-serving changes in the abundance of only one species, which is the most sensitive to pollution.

Application of both methods in areas of offshore oil and gas production in the North Sea revealed a clear and strong negative cor-relation between the measured biological parameters and the content of oil in the sediments (see Figures 72–74). This undoubtedly indi-cates the existence of a stress impact of oil pollution on benthic biota [Mair et al., 1987; Davies, Kingston, 1992]. At the same time, making quantitative estimates based on these methods is complicated. First, the range of oil hydrocarbon concentrations in sediments shown to reduce zoobenthos diversity is very wide, from 10 mg/kg to 100 mg/kg [Reiersen et al., 1989; Davies, Kingston, 1992]. Besides, re-duced diversity can correlate not only with the total content of oil hydrocarbons in sediments but with the presence of individual frac-tions of oil (e.g., naphthalenes) [Davies et al., 1984; Moore, 1983] and other components of drilling wastes. In particular, Table 57 summa-rizes data on the sediment levels of individual PAHs that cause dif-ferent biological effects in benthos. These materials, along with the data discussed in Chapter 4 (see Table 30), can be useful for inter-preting the monitoring results and assessing the possible ecological effects in areas of the offshore activities.

At the same time, the complexity of composition and toxic properties of both drilling wastes and contaminated sediments could make the environmental assessments based on the data for sediment levels of oil inadequate. This explains why some researchers prefer to study effects in benthic communities depending on the distance from the place of discharge [Daan et al., 1994]. In this case, the re-sponses of the most sensitive (indicator) species are analyzed with the use of methods of multifactorial statistics. This can provide a more detailed picture of biological processes, responses, and restruc-turing that take place in communities of bottom sediments subjected to the impacts of the oil and gas industry [Gray et al., 1990; Warwick, Clarke, 1993].

Practically all environmental assessments raise the question about the possible sizes of the areas of biological effects. Most of these studies reveal such effects within maximum radiuses of 250–2,000 m from drilling sites (see Table 58). Significantly, the zones of elevated

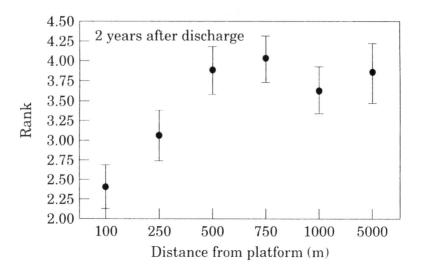

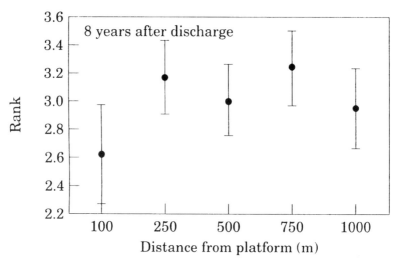

Figure 72 Relative macrofauna abundance along the residual current transect in the vicinity of a production platform in the Dutch sector of the North Sea in 1986 (top) and 1992 (bottom) [Daan, Mulder, 1996].

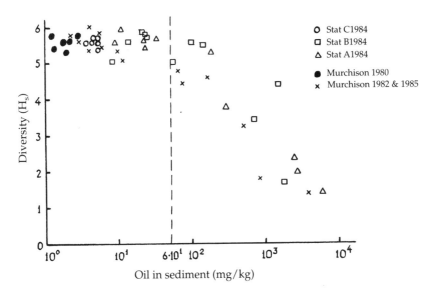

Figure 73 Relationship between diversity (H$_s$) and sediment oil levels at the Stratfjord Field [Mair et al., 1987].

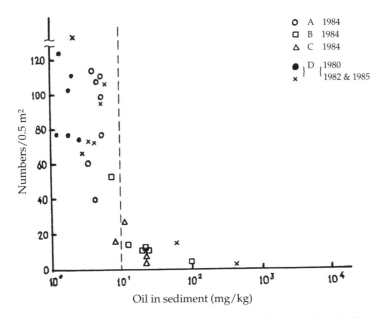

Figure 74 Relationship between abundance and sediment oil levels for an example of a sensitive species of benthic worm *Aonides paucibranchiata* from four North Sea oil production platforms between 1980 and 1985 [Davies, Kingston, 1992].

Table 57

Threshold Levels (mg/kg of dry matter) of Individual PAHs in
Bottom Sediments [Long, Morgan, 1990]

Polycyclic Aromatic Hydrocarbons	Lower Threshold*	Upper Threshold**
Naphthalene	0.340	0.50
2-methylnaphthalene	0.065	0.30
Acenaphthene	0.150	0.15
Fluorene	0.035	0.35
Phenanthrene	0.225	0.26
Anthracene	0.085	0.30
Fluoranthene	0.600	1.00
Pyrene	0.350	1.00
Benzo(a)anthracene	0.230	0.55
Benzo(a)pyrene	0.400	0.70
Dibenzo(a)anthracene	0.060	0.10

Notes: *—Concentration that causes primary biological responses with 10%-probability.
**—Concentration that causes statistically significant effects with 100%-probability.

hydrocarbon concentrations in sediments can spread considerably
further. For example, studies on the Norwegian shelf of the North Sea
revealed the hydrocarbon presence in sediments at the distance of up
to 12,000 m from the development wells where oil-based drilling
muds were used [GESAMP, 1993]. With increasing distance from the
place of waste discharge (in this case, from drilling rigs) and decreasing pollution levels, the reliability of assessing the stress effects
in the benthic communities at the background of their natural variability certainly decreases. Sometimes this provokes hot debates
about the methodology of such assessments [Gray, 1988; 1991].

According to a number of publications [Moore, 1983; Reiersen
et al., 1989; Gray et al., 1990; Kingston, 1992; Daan et al., 1994], stress
impacts on zoobenthos can be reliably detected up to 3,000–5,000 m
from the drilling platforms by methods of biological indication. This
approach has been used for a long time in monitoring the state of polluted water ecosystems. It takes into account the species differences
in sensitivity to hazardous factors. For example, some species of
polychaetes (especially *Capitella capitata*) have a higher resistance to
oil pollution. Due to that ability, they usually dominate in sediments
containing elevated levels of oil. Figure 75 clearly shows a direct correlation between the domination of *C. capitata* and the content of aromatic hydrocarbons in bottom sediments.

Table 58
Area of Seabed Around North Sea Drilling Sites Affected by
Oil-based Drilling Muds [GESAMP, 1993]

Sector	Well Sites Examined (Number of Sites for U.K. Area Only)	Average Size/ Shape of Zone
United Kingdom:		
Major biological effects	40 development	500 m radius
	380 single	250 m radius
Subtle biological changes	40 development	2,000 × 1,000 m ellipse
	380 single	1,000 × 500 m ellipse
OBM hydrocarbons present	40 development	4,000 × 8,000 m ellipse
	380 single	1,000 × 2,000 m ellipse
Norway:		
Major biological effects	Development	1,000 m radius
	Single	500 m radius
Minor	Development	3 to 5,000 m radius
	Single	1,000 m radius
Hydrocarbons present	Development	5,000 × 10,000 to 12,000 m ellipse
	Single	2,000 × 4,000 to 6,000 m ellipse

Along with species that are resistant to pollution, some benthic organisms are more vulnerable to stress impacts. For example, long-term studies in areas of oil and gas production in the shelf zone of the Netherlands [Daan et al., 1994] revealed a number of organisms, including amphipods *Harpinia antennaria*, echinoderms *Echinocardium cordatum*, and clams *Montacuta ferruginosa*, that had a higher sensitivity to the impact of the oil-containing drilling wastes. Observations showed a considerable decrease in their abundance (even complete disappearance in samples) within radiuses up to 1,000 m and more from the platforms. This occurred even though the volumes of discharged wastes in these locations were several times less

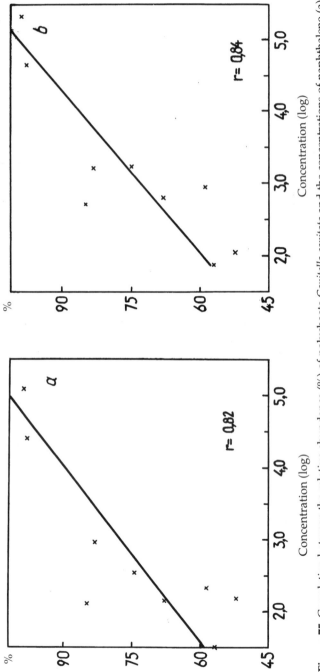

Figure 75 Correlation between the relative abundance (%) of polychaete *Capitella capitata* and the concentrations of naphthalene (a) and anthracene and phenanthrene (b) in bottom sediments in the area of an oil platform in the North Sea [Moor, 1983].

than in other areas of the North Sea. This and other similar studies prove the high sensitivity of biological indication.

The program of ecological monitoring around oil platforms on the U.S. shelf in the Gulf of Mexico uses methods of assessing populations of crustaceans (shrimps and crabs) based on their ecophysiological parameters. These include size structure, reproduction rates, pathological effects, and some other characteristics [Wilson-Ormond et al., 1994]. The system of such parameters proved to be rather sensitive. It revealed chronic adverse effects in the populations of crustaceans at the distance of up to 3 km from the platforms.

An important, although still poorly studied, issue is the rate of recovering of benthic environment and communities after removal of the stress impacts. According to some studies [Mair et al., 1987; Davies et al., 1989; Grahl-Nielson, 1987], the signs of an improved ecological situation on the bottom start revealing themselves 1–2 years after the termination of drilling operations. This happens due to the processes of oil biodegradation and secondary colonization of bottom sediments by larvae of benthic fauna [Gray et al., 1990]. An important condition affecting the success of this colonization is the decontamination of the surface sediment layer, even if the deeper mass of sediments has elevated levels of oil [Blackman et al., 1985].

In areas where water-based drilling fluids were used, benthos recovering goes considerably faster. For example, 2–14 months after the termination of such activities on the Norwegian shelf, the disturbances of benthic communities were not detected even at distances of about 25 m from the platforms [Daan, Mulder, 1994]. At the same time, oil-contaminated sediments are often characterized by prolonged deterioration of the oxygen regime and developing anaerobic conditions [Blackman et al., 1985]. Even relatively low (30–40 $\mu g/l$) levels of oil pollution in near bottom environment are shown to suppress the larvae settling for some dominating species of zoobenthos (especially amphipod) [Bonsdorff et al., 1990].

A detailed study in the southern North Sea conducted three years after the termination of drilling operations [Kroencke et al., 1992] found that along with noticeable reestablishment of benthic communities, some biological parameters (the number of species, total biomass, and abundance) still clearly correlated with the distance from the place of former discharges. These gradient changes were observed within a radius of up to 1,000 m from the platforms [Kroencke et al., 1992]. Figure 76 gives another illustration of relative persistence of chemical and biological effects in bottom sediments one year after the termination of drilling operations with oil-based fluids.

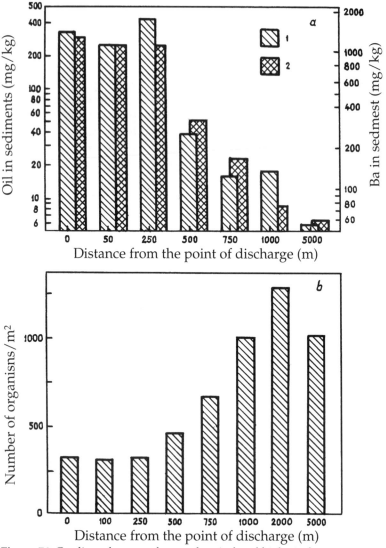

Figure 76 Gradient changes of some chemical and biological parameters in bottom sediments in the southern shelf of the North Sea (Netherlands) a year after the completion of exploratory drilling in 1987 [Stebbing, Dethlefsen, 1991]: a—concentrations of barium (1) and oil (2) in the 10-cm layer of sediments; b—the total abundance of macrofauna at different distances from the point of discharge.

7.3. Accidental Oil Spills

The scope of literature on the environmental effects of accidental oil spills in the sea is incredibly wide. Even a short review of this topic would take too much space. Chapters 3 and 4 already discussed the causes, nature, and scale of accidents at different stages of oil production and transportation. This section will try to cover the most typical and widespread ecological effects of large accidental oil spills.

From the ecotoxicological perspective, each spill at its initial stage resembles an acute experiment when high concentrations of a toxicant are introduced into a testing system. Such conditions provoke rapid and obvious biological effects that are easy to notice and measure, for example by organism mortality. This analogy, however, soon becomes inadequate. In several hours or days, a rapid separation of oil into fractions, decrease in its toxic properties, accumulation in bottom sediments, and many other processes take place (see Chapter 4). The variety of conditions and factors involved lead to practically unlimited scenarios that occur after any accidental oil spill in the sea.

At the same time, all these scenarios have a critical moment that ultimately defines the nature and severity of the possible consequences. The oil slick can either stay in the open sea or enter the coastal waters and contact the shoreline. In the first case, due to wind, currents, turbulent mixing, and other hydrodynamic impacts, the oil is usually dispersed, emulsified, and spread over extensive areas. These processes can decrease oil concentrations and bioavailability relatively rapidly. Biological effects will be limited to local, quickly reversible disturbances in the water column and on the sea surface. In the second case, however, the consequences are more diverse, severe, and persistent. The data in Table 59 gives a short description of possible oil's behavior when it washes ashore. This data also classifies the types of shoreline depending on their vulnerability to the impact of heavy oil pollution. When moving from open rocky beaches to sandy, gravel ones and then to sheltered bays, oil's persistence, and hence its hazardous impacts, radically increase.

Numerous (over 500) publications, including detailed reviews, give the environmental analysis of hundreds of accidental and experimental oil spills in the sea [e.g., Baker et al., 1990; Baker et al., 1991; GESAMP, 1993; Neff, 1993; Burger, 1994]. These extensive studies indicate that the situations when oil slicks do not get in contact with the shoreline and stay in the open waters are rather rare. Examples of such events include the rupture of underwater pipelines in

Table 59
Types of Shorelines Given in the Order of Increasing Their
Ecological Vulnerability to Oil Spills (based on [GESAMP, 1993])

Index of Vulnerability	Type of Shoreline	Notes
1	Open rocky shoreline	Wave action limits the oil washing ashore; no need in cleaning the coast exists
2	Flat rocky shoreline	Oil is removed in several weeks due to wave action and other natural processes
3	Fine sand beaches	Oil usually doesn't penetrate deep into sand; oil can be removed by mechanical means; oil pollution stays for several months
4	Coarse sand beaches	Oil is quickly accumulated in sediments that complicates cleaning; under favorable weather conditions, oil pollution of sediment surface disappears in several months
5	Open shallow tidal areas and packed sand bars	Main part of oil is removed by wave action and other natural factors; usually no need in cleaning exists
6	Sand, pebble, and gravel beaches	Oil rapidly penetrates into deeper layers; oil pollution can persist for years
7	Gravel beaches	Oil rapidly penetrates into gravel; oil pollution can persist for years; under heavy pollution, oil can form asphalt crust
8	Sheltered rocky shore and bays	Oil can stay for years due to weak wave action; cleaning is not recommended except for conditions of very heavy pollution

the North Sea in 1980 and 1986 [Haldane et al., 1992] and a large accident on the IXTOC-1 exploration well off the coast of Mexico in 1979 [Gourlay, 1988]. Due to long-term oil gushing after the accident in the Gulf of Mexico, oil losses totaled hundreds of thousands of tons. In spite of the impressive scales of these spills, they did not lead to some significant and long-term ecological consequences. The main part of the oil was dispersed, emulsified, and then decomposed due to the biological and chemical processes of self-purification in the open waters. Unfortunately, such happy endings occur extremely rarely. Most accidental spills affect the coastal zone and shoreline to some extent. Besides, the frequency of accidents in the coastal zone is higher than in the open waters. As mentioned previously, the ecological consequences of such spills are especially diverse and severe.

Acute (lethal) effects of oil are usually observed at the first stages of the oil's distribution. Oil mostly affects the organisms that directly contact the slick. Marine and coastal birds and mammals can die not only because of oil intoxication but also as a result of disturbed thermal regulation and the loss of the abilities to migrate and feed. The data on the actual scales of such events are inconclusive. Assessing the number of organisms that died from the oil's impact as compared with those that died naturally is difficult, especially due to the high variability of natural mortality. At the same time, it is widely documented that the direct contact of colonial populations of marine birds and mammals with large oil slicks can have fatal effects. For example, only about 50% of sea otters are estimated to survive when they contact crude oil [Waldichuk, 1990]. The birds most vulnerable to direct oil contact include those species that live in dense flocks and spend most of their time swimming on and under the sea surface. Cleaning oiled wildlife with different solvents and agents usually is not very effective.

Large losses of mammals and birds often occur during spills in the northern, particularly Arctic, regions. Here, numerous colonies of marine mammals and birds are especially vulnerable to oil pollution because their breeding, feeding, and migrations take place in locations that are the most often affected by oil spills (coastal zone, shoreline, surface waters, tidal flats, and others). Besides, at low temperatures, especially under conditions of ice cover, the persistence of oil is higher and natural degradation may take several decades. Nevertheless, most available publications, including the ones on the Arctic areas [Baker et al., 1991], indicate that affected populations of marine birds and mammals can restore the optimal population abundance several years after a spill.

In contrast with birds and mammals, the majority of fish are able to leave the zone of heavy pollution and thus avoid acute intoxication. In any case, the evidence indicates that populations of free-swimming fish are not damaged by oil spills in the open sea [Baker et al., 1991; Squire, 1992; GESAMP, 1993]. At the same time, juvenile fish, floating eggs, and larvae of many species can be killed when contacted by the oil. In the coastal shallow waters with slow water exchange, oil spills pose a similar threat to benthic fish and invertebrates as well as to organisms cultivated at marine farms. When the zones of spilled oil and marine aquaculture overlap, the objects of aquaculture either die from oil impact or accumulate hydrocarbons in their organs and tissues. They often acquire oil odor and flavor, thus losing commercial value. This happened, for example, after the tanker *Braer* accident near the Shetland Islands in 1993 when oil spill damaged local salmon-cultivating industry [MLA, 1993b].

Data on oil spill impact on the commercial species in the coastal and estuarine zones are of special interest and importance. In particular, some concern is expressed regarding a possibility of adverse effects of oil spills on the populations of spawning fish [Keiso, Kendziorek, 1991]. Despite the methodological difficulties of assessing such effects [Norcross, 1992], observations of spawning salmon and herring and their juveniles during and after the large oil spill near the Alaskan shore in 1989 did not reveal any noticeable reduction in the stock, returns, and commercial catch of these fish [Baker et al., 1991; McGurk et al., 1993]. At the same time, the fishing activity in that period was considerably limited, and local fishermen suffered considerable losses [Picou et al., 1992].

The most serious and long-term consequences of oil spills are found in situations when oil accumulates in the sediments of shallow coastal zones with slow water exchange, especially in the ecosystems of coastal mangroves, salt swamps, and coral reefs. In such cases, both acute (lethal) impacts on the local flora and fauna (especially on benthos) and chronic stress on organisms that survive inevitably occur. Significantly, the composition and structure of the surviving communities greatly differ from the original parameters. As time passes, these differences gradually disappear. However, some recovering communities (especially the benthic ones) can manifest the specific adaptive effects to oil stress for a rather long time.

The nature and rate of recovering marine communities affected by an oil spill depend on many parameters (spill size, oil type, season, weather conditions, shoreline type, and so on). The critical role belongs to those factors that define the processes of secondary colonization of the oil-polluted biotopes, primarily bottom sediments. Since each situation is unique, the processes of biota recovering in different areas can take very different times—from a few months to decades. For example, the ecological situation in the area of an oil spill off the California coast was recovered twenty days after the ac cident. It could be explained both by to the high self-purification ability of the marine environment and the effectiveness of the cleaning procedures [Francis et al., 1992]. A situation of another kind developed at the Atlantic coast off Massachusetts. Twenty years after the spill, the oil residuals, in the form of aromatic hydrocarbons and cycloalkanes, were found in deposits of coastal salt marshes 5–15 cm from the sediments' surface [Teal, 1993]. This residual oil caused weak induced biochemical responses in the cytochrome P4501A system in the tissues of local bottom-dwelling fish, although the general ecological situation in this zone had been normalized.

The very slow rates of detoxification processes and the corresponding long-term nature of stress caused by oil pollution typically occur in coastal mangrove ecosystems. Such areas actually trap the spilled oil. This was shown in a long-term observation of a mangrove ecosystem in one of the shallow bays of Panama. Twenty years after a large oil spill, both high levels of aromatic hydrocarbons in sediments and an increased number of damaged roots of mangrove *Rhizophora mangle* were found [Burns et al., 1994].

Species composition and trophic relationships in communities that recover after oil impacts may differ from the original parameters for a long time. The highest recovery rates were shown for populations of some species of algae, worms (especially polychaetes), and bivalve mollusks (especially mussels). Recovering populations of macrobenthic crustaceans (especially amphipods, ostracods, and decapods) usually takes more time.

One of the key factors affecting the self-purification of the marine environment from oil is temperature (see Chapter 4). Although some data indicate a rather high rate of oil degradation under low temperatures (e.g., in the wintertime in the North and even Barents Seas [Bruns et al., 1993; Gray, 1993; GESAMP, 1993]), this process goes much faster at high temperatures. It was shown, in particular, during observations of an extraordinary ecological catastrophe—probably the largest ever oil pollution—which developed as a result of hostilities during the 1991 Gulf War. Inputs of crude oil released into the waters of the Persian Gulf were estimated at 1 million tons. This was combined with the impacts of 67 million tons of oil combustion products in the form of aerosol residuals that partially fell on the sea surface. Over 500 km of the coastline of Saudi Arabia was covered by thick oil slicks [McKinnon, Vine, 1991]. By taking into consideration the scale of the impact that caused the death of large numbers of birds, damage to mangrove trees, and other acute effects of oil stress, many expected an ecological disaster with long-term consequences at the regional scale. However, the ecological situation in the area took a somewhat unexpected turn.

Prompt studies in the area of the Persian Gulf [Fowler et al., 1993] and adjoining areas of the Arabian Sea [Gupta, 1991] showed that 3–4 months after termination of the oil inputs into the environment, the concentrations of oil hydrocarbons, including PAHs, in bottom sediments and bivalve mollusks everywhere dropped to the levels observed before the hostilities. These levels did not exceed the background concentrations typical for other marine regions, for example the Baltic and North Seas and the coastal zones of North America and Europe. These preliminary conclusions were supported by later observations, including a detailed study conducted by an international expedition under the UN umbrella in 1993 [Price, Robinson, 1993]. The results indicated an amazingly high self-purification ability of the marine environment in the Persian Gulf. This must be typical for other regions with similar natural conditions. Most likely, this was a result of the radical acceleration of oil decomposition at high temperatures. Rapid development of hydrocarbon-degrading microorganisms in these waters also increased the rate of oil biodegradation. Besides, intense solar radiation could intensify the photooxidation of PAHs and other aromatic hydrocarbons in the marine environment.

Figure 77 gives a simplified diagram that emphasizes the variety of forms that stress oil effects take in marine organisms and ecosystems. These effects vary from acute intoxication in zones of high oil concentrations to chronic stress. The nature and consequences of this stress are extremely changeable and depend on many factors.

In conclusion, it should be remembered that regular discharges of oil-containing wastes raise more serious concerns than random and short-term accidental oil spills. The relatively small technological ("normal") discharges occur regularly in many regions during oil and gas production activities, tanker oil transportation, and shipping (see Chapter 3). They form areas of persistent oil pollution in the coastal zone and can cause long-term adverse effects in all groups of marine biota. For example, regular observations in the coastal zone of the Netherlands, Denmark, and Germany [Dahlmann et al., 1994] as well as on the shelf of Patagonia in Argentina [Esteves, Commendatore, 1993] revealed the existence of chronic oil stress for marine biota. The adverse effects on populations of marine birds, mammals, and invertebrates were associated with uncontrolled discharges of oil-containing wastes on the shelves of these countries. In the Gulf of Mexico, hundreds of thousands of young turtles died as

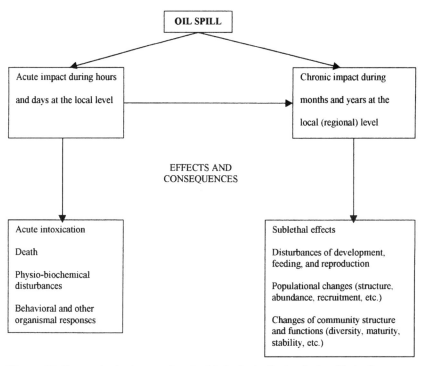

Figure 77 Conceptual scheme of major biological effects of oil spills in the sea.

a result of incidental ingestion of tar balls from the water surface during their first migrations [Ambrose, 1994]. At the same time, quantitative estimates of the scale of adverse biological effects caused by oil pollution (both from spills and from waste discharges) are complicated. These assessments must take into consideration other circumstances affecting the populations of marine biota, including extreme weather conditions, lack of food, epidemic diseases, and other factors that control natural mortality.

7.4. Chronic Contamination and Long-term Environmental Effects

The analysis of the general ecological situation in the World Ocean, given in Chapter 2, leaves no doubts about the global nature of anthropogenic impact on the marine environment. A leading factor of this impact is pollution which scale has become comparable with the natural cycles of matter and energy in the biosphere. The worldwide scientific community expresses concern about the possible long-term environmental consequences that spread beyond the areas of direct impact. These large-scale chronic disturbances are much less obvious and usually more difficult to reveal than, for example, effects caused by accidental oil spills or drilling discharges discussed above.

The issues of chronic contamination and long-term environmental effects first emerged as a topic of discussion back in the 1970s [Goldberg, 1970; Patin, 1979]. Since that time, the significance of these problems has increased tremendously. The coastal-shelf zone has become an arena for developing many dynamic and large-scale economic activities [Aibulatov, 1994]. Offshore resources have been increasingly exploited. The ecological situation and quality of the marine environment in many areas have been deteriorated [Patin, 1982; Izrael et al., 1993]. It has been recognized that marine pollution has become a problem requiring international control efforts. New preventive approaches specifically addressing the matter have been introduced in the national and international environmental practice [Wells, Bewers, 1992]. More attention has been devoted to the possibility of subtle (nonobvious) effects of low levels of chronic contamination in the marine environment. Such effects may emerge as a result of gradual cumulative changes caused by continuous inputs of anthropogenic substances, including routine discharges of different wastes into the sea.

Humankind has already dealt with alarming lessons from ignoring gradual and inconspicuous alternations in the nature. For example, possible global climatic changes and disturbances of the ozone layer that may occur in the near future have emerged and developed as a result of the combined impacts of local sources. Each of them was too weak, insignificant, and hardly noticeable to be taken into serious consideration. Combined together as time passed (only about 100 years), these local changes are causing a global effect.

Compared with the atmosphere, the World Ocean is certainly more conservative and slow to respond. It has a longer latent period before revealing nonobvious (subtle) effects. The complexity and potential tragedy of the situation lie with the fact that when the global changes in the hydrosphere do happen, it will be too late or impossible to do anything. The grown awareness of the critical significance of this problem has resulted in developing new preventive strategies to protect the ocean [MacGarvin, 1995].

To what extent is the offshore oil and gas industry responsible for the environmental disturbances in the seas and oceans? Could the consequences of its activity cause large-scale and persistent effects in marine ecosystems?

These questions in the context of specific situations or in relation to methodological approaches to this problem are discussed in many scientific publications. At the same time, the general agreement of opinions on these issues does not exist. This fact should not be very surprising if we consider the complexity of marine ecosystems and the difficulty of identifying slight anthropogenic effects in these ecosystems. Besides, the consensus on this matter is difficult to reach due to the major practical significance of these questions. The answers would directly affect strategic economic policies as well as the ways of their implementation and, hence, would define the expenses required for environmental protection measures.

A number of studies [Reid, Steimle, 1978; Clark, 1987; Payne, 1989; Baker et al., 1990; Neff, 1993; Kriskunov et al., 1993] deny, question, or ignore the possibility of long-term ecological disturbances (or their reliable detection) beyond the areas of direct impacts of the offshore oil and gas development. The most common evidence used to support this position include:

- the total contribution of offshore oil production into the global flow of oil hydrocarbons into the World Ocean does not exceed several percent of land-based discharges and all other sources of oil pollution;
- oil and gas belong to the group of natural organic substrates that are always present in the marine environment (sometimes in high concentrations); in many situations, they may degrade and lose their toxic properties relatively easily and quickly;
- environmental impact of the offshore industry has a local nature and covers insignificant areas; these include not more than 1–2% of an oil- and gas-bearing region (for example, the North Sea or the Gulf of Mexico); and
- even in case of strong impacts, for example during oil spills or explosive operations, only a part of the marine populations is damaged; ultimately, the high reproduction rate of most species and other regulatory mechanisms help reestablish the affected communities.

All of these considerations are mostly quite valid. Previous chapters have already partially discussed them. However, they still cannot be the foundation for complete excluding the possibility of cumulative ecological disturbances and subtle effects due to chronic low-level impacts on the marine environment during oil and gas

developments on the shelf. Some summarized ecotoxicological studies and field observations in areas of persistent oil pollution (given in
Table 60) support this statement. In particular, numerous research
suggest the following conclusions:

1. Practically everywhere where excessive (compared with natural background) amounts of oil hydrocarbons appear in the marine
environment, rapid development of hydrocarbon-degrading microflora starts. The portion of these organisms in bacterioplankton reaches
10% and more. This indicates that structural changes in so-called heterotrophic microplankton occur. This biotic component plays an important role in the transformation and distribution of organic matter in
the marine ecosystems [Sorokin, 1975]. Such microplankton restructuring was noted in the North Sea [Bruns et al., 1993; Minas, Gunkel,
1995], Gulf of Mexico [Lizarraga-Partida et al., 1991], and other areas
of intensive oil and gas production. The increased abundance of
hydrocarbon-degrading microorganisms and restructuring of microbial communities are found also in bottom sediments both in chronically polluted areas and in areas affected by an accidental oil spill
several years after the event [Braddock et al., 1995]. These circumstances support the possibility of large-scale and persistent ecological
changes in the marine environment caused by the oil and gas industry's activity. This certainly does not mean that such responses at the
bacterioplankton and bacteriobenthos levels can cause adverse effects in the whole ecosystem. However, such kind of large-scale
changes can not be ignored.

2. One of the most widespread features of contaminants' behavior in the marine environment is their localization in the surface
microlayer (at the water-air interface) and bottom sediments (at the
water-seafloor boundary). This is especially typical for oil hydrocarbons. Their concentrations at these boundaries can be hundreds
and thousands of times higher than in the water column. The water-
atmosphere interface is known to be the biotope for hyponeuston,
where eggs and larvae of many fish and invertebrates develop. The
ecological risk of exposing sensitive hyponeuston organisms to elevated levels of pollutants in the surface microlayer have been
stressed repeatedly for a long time [Zaitsev, 1970; Patin, 1979]. However, direct proof of actual threat was obtained only recently. Figure 78 gives such piece of evidence. It clearly shows a significant rise
of fish larvae mortality with increasing concentrations of PAHs in
water samples taken from the surface microlayer in the coastal zone
of the English Channel. In the most polluted places, the concentra-

Table 60
Summarized Results of Studies Indicating the Possibility of Long-term Effects in Marine Ecosystems Under Chronic Oil Contamination

Objects and Conditions of the Studies	Effects and Changes	Scale
Bacterioplankton exposed to oil hydrocarbons	Domination of hydrocarbon-degrading bacteria, reconstruction of microbial communities, changes in heterotrophic microplankton	Local, regional
Hyponeuston in the surface microlayer of seawater	Localization of PAHs in the surface microlayer, lethal and sublethal damages of hyponeustonic forms at embryonic and larval stages of development, changes in hyponeuston composition	Local, regional
Organisms and communities of different systematic and ecological groups under experimental and field conditions	Sublethal effects and elimination of sensitive species and forms under low levels of oil hydrocarbons in seawater (up to 10 μg/l) and bottom sediments (up to 10 mg/kg)	Local, regional (?)
Bottom fish and bivalves under background contamination of sediments by PAHs	Biological responses at the cellular level (induction of enzyme systems such as P-4501 A), diseases, histopathological changes	–"–
PAHs in the surface layer of seawater affected by solar radiation	Photooxidative processes and reactions forming high-molecular-weight compounds with mutagenic and carcinogenic properties	Local, regional(?)
PAHs in seawater in combination with chlorinated hydrocarbons	Synergetic increase in toxic properties of pollutants	Local

tions of PAHs in this microlayer were 10,000 times higher than at 0.5 m deep, and the larval mortality reached 100% [NERC, 1994].

A similar relationship was found between benthic fauna mortality and oil levels in bottom sediments [Jong et al., 1991]. Figure 79 shows a direct correlation between lethal effects in *Amphiura filiformis* and oil content in sediments in areas of oil-containing discharges

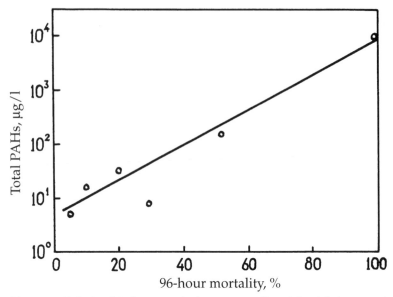

Figure 78 Relationship between the larvae mortality of *Scophthalmus maximus* and the PAH concentrations in the samples of the surface microlayer of water collected in the coastal waters near Plymouth [NERC, 1994].

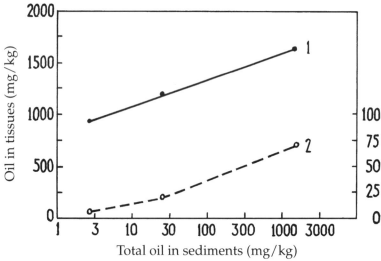

Figure 79 Dependence of oil accumulation in tissues (1) and mortality (2) of *Amphiura filiformis* from oil concentrations in bottom sediments in the areas of drilling discharges [De Jong et al., 1991].

(mainly drilling cuttings). In the North Sea, the total area of oil-polluted sediments reaches 8,000 km². The elevated oil concentrations in sediments can be found six and more years after the termination of drilling operations [ICES, 1992].

3. Summarized ecotoxicological materials, discussed in Chapter 4, show the possibility of hazardous effects of oil products when they are present in relatively low amounts in seawater and bottom sediments—from 10 $\mu g/l$ and 10 mg/kg, respectively. For individual PAHs, these concentrations are at least one order lower (see Table 51). Such levels, repeatedly recorded in the coastal zone, including areas of offshore oil production and transportation (see Chapter 4), can be the source of subtle toxic effects. These effects are difficult to reveal and measure. However, they have a long-term (chronic) nature and cover vast areas. Possible signs of these effects may include the following processes, which have been regularly observed lately:

- induced biochemical responses of enzyme systems (especially cytochrome P-450) in fish and invertebrates exposed to slightly elevated concentrations of PAHs found in the North Sea [Stebbing et al., 1992], on the Atlantic shelf of the United States and Canada [Addison, 1992; ICES, 1994], and in a number of other regions [GESAMP, 1995];
- increased frequency of histopathological changes and diseases of bottom fish and invertebrates that live in contact with sediments polluted by PAHs [Ilnitski et al., 1993]; these include areas of offshore oil production [ICES, 1994];
- degradation of ichthyoplankton communities in oil-polluted areas. For example, in the North Sea, high frequency of egg and larvae malformations in some common species of fish have been reported [Anonymous, 1990; Cameron et al., 1992; MacGarvin, 1995]; in the coastal waters of Australia, morphological anomalies at the embryonic stages of development reach 30–80% as compared with 10% for the background level in the clean areas [Klumpp, Westernhagen von, 1995]; in the Novorossiysk Bay of the Black Sea over the last 10 years, the ichthyoplankton abundance dropped 4–8 times as a result of chronic oil pollution [Bolgova, 1994]; in the Lower Volga, anomalies in embryonic development of sturgeon reach 50–100% [Shagaeva et al., 1993].

4. The ecological risk of chronic oil contamination is increased considerably by the mutagenic and carcinogenic properties of some oil compounds. Numerous examples of tumors developing among fish and invertebrates are often explained by the presence of high-molecular-weight PAHs (like benzo(a)pyrene), and the products of their photochemical and biochemical degradation (see Chapter 4) in the marine environment [Anderson, 1990; Ilnitski et al., 1993]. Mutagenic effects caused by chronic oil pollution of the coastal zone were found in mangrove ecosystems [Klekowski et al., 1994]. Mutagenic changes in the structure and sex ratio were revealed in fouling populations of bivalve mollusks from offshore drilling platforms [Dolgov, 1991].

5. The combination of PAHs and organochlorines in the water environment can significantly increase the severity of hazardous effects. The results of long-term experiments [Altufiev, 1994] suggest that such synergetic impacts could cause pathological changes in tissues and organs of Russian sturgeon. These effects lead to degradation of whole populations of these unique fish in the Caspian Sea and the Sea of Asov.

The list of such observations is far from being complete. However, the information presented above suggests that even relatively low levels of chronic contamination can cause large-scale subtle effects in the marine biota, especially in the coastal zone. To a certain extent, these effects could be a result of chronic contamination from offshore oil and gas development. Such possibility should become a subject of special and comprehensive studies. Some methodological approaches have been proposed to solve this very complicated problem [Howells et al., 1990]. However, the current level of knowledge in marine biology and ecology is insufficient to reveal and assess the cause and effect relationships of these subtle anthropogenic disturbances.

The offshore oil and gas industry, certainly, is not the sole contributor to the ecological problems on the shelf. On a global scale, only about 2% of the total input of oil pollution into the marine environment come from this kind of activity. At the same time, it must be remembered that at the regional level (for example, in the North Sea), this contribution can reach up to 30% (see Chapter 4). Discharges of other pollutants, seismic surveys, and other numerous impacts of the offshore oil development also affect the marine environment, as discussed in detail in Chapter 3. Finally, the most important fact is that all these impacts take place in the coastal zone that is already subjected to anthropogenic stress from other kinds of human activities. The intensity of this stress reaches a critical point in some coastal areas and requires immediate measures to control it.

7.5. Biological Resources, Fisheries, and Offshore Oil and Gas Industry

The impact of the offshore oil and gas industry on biological resources of the sea has repeatedly emerged as a topic in the pages of this book. This is quite understandable because any impact affects the environmental conditions and habitats of commercial species. Seas and oceans have been an important source of human food since the remotest times. Since World War II the production of fish and fish products has increased faster than that of any other major food. At the same time, dynamic offshore oil and gas developments over the last 30 years pose a certain risk of adverse effects on marine biological resources. Figure 80 gives the general schema of environmental stresses caused by the offshore oil and gas activity and possible consequences for commercial species.

The physical impacts of the offshore industry on biological resources and commercial fisheries were discussed in Chapter 3. These impacts include seismic surveys, explosive activities, platform installations, pipeline construction, and others. This section is going to focus mainly on the effects of oil pollution for commercial species and the fishing industry.

7.5.1. Oil Impacts on Fisheries Resources

Fish is a rather well-studied component of marine ecosystems. They display high sensitivity to stress impacts and ability to respond rapidly to changes in the quality of the marine environment. This made possible the development of ichthyological monitoring as a way to control and assess the ecological situation in the hydrosphere [Mazmanidi, Kotov, 1991; Lukyanenko, 1993].

Numerous data, including those discussed in this book (Chapters 3, 4, and 7), suggest that the acute (lethal) impact of oil, even during catastrophic spills, does not pose any serious threat to populations of free-swimming fish. It was clearly shown in a study devoted to the ecological situation during and after a large spill in the Santa Barbara Strait off the California shore in 1969 [Squire, 1992]. This spill resulted from long-term open gushing of an oil well. A thick oil slick covered a vast area, rich in fish resources, for several months. Detailed observations of the distribution, migration, and abundance of local pelagic fish (mainly anchovy, sardine, and mackerel) before and during the period of heavy oil pollution as

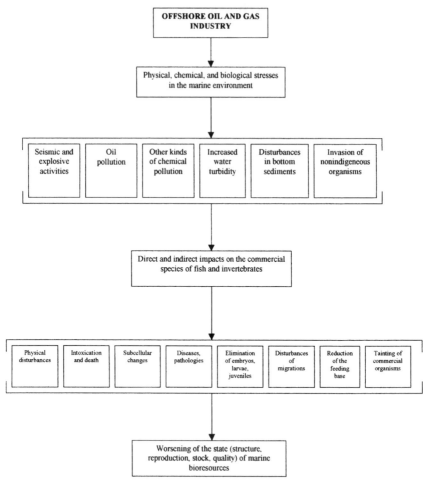

Figure 80 Scheme of complex impact on the marine biological resources during offshore oil and gas development.

well as in the following years did not reveal any changes in fish stock and catch.

This and many other similar studies, conducted in different regions, suggest that local oil spills cannot undermine the stock of pelagic species [Clark, 1987; Payne, 1989; Baker et al., 1990; Neff, 1993]. Although sublethal and lethal effects on some fish, especially at the early stages of development, could occur in affected areas, most pelagic adult fish are able to swim away from oil spills and are rarely at risk in the open sea. This does not mean, of course, that commercial fisheries are not affected in these situations. Usually, fishing activity in the spill areas is interrupted. The fishermen have to avoid the polluted areas to prevent fouling boats and gear. The losses for commercial fisheries can be significant due to the interruption of fishing activity, a reduction in the abundance of important species, or tainting and contamination of commercial organisms [Keiso, Kendziorek, 1991; Picou et al., 1992]. Besides, localized populations of commercial fisheries resources living in the shallow coastal zones with limited water circulation (e.g., menhaden, shrimps, blue crabs, and oysters) as well as objects of aquaculture are repeatedly shown to suffer from acute stress and high mortality after a number of catastrophic oil spills. At the same time, no significant cumulative effects from oil spills on commercial fisheries or irreversible undermining of fish stock were found [GESAMP, 1993; Neff, 1993].

Although these data seem to be reassuring, the negative effect of oil pollution on commercial ichthyofauna and biological resources in general undoubtedly exists. In particular, pollution can affect the feeding base of demersal and semidemersal fish. As previously discussed, the adverse effects of oil pollution (both in the zones of offshore developments and coastal oil spills) are often observed in benthos. At the same time, benthic fauna is a major feeding source for many commercial bottom-dwelling fish. Thus, pollution impact can lead to a reduction of the feeding base of these fish. Evident and long-term effects of oil pollution on benthic organisms were observed, for example, after the famous *Amoco Cadiz* oil spill in the English Channel in 1978. Eight years after this catastrophe, the benthos biomass in some coastal areas was still 40% less than that before the spill [Page et al., 1989]. Populations of important objects for fish feeding (especially amphipods) were depressed [Dauvin, 1987]. Bottom sediments in the areas of oyster reproduction had elevated concentrations of aromatic hydrocarbons. In addition, tumors in the digestive tract and gonads, which usually develop under the impact of carcinogenic PAHs, damaged the oysters themselves [Berthou et al., 1987].

In the long run, the most serious concerns arise regarding the possibility of sublethal effects in commercial species exposed to chronic contamination of the marine environment. The fish sublethal responses include a wide range of compensational changes. These start at the subcellular level and first have a biochemical and molecular nature. However, if the hazardous impacts persist, stress effects may manifest themselves later in a form of histological, physiological, behavioral, and even populational responses, including impairment of feeding, growth, and reproduction. These can affect the abundance of commercial populations.

Of course, such hazardous consequences of the long-term chronic impact of the offshore oil and gas industry on fish and other commercial species are only a general possibility. No quantitative assessments of these effects in different regions are available yet. At the same time, the problem attracts more and more attention. Many recent studies have been devoted to this issue. The available data suggest that primary sublethal responses of marine biota (including commercial species) in the form of biochemical changes, histological anomalies, diseases, and some other effects of chronic stress have become global phenomena in the marine coastal zone. An early diagnostic of these responses is presently an important and challenging task, especially in areas directly affected by anthropogenic impact. These areas undoubtedly include places of the offshore oil and gas developments.

7.5.2. Deterioration of the Quality of Fish and Other
Commercial Organisms

The contact of any water organism with oil most often results in the appearance of oil odor and flavor in their tissues. In case of commercial species, this certainly means the loss of their value and corresponding fisheries losses. In spite of the widespread and even routine character of such incidents and their economic implications, scientific information about this issue is rather limited and controversial.

Experimental studies [Brandal et al., 1976; Volkman et al., 1994; Davis et al., 1992; Neff, 1993] show that the range of water concentrations of oil causing the taint in fish, crustaceans, and mollusks is very wide. Usually, these concentrations vary between 0.01 mg/l and 1.0 mg/l depending on the oil type, composition, form (dissolved, slick, emulsion), duration and conditions of exposure, kind of organism, and other factors. An analysis of these extremely heterogeneous experimental data leads to the following conclusions:

- Accumulation of oil and its fractions in water organisms occurs within several hours or days. The rate of this process directly depends on the oil concentration in the environment.
- Fish can accumulate considerable amount of oil, which is easily detected organoleptically, without any obvious signs of acute intoxication.

- The rate of fish decontamination from oil varies within a wide range—from days to several weeks. One of the important factors is the lipid and fat content in fish tissues. With increasing lipid content, the rate of oil excretion from the organism declines while the rate and level of its accumulation, in contrast, increase.

- Among all oil hydrocarbons, water-soluble monoaromatic compounds like benzene, toluene, xylene, and their derivatives produce the strongest persistent oil flavor and odor in fish tissues. In contrast, poorly soluble aliphatic hydrocarbons (alkanes) practically do not affect organoleptic parameters. Data indicate that the tainting effect is more intense in the presence of such oil components as kerosene, naphthalene, and their derivatives.

Some studies [Swan et al., 1994] give the following estimates of threshold water levels of oil products that could lead to tainting of fish and other organisms (μg/l): cresol—70; ethylbenzene—250; kerosene—100; kerosene and kaolin—1,000; naphtha—100; naphthalene—1,000; 2-naphthol—300; phenol—1,000–10,000; styrene—250; toluene—250.

In the areas of offshore oil and gas field developments, the highest risk of oil accumulation and associated tainting effects most likely exist for the bottom-dwelling commercial species (e.g., saithe, turbot, sole, and other fish). The data in Figure 81 represent the rate of oil hydrocarbon accumulation in fish living in contact with oil-polluted drilling cuttings in bottom sediments.

Despite the high variability of the results of studies devoted to the tainting of commercial organisms in oil-polluted areas [McIntyre, 1982; Tidmarsh et al., 1985; McGill et al., 1987; McIntosh et al., 1990; Davies, Kingston, 1992], it is possible to make the following conclusions:

- The contact of commercial fish and invertebrates with the oil during accidental oil spills practically always leads to accumulation of oil hydrocarbons in their tissues and organs (usually within the range 1–100 mg/kg). In most cases, the organisms acquire oil odor and flavor. This fact is the main reason for closing fisheries in the affected areas.

- Coastal farms for cultivating marine fish and invertebrates can be exposed to the severe impact of accidental oil spills. Observations showed that several months after the spill, the fish (salmon) cultivated on marine farms still had elevated concentrations of oil hydrocarbons in their tissues and suffered diseases and increased mortality [MLA, 1993a].

- Limited data available on fish tainting in areas of the offshore oil production are rather controversial. Some publications [Davies et al., 1989; Picken, 1989; McIntosh et al., 1990; Rasmussen et al., 1992] document the absence of any tainting in benthic and pelagic fish caught 1–3 km and more from the platform. At the same time, other researchers [McGill et al., 1987; Parker et al., 1990] observed oil taint in fish caught within 5 km from platforms. The degree of tainting directly correlated with the distance from the platforms. The tainting effect is most often found in the bottom-dwelling fish. This is quite understandable considering the accumulation of oil hydrocarbons in bottom sediments, especially when they are polluted by oil-based drilling muds and cuttings (see Figure 81).
- Most studies reveal the difficulties of sensory evaluation of seafoods and the absence of a clear correlation between the results of such evaluation and the data of chemical analyses.

Many authors of the previously cited publications do not consider the problem of the tainting of marine commercial organisms urgent. They claim that this problem does not pose any real threat to the fishing industry, at least in oil production areas in the North Sea. This is probably true for the North Sea, where rather rigid regulations and international control of the oil and gas production activities are practiced. Although even for this region, some data (as previously described) could cool down such an optimistic outlook. By considering this issue in a broader geographic aspect, we have to admit that in some regions, fish tainting can become a real problem, especially for the coastal fishing and aquaculture. The absence of regular observations and any statistics as well as difficulties of quantitative assessment of taint could lead to underestimation of the scale of the effect. Thus, the deterioration of the seafood quality due to oil odor and flavor may be much more widespread than can be concluded from the rather limited information available at present [GESAMP, 1993].

Conclusions

1. The ecological effects of the offshore oil and gas industry involve many complex factors. They affect all marine biotopes and can manifest themselves at all levels of biological hierarchy in the sea—from responses at the cellular and subcellular levels to populational disturbances and changes in the structure and functions of communities and ecosystems. Modern methodology to reveal and

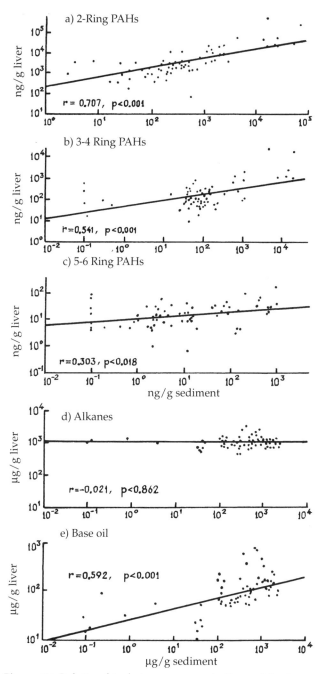

Figure 81 Relationship between concentrations of hydrocarbons in the liver and in sediment in a multi-tank experiment to investigate the effects of oil-based mud drill cuttings on the dab (*Limanda limanda*). Each point is the mean of a pooled liver sample from five fish killed 5, 15, and 30 days after the start of exposure and a sediment sample taken at the same time. Lines were fitted by least squares regression analyses [MLA, 1993].

measure such effects on the background of the very complex natural dynamics of the ecosystem is still searching for adequate approaches and procedures. The most promising ones are the methods of early diagnostics of subtle anthropogenic effects, including sensitive biological tests and biomarkers to reveal the primary stress responses at the molecular and biochemical levels.

2. Typical characteristics of the ecological situation in the areas of offshore activities include the gradual changes of chemical and biological parameters with distance from the offshore developments. In the water column, these changes usually rapidly attenuate due to the intensive dilution of discharged waters. The environmental parameters usually reach background levels at a distance of few hundred meters (maximum—up to 1,000 m) from the platforms. In contrast, the oil pollution of bottom sediments may extend as far as five and more kilometers from the source of oil-containing discharges. Its biological effects in the form of structural changes in benthic communities can be found up to 1 km from the platforms. Recovering of benthic communities after terminating discharges of oil-containing wastes occurs at different rates, depending on many factors, and often takes several years.

3. The ecological consequences of accidental oil spills are characterized by an extremely wide range of possible effects and stresses. Acute manifestations and chronic effects should be distinguished. Acute manifestations take place at the initial stages of oil spreading (the first hours and days after the accident). They include the death and heavy intoxication of marine organisms, especially birds and mammals. Chronic effects emerge due to the oil accumulation in bottom sediments and may manifest themselves for 20–30 years after the spill. The most hazardous consequences of oil spills occur when oil enters shallow coastal zones and contacts the shoreline. In contrast, in the open sea, oil slicks usually quickly disperse without any significant ecological effects.

4. The biggest concern regarding the ecological consequences of the offshore oil and gas activity in the coastal zone is connected with the possibility of long-term effects caused by low levels of chemical (mainly oil) pollution. In spite of the difficulties of revealing such responses, more and more studies prove the existence of nonobvious (subtle) long-term consequences of chronic contamination. The evidence include regional structural changes in marine bacterioplankton (e.g., growing abundance and activity of hydrocarbon-degrading microorganisms); deterioration of ichthyoplankton (e.g., high mortality, morphological anomalies); symptoms of chronic stresses at the

subcellular and cellular levels (e.g., induced biochemical responses of enzyme systems in fish and invertebrates); and increased frequency of mutagenic and carcinogenic effects and diseases (especially among benthic organisms).

5. Pelagic species of commercial fish in the open waters are least vulnerable to the hazardous impacts of oil even under conditions of heavy oil pollution. In contrast, bottom-dwelling fish and invertebrates as well as the objects of aquaculture in the coastal zone are subjected to the higher risk of damage both during accidental oil spills and under chronic contamination. Direct negative impacts of offshore oil production on fisheries include displacing the traditional areas of commercial fishing, interfering physically with fishing activities, deteriorating the quality of the marine environment in the areas of marine aquaculture, interrupting fishing during accidental oil spills, and reducing the quality and commercial value of the seafood organisms.

References

Abakumov, ed. 1991. *Ecological modifications and criteria for establishing the ecological standard.* Leningrad: Gidrometeoizdat, 384 pp. (Russian)

Addison, R. F. 1992. Detecting the effects of marine pollution. In *Science review 1990–1991.*—Dartmouth, Nova Scottia, pp.9–12.

Aybulatov, N. A. 1994. Anthropogenic expansion in the coastal-shelf zone. *Vestnik RAN* 64:4:340–348. (Russian)

Al-Hadhrami, M. N., Lappin-Scott, H. M., Fisher, P. J. 1995. Bacterial survival and n-alkane degradation within Oman crude oil and a mousse. *Mar.Pollut.Bull.* 30(6):403–408.

Altufiev, U. V. 1994. Morphofunctional state of the muscle tissue and liver of the juvenile Russian sturgeon in the experiments on chronic intoxication. *Voprosy ikhtiologii* 43(1):135–138. (Russian)

Amrose, Ph. 1994. Tarred loggerhead turtles. *Mar.Pollut.Bull.* 28(5):273.

Anderson, D. P. 1990. Immunological indicators: effects of environmental stress on immune protection and disease outbreaks. In *Biological indicators of stress in fish: Amer.Fish.Soc.Symp.8.*—Maryland, pp.1–8.

Anonymous. 1990. Deformed fish embryos in the North Sea. *Mar.Pollut.Bull.* 21(3):106.

Armstrong, H. W., Fuick, K., Anderson, J. W., Neff, J. M. 1979. Effects of oilfield brine effluent on sediments and benthic organisms in Trinity Bay, Texas. *Mar.Environ.Res.* 2:55–69.

Baker, J. M., Clark, R. B., Kingston, P. F. 1990. *Two years after the spill: environmental recovery in Prince William Sound and the Gulf of Alaska.* Edinburgh: Institute of Offshore Engineering, Heriot-Watt University, 31 pp.

Berthou, F. G., Balouet, G., Bodennec, G., Marchand, M. 1987. The occurrence of hydrocarbons and histopathological abnormalities in oysters for seven years following the wreck of the Amoco Cadiz in Brittany (France). *Mar.Environ.Res.* 23:103–133.

Blackman, R. A. A., Fileman, T. W., Law, R. J., Thain, J. E. 1985. *The effects of oil-based drill-muds in sediments on the settlement and development of biota in a 200-day tank test. International Council for the Exploration of the Sea.* ICES CM 1985/E:23.

Blackman, R. A. A., Law, R. J. 1981. *The oil content of discharged drill-cuttings and its availability to benthos. International Council for the Exploration of the Sea.* ICES CM 1981/E:23, 14 pp.

Blackman, R. A. A., Law, R. J. 1986. *The effects of new oil-based drill-muds in sediments on settlement and development of biota in an improved tank test. International Council for the Exploration of the Sea.* ICES CM 1986/E:13.

Bolgova, L. V. 1994. *Ichthyoplankton of the Novorosiisk Bay under conditions of anthropogenic impact. Theses of dissertation.* Moscow, 23 pp. (Russian)

Bonsdorff, E., Bakke, T., Pedersen, A. 1990. Colonization of amphipods and polychaetes to sediments experimentally exposed to oil hydrocarbons. *Mar.Pollut.Bull.* 21(7):355–358.

Boothe, P. N., Presley, B. J. 1989. Trends in sediment trace element concentrations around six petroleum drilling platforms in the northwestern Gulf of Mexico. In *Drilling wastes.*—London: Elsevier Applied Science, pp.3–21.

Bourne, W. R. P. 1990. Dutch beached bird surveys. *Mar.Pollut.Bull.* 21(7):360–361.

Braddock, J. F., Lindstorm, J. E., Brown, E. J. 1995. Distribution of hydrocarbon-degrading microorganisms in sediments from Prince William sound, Alaska, following the Exxon Valdez spill. *Mar.Pollut.Bull.* 30(2):125–132.

Brandal, P. E., Grahl-Nielsen, O., Neppelberg, T., Palmork, K. H., Westrheim, K., Wilhelmsen, S. 1976. *Oil-tainting of fish, a laboratory test on salmon and saithe. International Council for the Exploration of the Sea.* ICES CM 1976/E:33.

Bruns, K., Dahlmann, G., Gunkel, W. 1993. Distribution and activity of petroleum hydrocarbon degrading bacteria in the North and Baltic Seas. *Hydrogr.Z.* 6:359–369.

Burger, J., ed. 1994. *Before and after an oil spill: the Arthur Kill.* New Brunswich (New Jersey): Rutgers University Press, 305 pp.

Burns, K. A., Garrii, S. D., Jorissen, D., MacPherson, J., Stoelting, M., Tiemey, J., Yelle-Simmons, L. 1994. The Galeta oil spill. 2. Unexpected persistence of oil trapped in mangrove sediments. *Estuar.Coast.Shelf.Sci.* 38(4):349–364.

Cameron, P., Berg, J., Dethlefsen, V., Westrnhagen, H. von 1992. Developmental defects in pelagic embryos of several flatfish species in the southern North Sea. *Neth.J.Sea.Res.* 29:239–256.

Chapman, P. M., Power, R. N., Dexter, R. N., Andersen, H. B. 1991. Evaluation of effects associated with an oil platform, using the sediment quality triad. *Environ.Toxicol. and Chem.* 10:407–424.

Clark, R. B. 1987. Summary and conclusions: environmental effects of North sea oil and gas developments. In *Environmental effects of North Sea oil and gas developments.*—London: Phil.Trans.R.Soc. B316, pp.587–602.

Clarke, K. R. 1993. Non-parametric multivariate analyses of changes in community structure. *Aust.J.Ecol.* 18:117–143.

Daan, R., Mulder, M. 1994. Biological effects of drilling activities in the North Sea. *Netherlands Institute for Sea Research, Publ.Ser.* 22:91–93.

Daan, R., Mulder, M., Leewen, A. V. 1994. Differential sensitivity of macrobenthic species to discharges of oil-contaminated cuttings in the North Sea. *Netherl.Journ. of Sea Res.* 33(1):113–127.

Daan, R., Mudler, M. 1996. On the short-term and long-term impact of drilling activities in the Dutch sector of the North Sea. *ICES J. of Mar.Sci.* 53(6):1036–1044.

Dahlmann, G., Timm, D., Averbeck, Chr., Camphuysen, C., Skov, H., Durinck, J. 1994. Oiled seabirds—comparative investigations on oiled seabirds and oiled beaches in the Netherlands, Denmark and Germany (1990–93). *Mar.Pollut.Bull.* 28(5):305–310.

Dauvin, J. C. 1987. Long-term evolution 1987–1988 of amphipod populations of the fine sands of Pierre Noire Bay of Morlaix, Western English Channel following the Amoco Cadiz disaster. *Mar.Environ.Res.* 21(4):247–274.

Davies, J. M., Addy, J. M., Blackman, R. A. A., Blanchard, J. R., Ferbrache, J. E., Moore, D. C., Somerville, Hh. J., Whitehead, A., Wilkinson, T. 1984. Environmental effects of the use of oil-based drilling muds in the North Sea. *Mar.Pollut.Bull.* 15(10):363–370.

Davies, J. M., Bedborough, D. R., Blackman, R. A. A., Addy, J. M., Appelbee, J. F., Grogan, W. C., Parker, J. G., Whitehead, A. 1989. Environmental effects of oil-based mud drilling in the North Sea. In *Drilling wastes.*—London: Elsevier Applied Science, pp.59–89.

Davies, J. M., Kingston, P. F. 1992. Sources of environmental disturbances associated with offshore oil and gas developments. In *North Sea oil and the environment. Developing oil and gas resources, environmental impacts and responses.*—London–New York: Elsevier Applied Science, pp.417–440.

Davis, H. K., Geelhoed, A. W., MacRae, A. W., Howgate, P. 1992. Sensory analysis of tainting of trout by diesel fuel in ambient water. In *Proceedings of the Third International Symposium on off-flavors in the aquatic environment, 3–8 March 1991, Los Ageles, California, USA.*

Dolgov, L. V. 1991. Sex expression and environmental stress in a mollusk, *Pictanda margaritifera. Invertebr.Reprod.Dev.* 20(2):121–124.

Esteves, J. L., Commendatore, M. J. 1993. Total aromatic hydrocarbons in waters and sediments in a coastal zone of Patagonia, Argentina. *Mar.Pollut.Bull.* 28(6):341–342.

Fowler, S. W., Readman, J. W., Oregioni, B., Vieneuve, J.-P., Mckay, K. 1993. Petroleum hydrocarbon and trace metals in nearshore Gulf sediments and biota before and after the 1991 war. *Mar.Pollut.Bull.* 27(3):171–182.

Francis, R. D., Javanfar, A., Brenner, J. P. 1992. Rapid disappearance of oil from a moderate-size spill: cleanup effort coupled with natural degradation processes. *Mar.Technol.Soc.J.* 26(3):26–33.

GESAMP. 1993. *Impact of oil and related chemicals and wastes on the marine environment. GESAMP Reports and Studies No.50.* London: IMO, 180 pp.

GESAMP. 1995. *Biological indicators and their use in the measurements of the condition of the marine environment. GESAMP Reports and Studies No.55.* UNEP, 56 pp.

Goldberg, E. D. 1970. *Chemical invasion of ocean by man.* McGraw-Hill Yearbook Science and Technology.

Gourlay, K. A. 1988. Chapter 3. The black death: oil. In *Poisoners of the seas.*— London: Zed Books Ltd, pp.133–188.

Grahl-Nielson, O. 1987. Hydrocarbons and phenols in discharge water from offshore operations. Fate of the hydrocarbons on the recipient. *Sarsia* 72(3–4):375–382.

Gray, J. S. 1988. Environmental politics and monitoring around oil platforms. *Mar.Pollut.Bull.* 19(11):549–550.

Gray, J. S. 1991. Anthropocentric or ecocentric? *Mar.Pollut.Bull.* 22(11):529–530.

Gray, J. S., Clarke, K. R., Warwick, R. M., Hobbs, G. 1990. Detection of the initial effects of pollution on marine benthos: an example from Ecofisk and Eldfisk oilfields, North Sea. *Mar.Ecol.Progr.Ser.* 66:285–299.

Gundlach, E. R., Hayes, M. O. 1978. Vulnerability of coastal environment to oil spill impact. *Mar.Tech.Soc.Jour.* 12:18–27.

Gupta, R. S. 1991. Gulf oil spill and India. *Mar.Pollut.Bull.* 22(9):423–424.

Haldane, D., Reuben, R. L., Side, J. C. 1992. Submarine pipelines and the North Sea environment. In *North Sea oil and the environment: developing oil and gas resources, environmental impacts and responses.*—London and New York: Elsevier Applied Science, pp.481–522.

Hinshery, A. K., Baghdadi, A. L., Kumar, N. S. 1991. Petroleum hydrocarbon levels near offshore water of Libya. Proceedings of UNESCO Symposium on Marine Chemistry in the Arab Region (Suez, Egypt, April 1991). *Bulletin of National Institute of Oceanography and Fisheries* 16(3):229–238.

Howells, G. D., Calamari, D., Gray, J. S., Wells, P. G. 1990. An analytical approach to assessment of long-term effects of low levels of contaminants in the marine environment. *Mar.Pollut.Bull.* 21(8):371–375.

ICES. 1992. *International Council for the Exploration of the Sea. Report of the Study Group on ecosystem effects of fishing activities.* ICES C.M.1992/ G:11, 144 pp.

ICES. 1994. *International Council for the Exploration of the Sea. Report of the Joint meeting of the Working Group on Marine Sediments in relation to pollution and the Working Group on Biological Effects on Contaminants (Nantes, France, 21–22 March 1994)*. ICES C.M.1994/ENV:2. 19 pp.

Ilnitski, A. P., Korolev, A. A., Khudolei, V. V. 1993. *Carcinogenic substances in the water environment*. Moscow: Nauka, 220 pp. (Russian)

Izrael, U. A. 1984. *Ecology and control of the state of the environment*. Moscow: Gidrometeoizdat, 560 pp. (Russian)

Izrael, U. A., Tsiban, A. V., Panov, G. V. 1993. On ecological situation in the Russian seas. *Meteorologia i gidrologia*. 8:15–21. (Russian)

Jong, S. A. De, Zevenboom, W., Van Het Groenewoud, H., Daan, R. 1991. Short and long-term effects of discharge OBM cuttings, with and without previous washing, tested in field and laboratory studies on the Dutch Continental Shelf, 1985–1990. In *Proceedings of the First International Conference on Health, Safety and Environment in Oil and Gas Exploration and Production (The Hague, 1991)*.— Pp.329–338.

Keiso, D. D., Kendziorek, M. 1991. Alaska's response to the Exxon Valdez oil spill. *Environ.Sci.Technol.* 25(1):16–23.

Kingston, P. F. 1987. Field effects of platform discharges on benthic microfauna. *Phil.Trans.R.Soc.Lond.* B316:545–565.

Kingston, P. F. 1991. The North Sea oil and gas industry and the environment. In *Proceedings Financial Times Conference on North Sea Oil and Gas (London, July 1991)*.—London: Financial Times, pp.19.1–19.6.

Kingston, P. F. 1992. Impact of offshore oil production installations on the benthos of the North Sea. *ICES J.Mar.Sci.* 28(3):166–169.

Klekowski, E. J., Corredor, J. E., Morrel, J. M., Del Castillo, C. A. 1994. Petroleum pollution and mutations in mangroves. *Mar.Pollut.Bull.* 28(3):166–169.

Klumpp, D. W., Westernhagen, H. von.1995. Biological effects of pollutants in Australian tropical coastal waters: embryonic malformations and chromosomal aberrations in developing fish eggs. *Mar.Pollut.Bull.* 30(2):158–165.

Konstantinov. A. S. 1986. *General hydrobiology*. Moscow: Vysshaya shkola, 470 pp. (Russian)

Kriskunov, E., Polonski, U., Vekilov, E. 1993. Seismic surveys taking into consideration the nature protection issues. *Neftyanik* 4:11–14. (Russian)

Kroencke, I., Duineveld, G. C. A., Raak, S., Rachor, E., Daan, R. 1992. Effects of a former discharge of drill cuttings on the microfauna community. *Mar.Ecol.Progr.Ser.* 91(1–3):277–288.

Law, R. J., Blackman, R. A. A 1981. *Hydrocarbons in water and sediments from oil-producing areas of the North Sea. International Council for the Exploration of the Sea*. ICES C.M. 1981/E:16, 18 pp.

Law, R. J., Blackman, R. A. A, Filman, T. W. 1982. *Surveys of hydrocarbon levels around five North Sea production platforms in 1981. International Council for the Exploration of the Sea*. ICES C.M. 1982/E:14, 18 pp.

Law, R. J., Hudson, P. M. 1986. *Preliminary studies of the dispersion of oily water discharges from the North Sea oil production platforms. International Council for the Exploration of the Sea.* ICES C.M. 1986/E:15, 12 pp.

Lein, T. E., Hlohlman, S., Fossaa, J. H., Aarrestad, K., Mortensen, P. B. 1993. *Oil pollution consequences on hard bottom communities. Rocky shores and kelp forests in mid-Norway.* Bergen (Norway): Dep.Fish.Mar.Biol., Univ.Bergen, 72 pp.

Lizarraga-Partida, M. L., Izquierdo-Vicuna, F. B., Wong-Chang, I. 1991. Marine bacteria on the Campeche Bank oil field. *Mar.Pollut.Bull.* 22(8):401–405.

Lukyanenko, V. I. 1993. Ichthyotoxicological monitoring—the most important tool for assessing the quality of the water environment and the state of the natural fish populations. In *Biological and economic issues of the world fisheries.*—VNIERH, Vip.5, pp.1–5. (Russian)

MacDonald, D. D. 1993. *Development of an approach to the assessment sediment quality in Florida coastal waters. Report prepared by MacDonald Environmental Sciences, Ltd.* Ladysmith, British Columbia for the Florida Department of Environmental Regulation. Florida: Tallahassee, 133 pp.

MacGarvin, M. 1995. The implication of the precautionary principle for biological monitoring. The challenge to marine biology in a changing world. Proceedings of the International Symposium (Helgoland, 13th–18th September 1992). *Helgolander Meeresuntersuchungen* 49(1–4):647–662.

Mair, L. McD., Matheson, I., Appelbee, J. F. 1987. Offshore macrobenthic recovery in the Murchison Field following termination of drill-cutting discharges. *Mar.Pollut.Bull.* 18(12):628–634.

Massie, L. C., Ward, A. P., Bell, J. S., Saltzmann, H. A. 1981. *The levels of hydrocarbons in water and sediments in selected areas of the North Sea and the assessment of their biological effects. International Council for the Exploration of the Sea.* ICES CM 1991/E:44, 25 pp.

Matheson, I., Kingston, P. F., Johnson, C. S., Gibson, M. J. 1986. Statfiord field environmental study. In *Proceedings of Conference on Oil-based Drilling Fluids. Cleaning and environmental effects of oil contaminated drill cuttings.*—Trondheim (Norway), pp.3–16.

Mazmanidi, N. D. 1991. Fish as objects of monitoring of the ecological situation in the Black Sea. In *Theses of the Second All-Union Conference on the Fisheries Toxicology.*—Spb., pp.22–23. (Russian)

Mazmanidi, N. D., Kotov, A. M. 1991. On some mechanisms of pathogenesis of poisoning by organic toxicants in the fish of the Black Sea. In *Theses of the Second All-Union Conference on the Fisheries Toxicology.*—Spb., pp.24–25. (Russian)

McGill, A., Mackie, P. R., Howgate, P., McHenry, L. G. 1987. The flavor and chemical assessment of dabs (*Limanda limanda*) caught in the vicinity of the Beatrice oil platform. *Mar.Pollut.Bull.* 18(4):186–189.

McGurk, M. D., Warbuton, H. D., Parker, T. B., Litke, M., Marliave, J. B. 1993. Effects of the Exxon Valdez oil spill on Pacific herring eggs and viability of their larvae. *Can.Tech.Rep.Fish.Aquat.Sci.* 1924:255–257.

McIntosh, A. D., Massie, L. C., Davies, J. M. 1990. *Assessment of fish from the northern North Sea for oil taint. International Council for the Exploration of the Sea.* ICES CM 1990/E:24, 14 pp.

McIntosh, A. D., Ward, A. P. 1982. *The levels of hydrocarbons in the water and sediments in areas around Sullom Voe and Beatrice oil field. International Council for the Exploration of the Sea.* ICES CM 1982/E:42, 21 pp.

McIntyre, A. D. 1982. Oil pollution and fisheries. In *Philosophical transactions of the Royal Society, London.*—B297:401–411.

McKinnon, M., Vine, P. 1991. *Tides of war. Eco-disaster in the Gulf.* London: Boxtree LTD, 192 pp.

Minas, W., Gunkel, W. 1995. Oil pollution in the North Sea—a microbiological point of view. The challenge to marine biology in a changing world. Proceedings of the International Symposium (Helgolander, 13th–18th September 1992). *Helgolander Meeresuntersuchungen* 49(1–4):143–158.

MLA. 1993a. *Marine Laboratory Aberdeen. Annual Review (1992–1993).* Agriculture and Fisheries Department. The Scotish Office, 68 pp.

MLA. 1993b. *Marine Laboratory Aberdeen. Interim report of the marine monitoring programmer on the Braer oil spill, March 1993.* Agriculture and Fisheries Department. The Scotish Office, 12 pp.

Moller, T. H., Dicks, B., Goodman, C. N. 1989. Fisheries and mariculture affected by oil spills. In *Proceedings of the 1989 International Oil Spill Conference (San Antonio, TX, USA).*—Amer.Petrol.Inst. Washington, DC. A.P.I.Public. N4479:389–394.

Moore, D. C. 1983. *Biological effects on benthos around the Beryl oil platform. International Council for the Exploration of the Sea.* ICES CM 1983/E:43, 19 pp.

Neff, J. M. 1988. *Bioaccumulation and biomagnification of chemicals from oil well drilling and production wastes in marine food webs: a review. Report to American Petroleum Institute.* Battelle Ocean Sciences, Duxbury, Mass.

Neff, J. M. 1993. *Petroleum in the marine environment: regulatory strategy and fisheries impact. Report to Exxon Company.* Huston, TX, 13 pp.

Neff, J. M. 1998. Fate and effects of drilling mud and produced water discharged in the marine environment. In *U.S.–Russian Government Workshop on Management of Waste from Offshore Oil and Gas Operation (April 1998, Moscow).*

Neff, J. M., Rabalais, N. N., Boesch, B. F. 1987. Offshore oil and gas development activities potentially causing long-term environmental effects. In *Long-term environmental effects of offshore oil and gas development.*—London: Elsevier Applied Science, pp.149–173.

NERC. 1994. *Natural Environment Research Council. Report of the Plymouth Marine Laboratory (1993–1994).* NERC, 70 pp.

Norcross, B. L. 1992. Responding to an oil spill: reflections of a fisheries scientists. *Fisheries* 17(6):4.

NSQSR. 1994. *North Sea Quality Status Report.* London.

Page, D. S., Foster, J. C., Fickett, P. M., Gilfillan, E. S. 1989. Long-term weathering of Amoco Cadiz oil in soft intertidal sediments. In *Proceedings of the*

1989 International Oil Spill Conference (San Antonio, TX, USA).—American Petroleum Institute, Washington, DC. A.P.I.Publ. N4479:401–405.

Parker, J. G., Howgate, P., Mackie, P. R., McGill, A. C. 1990. Flavor and hydrocarbon assessment of fish from gas fields in the southern North Sea. *Oil Chemical Pollution* 6:263–277.

Patin, S. A. 1979. *Pollution impact on the biological resources and productivity of the World Ocean.* Moscow: Pishchepromizdat, 305 pp. (Russian)

Payne, J. F. 1989. Oil pollution: a penny ante problem for fisheries (if it weren't for erroneous perception). In *Proceedings of the 1st International Conference on Fisheries and Offshore Petroleum Exploitation (Bergen, October 1989).*—Bergen: Chamber of Commerce and Industry, 21 pp.

Picou, J. S., Gill, D. A., Dyer, C. L., Curry, E. W. 1992. Distribution and stress in an Alaskan fishing community: initial and continuing impacts of the Exxon Valdez oil spill. *Ind.Crisis Q.* 6(3):235–257.

Picken, G. 1989. UK fisheries research around offshore oil installations. In *Proceedings of the 1st International Conference on Fisheries and Offshore Petroleum Exploitation (Bergen, October 1989).*—Bergen: Chamber of Commerce and Industry.

Price, A. R. J., Robinson, J. H., eds. 1993. The 1991 Gulf War: coastal and marine environmental consequences. *Mar.Pollut.Bull.* 27:320.

Rasmussen, T., Skara, T., Aabel, J. P. 1992. Oil tainting of cod. In *Proceedings of the Conference on Quality Assurance in the Fish Industry (August 1991, Lyngby, Denmark).*

Reid, R. N., Steimle, F. W. 1978. *Offshore oil production and the United States fisheries. International Council for the Exploration of the Sea.* ICES CM 1978/E:46, 11 pp.

Reiersen, L. O., Gray, J. S., Palmork, K. H., Lange, R. 1989. Monitoring in the vicinity of oil and gas platforms: results from the Norwegian sector of the North Sea and recommended methods for forthcoming surveillance. In *Drilling wastes.*—London: Elsevier Applied Science, pp.689–702.

Semenov, A. D., Pavlenko, L. F. 1991. Pollution of the marine environment during the development of the oil and gas fields on the shelves of the Black Sea and the Sea of Asov. In *Theses of the Second All-Union Conference on the Fisheries Toxicology.*—Spb., pp.159–160. (Russian)

Shagaeva, V. G., Nikolskaya, M. P., Akimova, N. V., Markov, K. P., Nikolskaya, N. G. 1993. Study of the early ontogeneses of Volga sturgeon under the anthropogenic impact. *Voprosy ikhtiologii* 33(2):230–240. (Russian)

Scholten, M. C. Th. 1995. Thirty years of studies at the TNO Laboratory for Applied marine Research, Den Helder (The Netherlands). The challenge to marine biology in a changing world. Proceedings of the International Symposium (Helgoland, 13th–18th September 1992). *Helgolander Meeresuntersuchungen* 49(1–4):689–702.

Somerville, H. J., Bennett, D., Davenport, J. N., Holt, M. S., Lynes, A., Mahie, A., McCourt, B., Parker, J. G., Stephenson, J. G., Watkinson, T. G. 1987.

Environmental effect of produced water from North Sea oil operations. *Mar.Pollut.Bull.* 18(10):549–558.

Sorokin, U. I. 1975. Geterotrophic microplankton as a component of marine ecosystems. *Jurnal obshchei biologii* 36(5):716–730. (Russian)

Squire, J. L., Jr. 1992. Effects of the Santa Barbara, Calif., oil spill on the apparent abundance of pelagic fisheries resources. *Mar.Fish.Rev.* 54(1):7–14.

Stebbing, A. R. D., Dethlefsen, V. 1991. *Interim Report on the ICES/ IOC Bremerhaven Workshop on Biological Effects Techniques. International Council for the Exploration of the Sea.* ICES CM 1991/E:6, 28 pp.

Stebbing, A. R. D., Dethlefsen, V., Carr, M. 1992. Biological effects of contaminants in the North Sea: results of the ICES/IOC Bremerhaven Workshop. *Mar.Ecol.Progr.Ser.* 91:361.

Swan, J. M., Neff, J. M., Young, P. C., eds. 1994. *Environmental implications of offshore oil and gas development in Australia—the findings of an independent scientific review.* Sydney: Australian Petroleum Exploration Association Limited, 696 pp.

Teal, J. M. 1993. A local oil spill revisited. *Oceanus* 36(2):65–70.

Tidmarsh, W. G., Ernst, R., Ackman, R. G., Farquharson, T. E. 1985. *Tainting of fisheries resources. Environmental Studies Research Funds Report No.021.* Ottawa, Canada, 174 pp.

Vinogradov, M. E., Shukshina, E. A., Bulgakova, U. V., Serobaba, I. I. 1995. Consumption of zooplankton by jellyfish *Mneopsis* and pelagic fish in the Black Sea. *Oceanologia* 35(4):569–573. (Russian)

Volkman, J. K., Miller, G. J., Revill, A. T., Connell, D. W. 1994. Oil spills. In *Environmental implications of offshore oil and gas development in Australia—the findings of an independent scientific review.*—Sydney: Australian Petroleum Exploration Association, pp.509–695.

Waldichuk, M. 1990. Sea otters and oil pollution. *Mar.Pollut.Bull.* 21(1):10–15.

Ward, A. P., Massie, L. C., Davies, J. M. 1980. *A survey of the levels of hydrocarbons in the water and sediments in areas of the North Sea. International Council for the Exploration of the Sea.* ICES CM 1980/E:48, 17 pp.

Warwick, R. M., Clarke, K. R. 1993. Comparing the severity of disturbance: a meta-analysis of marine macrobenthic community data. *Mar.Ecol.Progr.Ser.* 92:221–232.

Wells, P. G., Bewers, J. M., eds. 1992. Progress and trends in marine environmental protection. *Mar.Pollut.Bull.* 25(1–4):117.

Wilson-Ormond, E. A., Ellis, M. S., Powell, E. N. 1994. The effects of proximity to gas production platforms on size, stage of reproductive development and health in shrimps and crabs. *J.Shellfish.Res.* 13 (1):306.

Yunker, M. B., MacDonald, R. W. 1995. Composition and origin of polycyclic aromatic hydrocarbons in the Mackenzie River and on the Beaufort Sea shelf. *Arctic* 48(2):118–129.

Zaitsev, U. P. 1970. *Marine neustonology.* Kiev, 250 pp.

Chapter 8

Environmental Management and Regulation of the Offshore Oil and Gas Industry

The environmental issues of the offshore oil and gas development have been the focus of scientific and public attention all over the world since the industry's very beginning. The possible effects of the industry 's expansion on the unique and vulnerable environment of the continental shelf have been of considerable concern. The continental shelf is already subjected to the powerful press of many other kinds of human activity. The possible ecological disturbances caused by introducing additional large-scale impacts on this environment can rapidly spread over huge areas and across any political borders and economic zones. Thus, the international significance of environmental management, control, and regulation of offshore oil and gas development cannot be overestimated.

8.1. Strategic Principles and Approaches

The environmental strategy of developing offshore oil and gas resources takes into account many factors, including the balance of current and future interests, possibilities of using alternative sources of energy (e.g., nuclear power, solar energy), natural conditions, and many other ecological, technical, and economic considerations. These ultimately define the specific decisions in relation to industrial exploitation of offshore resources in individual regions. It should be mentioned that some experts oppose the increasing scale of developing nonrenewable sources of energy (including oil and gas) in remote areas, especially in the Arctic [Gorshkov et al., 1994]. However, their arguments are usually based on the concept of limited consumption, which does not receive wide support. Most countries assume that sustainable economic development involves continued exploitation of natural resources while ensuring environmental stability.

The following conceptual principles regarding environmental protection of the sea should be applied to the oil and gas activity on the shelf:

1. **Acknowledgement of socioeconomic stipulation and expediency of developing the offshore natural resources (including oil hydrocarbons) taking into account the priority of preserving living renewable resources.** This thesis is shared even by the fishery circles in a number of countries [Buchan, Allan, 1992] despite the fact that the fishing industry is the first to suffer from neighboring with offshore oil production. Unconditional rejection and protest against the oil and gas industry's expansion on the shelf is as constructive as, for example, protesting against the bad weather. The search for a balance of interests and cooperation among the oil producers, fishermen, and environmental protection circles in the process of solving ecological problems seems to be more beneficial [Hinssen, Van der Schans, 1994]. At the same time, this general thesis should not be applied blindly to a country such as Russia. Here, due to economic, political, and resource-energetic considerations [Gorshkov et al., 1994], the advisability of expanding the oil and gas activity from the land to the offshore areas requires extremely careful and weighted consideration.

2. **Using an ecocentric (in contrast with anthropocentric) approach, which ensures the stability of natural ecosystems and supports conditions for optimal self-renewal of biological resources.** Such an approach, close to the idea of the sustainable development [GESAMP, 1991; Sdasyuk, Shestakov, 1994], gives priority to protecting renewable resources to satisfy the needs of the present and future generations. The main condition of this approach applied to the hydrosphere is to maintain the main parameters of the water ecosystems within the range of their natural variability. Such an approach guarantees not only general stability of the ecological situation in the hydrosphere but also the protection of direct human interests during any kind of water use (water supply, recreation, fishing, aquaculture, and so on). The ecocentric approach is based on an obvious and proven fact—the only way to protect the quality of the water environment is to preserve the structure and functions of water ecosystems.

3. **Using the preventive principle to control and protect the water environment, which prioritizes early detection of subtle**

anthropogenic effects (instead of documenting the already-ob-
vious anomalies) and implies taking corresponding preventive measures at international, regional, and local levels.
Preventive principle is essential for protecting the World Ocean,
which has a relatively high inertia of response to stress impacts
in comparison with the fresh water systems. From the preventive
perspective, the subtle changes in the marine environment (e.g.,
large-scale chronic contamination by oil hydrocarbons, especially PAHs,) and nonobvious effects in biota (e.g., sublethal biochemical responses) indicate the need to introduce regulatory
measures regarding the impact sources. In a broader sense, this
principle implies adopting stricter environmental requirements.
Such requirements should be enforced even if no direct proof yet
exists of the cause-effect relation between a pollutant discharge
and observed biological effect [Bewers, Wells, 1992; Patin, 1995].

4. **Using a regional approach, which takes into account the specific features of different marine basins, including their diverse climatic, social, economic, and other characteristics.** The
strategy of the World Ocean protection should be conceptually
directed toward the global level and practically implemented at
the regional level [GESAMP, 1990]. The regional approach
should be used both in the international and national practice
of environmental protection. Applying this principle within a
large country (e.g., in Russia) requires developing regional programs of oil and gas activities that take into account the specific
conditions of each region.

These conceptual principles, especially preventive approach,
have recently become the focus of major scientific discussions in
many countries and at the international level [GESAMP, 1991; ICES,
1994a; Lukyanenko, 1994; McIntyre, 1995, Patin, 1995]. The industrial
circles often resist the acceptance of these principles and approaches
[Wynne, Mayer, 1993]. This is quite understandable because their
wide recognition and practical implementation would mean revising
the whole strategy of marine environmental protection and enforcing
rigid ecological requirements for many kinds of industrial activity.
This would include the offshore oil and gas industry. Nevertheless, in
spite of the mentioned opposition and discussions, some of these
principles are already used in national and international documents
and agreements regulating industrial activity in a number of regions.

This certainly does not mean that new strategies and policies
to protect the World Ocean have become widely accepted. Many

countries are not ready for them because of pure economic reasons, such as the absence of necessary financial and material resources. However, the process of revising the conceptual approaches in this area has started. Undoubtedly, it will lead to developing a profound scientific substantiation of the marine protection and a corresponding revising of legal and legislative basis.

8.2. General Systems of Environmental Management, Control, and Regulation

The theory and methodology of optimal exploitation of natural resources and environmental management have been developing for a long time all over the world. This is one of the areas where the fundamental and applied branches of science come in close contact. At the same time, many economic, social, and psychological circumstances (e.g., the lack of funds, need for industry restructuring, management inertia) complicate the immediate practical implementation of theoretical developments. This section reflects the environmental management and control requirements for the offshore oil and gas industry and gives some idea about the major problems in this area.

Figure 82 shows a scheme of complex environmental management and regulation of developing the hydrocarbon resources on the continental shelf. It is a modification of a more general system of environmental management developed by the Joint Group of Experts on Scientific Aspects of Marine Pollution [GESAMP, 1991]. The system is designed for solving strategic and long-term problems of marine environmental protection.

The suggested scheme distinguishes three main blocks of environmental protection (planning, assessment, and regulation). Each block is constituted from several stages. At the first stage, a general (strategic) goal of the environmental protection is formulated basing on the general principles described in Section 8.1. In this case, the strategic goal involves preserving the quality of the marine environment, maintaining stability of marine ecosystems, and protecting biological resources. At the second stage, this general goal is adopted according to specific conditions of concrete regions. The priority regional goals are defined. In this case, indisputable priority should be given both to protecting valuable fisheries resources (e.g., salmon, sturgeon, and other commercial organisms) and areas where these organisms reproduce, grow, and migrate. The third, last, stage of the first block (see Figure 82), involves selecting environmental requirements to ensure achieving the priority goal.

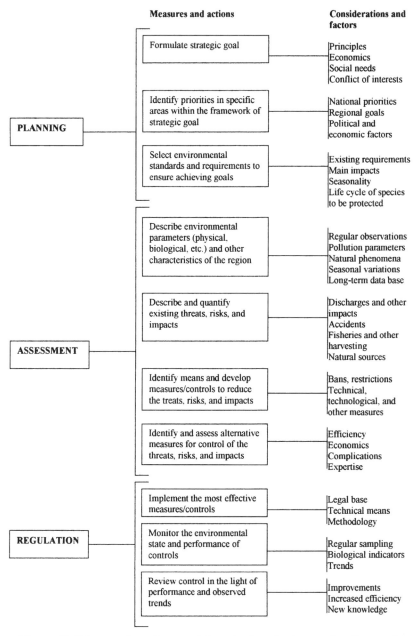

Figure 82 Scheme of environmental protection during exploration and development of the offshore oil and gas resources (from [GESAMP, 1991] with changes).

The second block of the scheme is devoted to assessing the impacts of the offshore oil and gas industry on the marine environment. It includes describing background characteristics of the region, analyzing environmental hazards and risks, developing measures to reduce these hazards and threats, and identifying alternative measures to control the adverse effects. This stage involves a wide range of activities—from conducting observations in the impact zone to developing practical measures to control the impact.

The third, concluding, block of the scheme includes implementing the most effective measures and assessing the effectiveness of the whole system of environmental control based on the results of the previous stages and monitoring data.

Other types of systems used to control and manage environmental impact during offshore oil and gas development include hazard and risk assessment. Figure 83 gives the schematic illustration of one such system. This general scheme was designed mainly to address environmental protection problems during acute impacts (e.g., oil well blowouts, pipelines ruptures, and other accidents) [Ramsay, Grant, 1992]. At the same time, this methodology has a potential to be used for assessing the hazards and regulating the risk of chronic impacts in areas of oil and gas development [Somerville, Shirley, 1992]. A possible application of risk analysis approach to ecological assessments of hazards resulting from chemical contamination of the water environment is given in Figure 84.

These briefly discussed schemes are certainly not the only possible systems used in environmental protection. However, they reflect the content and procedures of management, control, and regulation of the offshore oil and gas production. These schemes can be used as general guidelines and a framework for environmental impact assessments of the future oil and gas activities on the continental shelf. They can provide decision-makers with information needed to predict, assess, and manage possible impacts on the affected marine areas. These schemes can also be helpful in controlling, monitoring, and regulating current oil and gas activity in a concrete region.

8.3. Environmental Requirements

The central place in modern systems of environmental control and regulation belongs to ecological standards. Establishing ecological standards is a rather complicated task debated in many publications [Izrael, 1984; Patin, 1988; Lukyanenko, 1994; Kreneva, 1993]. Two

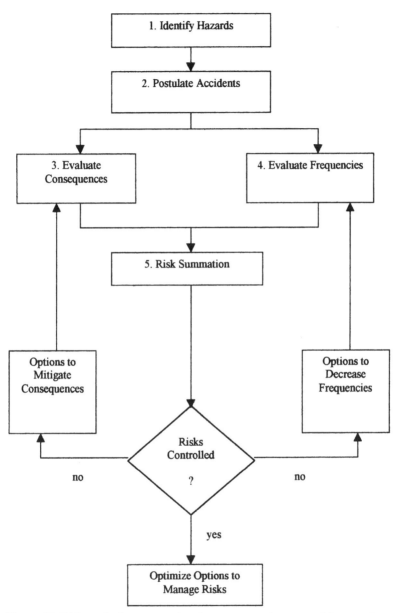

Figure 83 Risk analysis to control and manage environmental impact during offshore oil and gas development [Ramsay, Grant, 1992].

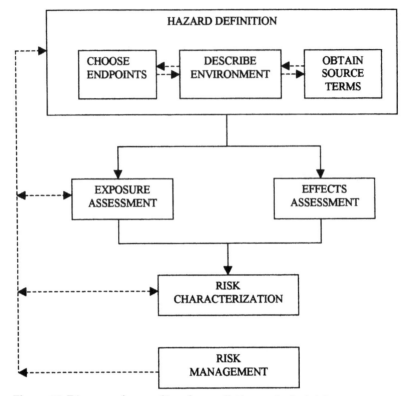

Figure 84 Diagram of a paradigm for predictive ecological risk assessments. The solid arrows represent the sequential flow of the procedure. The dashed arrows represent feedback and other constraints on one assessment process by other processes [Suter, 1994].

types of standards should be distinguished. The first type defines effluent guidelines. The second type sets water quality-based limits. Both approaches, based either on establishing a standard for either the composition and volumes of discharges or the quality of natural waters, are widely used in environmental control and regulation of pollution during offshore oil and gas production.

8.3.1. Standards and Requirements for Discharges

National and international standards and requirements for discharges from offshore industry set the limits of individual pollutants in discharges and the total volumes of major discharges. As an ex-

Table 61
Major Permitted Discharges and Potential Impact-Causing Agents
Associated with Offshore Oil and Gas Exploration and Production
[Neff et al., 1987]

Major Wastes	Permitted Discharges and Standards
Drilling cuttings	1,100 tons/explorations well, less for development well
Drilling fluids	900 tons/explorations well, 25% less for development well
Cooling water, deck drainage, ballast water	May be treated in an oil/water separator
Domestic sewage	Primary activated sludge treatment
Sacrificial anodes, corrosion, antifouling paints	May release small amounts of several metals (Al, Cu, Hg, Zn, etc.)
Produced water	Treated in oil/water separator to reduce total hydrocarbons to mean of 48 mg/l, daily maximum 72 mg/l

ample, Table 61 gives the requirements for discharges during the offshore oil and gas developments in the United States and Canada.

Establishing a standard for discharges takes into account the
performance of the best available technology and the dilution rates
of discharged wastes. Both technological capacities and dilution rates
may differ considerably in different situations. This is why such standards always have an element of some uncertainty and subjectivity.
They are usually revised with improved cleaning technologies, development of environmentally safe formulations, and obtaining new
data on the self-purification ability of the marine environment.

Current regulatory requirements for discharges of drilling
wastes into the sea, adopted in the United States, include the following main stipulations [Burke, Veil, 1995; Neff, 1998]:

- discharges to the sea require authorization and must comply
  with regulatory limits;
- concentrations of oil and oil products in discharges, determined
  during standard tests, should not exceed established standards;
- the LC_{50} values for discharge samples during 96-hour acute
  testing with mysids should not be less than 30 g/kg (a
  schematic reflection of standard procedure for mysid toxicity
  testing is given in Figure 85);

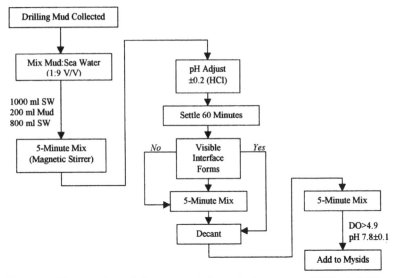

Figure 85 Preparation of the suspended particulate phase (SPP) of drilling mud for mysid toxicity testing [Dorn, Compernolle, 1995].

- the content of mercury and cadmium in the barite base of drilling fluids is restricted;
- no discharge of drilling wastes allowed in waters within 3 miles from shore (except Alaska);
- no discharge of diesel oil is allowed; and
- no discharge of free oil based on static sheen test is allowed.

Besides, at present in the United States, the possible revision of the discharge standards for drilling cuttings containing new environmentally safe synthetic-based fluids is discussed (see Chapter 6, Section 6.1.1) [Burke, Veil, 1995]. Permissible content of oil and oil products in offshore discharges (mainly in the produced waters) on the shelf of the United States and some other countries [Neff, 1993] varies from 7 mg/l to 100 mg/l. Lately, a clear tendency to enforce stricter norms has been observed. For example, in the United States, the average limit of oil concentration of 48 mg/l is going to be changed to a standard of no more than 7 mg/l for the average monthly oil content and to not more than 13 mg/l for the average daily oil concentration for discharges of produced waters within 4 miles from shore [Neff, 1993]. Such a revision of the effluent guidelines has become possible due to development of new effective technologies of waste treatment (e.g., with the help of membranes [Veil, 1992]). The final goal of the Environmental Protection Agency in the United States is to achieve zero oil content in discharged effluents or to prohibit discharges of oil-containing wastes [EPA, 1991].

This tendency to enforce stricter environmental requirements or even prohibit discharges during offshore oil and gas production (achieving the zero discharge) has been seen lately in many countries [Thorson, 1991; Johnston, 1993; Read, Reiersen, 1992; Reynaids, 1993]. In countries adjoining the North Sea, national requirements to oil production are often more rigid than international standards. For example, in Norway, the maximum permissible oil contents in discharges of produced waters and drilling cuttings were decreased to 25 mg/l and 10 g/kg, respectively, instead of regulatory limits of 40 mg/l and 100 g/kg stipulated by international agreements [Gramme, 1991; Melberg, 1991]. In the North Sea, the use of diesel fuel as a base for drilling fluids has been prohibited and the use of oil-based drilling fluids has been strictly regulated since the 1980s. Some countries (e.g., the United States, Canada, Norway) practice reinjection of oil-containing drilling wastes into the deep geological structures [GESAMP, 1993; Minton, Last, 1994]. Others (for example, India) require transportation of these wastes to coastal terminals [Marrah, Kamal, 1992]. At the same time, it must be remembered that in a few regions, waste discharges into the sea during oil and gas production are not subjected to any national or international regulation at all [GESAMP, 1993].

8.3.2. Standards and Requirements for Quality of the Marine Environment

In contrast with discharge limits, environment quality-based requirements consider site-specific environmental effects. Methods of establishing permissible limits of contaminants in the marine environment differ widely from country to country. In particular, in Russia, they involve studying the chronic toxic impacts on organisms of different trophic levels, assessing contaminant bioaccumulation and environmental persistence, and revealing their hazardous biological effects. Quantitative interpretation of combined results is used to estimate the permissible content of anthropogenic substances in the marine environment [Patin, Lesnikov, 1988].

Differences in approaches used to establish the limits and differences in criteria used to determine the contaminant hazards result in high variability of data and corresponding standards for the marine environment. The water quality-based limits are more variable than previously discussed requirements for discharges. Permissible limits established in different countries and for different purposes can vary within several orders of magnitude. Sometimes the adopted standards differ even within the same country. For example, in the United States, the water quality-based standard regarding the total oil hydrocarbon content varies from 15 μg/l to 10,000 μg/l in different states [Neff, 1993]. The limits for individual hydrocarbons also vary within a rather wide range, depending on the used criteria, as Tables 62 and 63 show. This shows once again that the very notion of a standard used in applied ecology is conditional and relative.

In Russia, two systems of water quality-based standards are used: sanitary-hygienic and fisheries maximum permissible concentrations (MPC). These regulate the chemical composition of the water environment from the perspective protecting human health and from the fisheries interests, respectively. Both systems are rather large, and at present, each of them includes over 1,500 standards. Although these systems are sometimes criticized [Volkov et al., 1993; Kreneva, 1993; Levich, 1994], they are widely used for water protection in Russia.

The officially adopted methodology of establishing the fisheries standards is based on a complex assessment of the environmental hazard of pollutants for all main biotic groups and trophic levels in water bodies [Patin, Lesnikov, 1988]. The fisheries MPC for the ma-

Table 62

National Water Quality Criteria for Aromatic Hydrocarbons Found in Crude and Refined Petroleum (μg/l) [Neff, 1993]

Chemical	Acute Criterion	Chronic Criterion	HHC*
Benzene	5,100	700	40
Toluene	6,300	5,000	424,000
Ethylbenzene	430	No value	328,000
Naphthalene	2,350	620	No value
Acenaphthene	970	500	No value
Pyrene	No value	10	No value
Fluoranthene	40	16	54
Total PAHs	300	No value	0.031**

Notes: *—HHC, Human Health Criterion based on fish consumption alone; **—Human Health Criterion for total PAHs (fish consumption) is based on the carcinogenicity of benzo(a)pyrene which is absent or only a trace component of the PAH fraction of crude oil.

Table 63
Preliminary No Observed Effect Concentrations (NOEC) of Some
PAHs in Seawater and Bottom Sediments [ICES, 1994]

Polycyclic Aromatic	Concentrations in Seawater (μg/l)	Concentrations* in Bottom Sediments (mg/kg)
Naphthalene	1–10	0.1–0.10
Anthracene	0.005–0.05	0.001–0.01
Fluoranthene	0.05–0.50	0.01–0.10
Benzo(a)pyrene	0.01–0.10	0.01–0.10

Note: *—Under C_{org} levels about 1% of the wet sediment weight.

jority of tested substances usually have lower thresholds of permissible content than the sanitary-hygienic standards. They can be used as ecological (ecofisheries) standards for the quality of natural waters.

Ecological standards are the most important element of regulating anthropogenic impact on the hydrosphere. Introducing these standards, in particular in areas of offshore oil and gas developments, includes a combined application of both types of standards discussed above (requirements for discharges and water-quality based limits). This approach makes establishing the limits of the maximum permissible discharges (MPD) possible.

Several very different methods are used to calculate the maximum permissible discharges (MPD). The most reliable ones are based on the concept of assimilation capacity of water ecosystems [Izrael, 1984]. They take into account not only the rate of discharge dilution (e.g., due to the currents, turbulent mixing, and water exchange) but other factors as well. These factors include microbial degradation, physicochemical transformations, bioaccumulation, sedimentation, and other processes that affect the pollutant distribution and the self-purification of the water environment.

In spite of certain criticism expressed regarding the concept of assimilation capacity and the maximum permissible discharges (MPD) [Jackson, Taylor, 1992; Laskorin, Lukyanenko, 1993; Bagotski, 1995], methods based on these notions are widely used in Russia and some other countries. They are supplemented by the MPC system to control and regulate anthropogenic impact on the water ecosystems. This methodology stipulates the gradual systematic decreasing of the MPD in accordance with technological progress and economic capabilities.

The previously discussed general principles and approaches are applied in environmental management and regulation of the offshore oil and gas industry's impact on the marine environment (see Section 8.2). At the same time, it should be noted that the systems of standards as well as their applications to offshore oil production need further development and unification both at the national and international levels.

Besides the water quality-based limits discussed above, it would be logical to include materials regarding the standards for bottom sediments. Their parameters reflect the consequences of offshore oil and gas production in a more reliable cumulative form than the seawater characteristics (see Chapter 7). Unfortunately, establishing ecological requirements for the quality of bottom sediments is presently only at its initial stage of development. No national or international standards of such kind have been established thus far. The high variability of the composition and properties of bottom sediments and the even higher variability of species composition, structure, and abundance of the benthic communities complicate developing requirements applicable to different conditions and regions.

Lately, some steps in this direction have been taken (including this book–see Chapter 4) [MacDonald, 1993; Thain et al., 1994; ICES, 1994b; GESAMP, 1995]. In particular, attempts have been made to estimate maximum levels of oil hydrocarbons and other pollutants in bottom sediments that do not cause toxic effects in benthic fauna [Long, Morgan, 1990; MacDonald, 1993]. Chapters 4 and 7 discussed these threshold levels. Figure 43, Tables 57 and 63 gave some idea about their magnitude.

8.3.3. Ecofisheries Requirements

Since ancient times, coastal fisheries and marine aquaculture have been supplying people with valuable food and raw materials. The loss (or tainting) of these products would cause irreparable harm for humankind and become a catastrophe for many coastal countries. At the same time, the fisheries are the first to suffer from the growing press of anthropogenic impact on the coastal zone. This is why fisheries interests are very close to the interests of protection of the marine ecosystems and maintaining the quality of the marine environment.

Ultimately, fisheries requirements for the offshore oil and gas industry (as well as for any other industrial activity) often coincide with environmental protection standards. Tables 64 and 65 attempt

Table 64
Main Ecofisheries Requirements to the Offshore Oil and Gas
Developments in Russia

Stage and Type of Activity	Nature and Content of Requirements	Notes Regarding Russian Practice
Giving permits and licenses for the rights to develop offshore oil and gas resources	Participation of fishing industry in making decisions about licensing the right to explore and develop offshore oil and gas, choosing the areas of seismic surveys, drilling, laying pipe-lines, etc.	Current procedures need improvements
	Limitations (up to banning) of activities in areas of high fisheries value and high ecological vulnerability	Ecofisheries mapping of the shelf is needed
Management planning process	Including assessments of potential hazards and impacts on biological resources and fishing in the plans of future developments; par-ticipation of fishing industry in planning process	Is not included in current procedures
Geophysical exploration	Coordination of areas and periods of explorations; fisheries inspection during seismic surveys	Is not included in current procedures or needs to be revised
Drilling and other oil production operations	Quality of the marine environment should meet the standard requirements at the distances beyond 250–500 m in any direction from the drilling sites	Is included in official documents
	Limitation or prohibition of discharges	–"–

Table 64 (*continued*)

Stage and Type of Activity	Nature and Content of Requirements	Notes Regarding Russian Practice
	Conducting of regular ecofisheries monitoring	Standardized methodological base is needed
	Compensation of fisheries damages and losses	Legal base needs improvement
Pipeline emplacement	Coordination of the places and periods of structure emplacements; the structures should not interfere with migrations of commercial species and with fishing	Standards are absent or need improvement
Termination of oil production	Removal of platforms, pipelines, and other objects interfering with fishing and shipping	Included in international agreements

to summarize these requirements taking into account the ecofisheries standards adopted in Russia and the considerations discussed above.

Broadly speaking, all of the ecofisheries guidelines shown in these tables can be divided into two groups. The first group includes organizational, procedural, and legal regulations. The second one unites the environmental standards. Ultimately, they all are aimed toward protection of water ecosystems and biological resources.

As Table 64 shows, a number of requirements are not supported by current procedural and legal regulations in Russia. Many decisions, which are critically important for the fishing industry, are often made without its participation (e.g., decisions regarding licensing, project expertise, coordination, developments, and so on). Such neglecting of fisheries priorities (which are often close to the environmental protection interests) may lead to long-term negative consequences beyond the limits of pure fisheries losses.

One radical way to solve the complex problem of balancing the interests of oil and fishing industries on the shelf could be ecofisheries mapping of the coastal areas. This would define the most productive and most valuable regions for fisheries. Development of oil and other resources should be restricted or totally prohibited in these regions.

In spite of the integrity of natural processes occurring in the coastal areas, the shelf is not homogeneous in its geomorphologic

Table 65
Ecofisheries Requirements to Surface Waters in Areas of Offshore Oil and Gas Development

Parameters	Standards and Requirements
Suspended substances	No more than 0.25–0.75 mg/l above natural concentrations at the distance of no more than 250–500 m from discharge point; discharges of suspense with sedimentation rate >0.25–0.4 mm/sec are prohibited
Floating components	Water surface should be free from oil slicks, tar balls, and other floating components
Color	Water should not have unnatural color
Odor, flavor	Water should not impart unnatural odors and flavors in fish and other marine organisms
Hydrogen-ion concentration	pH values should stay within 6.5–8.5
Levels of dissolved oxygen	In the winter (ice) period—no less than 4–6 mg/l, in the summer—no less than 6 mg/l
Water toxicity, biochemical oxygen demand (BOD)	At discharge site, sewage water should not cause acute toxic effects; water should not cause chronic toxic effects within 250 m from the site
Pollutants:	Concentrations should not exceed fisheries MPC beyond 250–500 m from the place of discharge:
Oil and oil products	0.05 mg/l
Surfactants	0.05–0.5 mg/l
Dispersants	0.005–0.25 mg/l
Other components	See Appendix

and biological characteristics. Different parts of the shelf can play different roles in maintaining the reproduction and abundance of commercial species. From the fisheries perspective, the most valuable areas include places where commercial species reproduce, develop, and migrate, zones of concentration of their feeding objects (especially in benthos), and locations with some other parameters defining the bioproductive potential of coastal ecosystems.

The methodology of distinguishing, assessing, and rating such shelf areas has not been developed yet. However, fishery science and marine ecology should consider this task as their priority. These disciplines have already accumulated numerous results from long-term observations of commercial organisms and their environment. These data should be used for ecofisheries mapping that provides carto-graphic reflection of summarized information regarding the state, distribution, and reproduction of biological resources.

A promising approach to this problem is the application of gen-eralized indexes describing the states of habitats and biota in differ-ent areas of the continental shelf. A similar method is used, for example, in economics, cybernetics, or global forecasting, for de-scribing the state of complex systems for practical decision making. Such an approach suggests creating a set of basic indexes, which in this case may be abundance and biomass of plankton and benthos, stock and catch of commercial species, primary and secondary pro-duction, and others. The procedure involves determining the scale of their natural variability using the estimates of limits for each index, rating the indexes according to their relative importance, and calcu-lating generalized indexes. These generalized indexes are based on all available data. In the integral form, they reflect all individual indexes for each area unit of the shelf outlined by a special grid on the map.

Mathematical procedures for calculating such indexes which use multidimensional statistical analyses and function of desirability are well known and widely used for solving similar tasks [Adler et al., 1976; Maksimov, 1991]. Implementation of such an approach could result in creating regional maps of the continental shelf. These would clearly display the relative ecofisheries value of different areas in the chosen scale of basic indexes. Such maps would help to justify fisheries and environmental protection requirements under any kind of anthropogenic impact on the shelf (including the areas of oil and gas developments). Lately, cartographic methods and approaches are becoming more common in environmental protection science and practice [Komedchikov et al., 1994; Smirnov, Shumova, 1994].

Other practical tasks connected with developing oil and gas re-sources on the Russian shelf include:

- development of methodological guidelines to assess the envi-ronmental impact in areas of oil and gas production;
- development of methods to calculate the fisheries damage and losses caused by different impacts of the offshore industry;
- substantiation of the general scheme and regional applications of ecofisheries monitoring in areas of developing oil and gas fields; and
- revision and improvement of requirements regulating the oil and gas activity in valuable fisheries areas.

The foundation for solving all these tasks has been already described in many publications. However, most of them are too general and their results have never been used with the purpose of environmental control of offshore oil production.

Table 65 shows some requirements for the quality of surface waters defined by official Russian documents, primarily the "Rules for the surface water protection" [1991]. Although most given standards were established based on experimental and field studies in freshwater, their extrapolation to the marine environment seems to be justified within the range of uncertainty and variability of modern estimates of ecological standards. At the same time, the maximum permissible concentrations (MPC) of many pollutants were specially determined for the marine environment.

8.4. Problems of Ecological Monitoring

8.4.1. General Features and Approaches

The concept of ecological monitoring has been the center of attention of many scientific disciplines for a rather long time [Izrael, 1984; Shvebs, 1993]. Such intent interest to this issue is quite understandable. Ecological monitoring plays (or should play) the same role in environmental science and practice as diagnostic plays in medicine. Such an analogy can be used with one stipulation. Diagnosing the health of natural ecosystems is still searching for the ways, approaches, and methods to assess and predict the symptoms of ecological stress.

This is true regarding the varieties of monitoring devoted to revealing anthropogenic changes in the marine ecosystems. From one side, the wide scope of research and practical activity at the national and international levels has resulted in developing numerous theoretical and methodological approaches—from detecting subcellular responses in marine organisms to remote observations of marine pollution (including systems of space control). From another side, many programs of monitoring are still not very effective. Often, they do not justify the expense to conduct them.

This section will outline those features of marine ecological monitoring that should be taken into consideration to solve monitoring problems in areas of offshore oil and gas development.

First, marine ecological monitoring is becoming much more than just a system to collect information about the changes in natural parameters. More and more often, it is considered a main element to control and manage those activities that affect (or could affect) the ecological situation. Monitoring turns into a valuable regulatory tool. It becomes a link in the chain of negative feedback between the intensity of anthropogenic impact and the state of the ecosystem. The results of monitoring have begun to directly define different managerial (technical, normative, legal, and so on) decisions. Obviously, such application of monitoring requires a clearly defined purpose.

The second feature of modern systems of marine monitoring involves the growing role and contribution of biological methods to assess the ecological situation in the sea [ICES, 1995; GESAMP, 1995]. Not long ago, chemical analyses dominated the array of monitoring tools. With all their advantages, they did not give answers to the main questions about the degree, nature, and mechanisms of stress processes in the ecosystems. Lately, biological methods have been actively introduced in many countries and international programs. Biological methods include both traditional hydrobiological techniques to assess the state of marine communities and new ecotoxicological methods. These new methods are based on experimental assessments of test organism responses to the presence of hazardous substances and stress factors in the marine environment. A combination (triad) of chemical, hydrobiological, and ecotoxicological methods increases the capabilities of ecological monitoring and makes it more informative. In some cases, for example if any obvious changes of biological parameters are absent, chemical control becomes unnecessary. This, of course, reduces the cost of monitoring programs.

Methods used for an early diagnostic of stress responses in marine organisms occupy a special place in biological monitoring. These methods are based on measuring molecular and cellular effects (e.g., biochemical, immune, histological) under low levels of impact that chemical analysis cannot determine. At present, they have already started playing a noticeable role in the practical implementation of preventive approach to protect marine ecosystems and biological resources [ICES, 1995]. Previous chapters have discussed some examples and results of using these methods for assessing the ecological situation, including in areas of offshore oil and gas production.

One of the latest innovations in ecological monitoring is supplementing it by controlling the sources of pollution. It uses not only traditional chemical analyses but new biological methods based on waste discharge testing. Controlling the sources of pollution helps to understand the cause of the actual or potentially possible adverse changes in the marine ecosystems and to develop preventive technological measures (especially at the stage of discharge treatment). It increases the role and efficacy of monitoring in regulation and control of the anthropogenic impact on the marine environment [Patin, 1991].

All these general features and characteristics are typical to some extent for very different programs of ecological monitoring (national, regional, international, and so on). These programs have been implemented almost everywhere, primarily in the marine coastal zone. The scale of such programs differs a lot—from local observations of isolated impacts to global monitoring.

Ecological monitoring in areas of offshore oil production is usually conducted at the local level. However, in some regions (e.g., the North Sea, the Gulf of Mexico, and the Persian Gulf), oil and gas developments have the large-scale and long-term nature. One can expect that in these areas, numerous local impacts can form zones of vast, regional disturbances. Hence, regional systems of monitoring seem to be quite justified in such situations.

Figure 86 shows the conceptual framework for ecological monitoring in the system of environmental protection. This scheme shows that the results of observations, conducted with the use of biological and chemical methods and criteria, serve as a foundation for developing regulatory measures in the industrial (technological) sphere to decrease the hazard from effluents and other discharges.

The process of monitoring includes several stages (see Figure 86):

- First, actual and potentially possible hazards from impact sources are identified. Both the information about the studied area and available data about similar situations in other places are taken into consideration.
- At the second stage, regular observations of marine biota are conducted. Their purpose is to reveal and quantitatively assess biological responses in organisms, populations, and communities.
- Next, the cause-effect relationships between documented biological effects and impact factors are described. At this stage, the combination of chemical and biological methods is especially important.
- The following stage involves assessing the total impact on the marine environment and biota, including the impacts on commercial species and biological resources in general.
- The final stage is based on all combination of monitoring results. It entails evaluating different regulatory measures oriented toward appropriate industrial-technological changes. These include correcting the standards for discharges and taking restrictive and preventive measures.

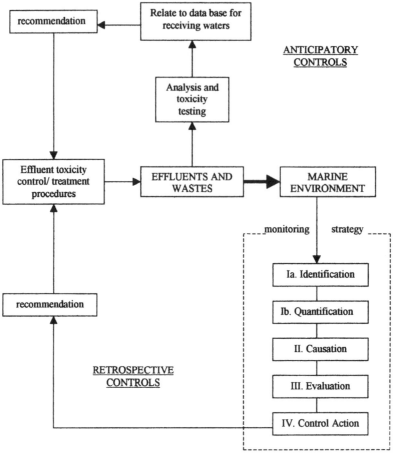

Figure 86 Conceptual framework for ecological monitoring to control and manage the quality of the marine environment [ICES, 1994].

8.4.2. Methodology and Methods

In spite of the fact that traditional chemical methods still dominate in marine monitoring, modern approaches give priority to biological methods for revealing anthropogenic changes. This is the latest strategic orientation of marine monitoring in the United States, Canada, and coastal European countries that implement large national and international programs of marine research and observations [ICES, 1995]. Such a methodological approach is quite valid due to a number of reasons. First of all, irrespective of the sources and factors of an environmental impact, the final and integral result of this impact is always manifested as different responses (effects) at the levels of marine organisms, populations, and communities. All ecological criteria ultimately have a biological basis, even though at present they are mainly expressed in chemical categories (MPC, norms of permissible discharges, and others).

A wide use of chemical parameters explains the overestimation of the importance of traditional chemical methods in the monitoring programs, in particular, their dominant position in the systems of discharge guidelines and environment quality-based requirements. At the same time, a simple comparison of the number of pollution components that actually enter the marine environment (tens of thousands of substances, including hundreds of components during oil and gas production) and the number of substances that can be reliably determined by chemical methods (not more than a hundred of the most common toxicants) shows that chemical control does not include the main portion of anthropogenic products causing ecological problems in the natural waters.

However, biological methods with all their advantages, also have drawbacks and limitations. Although sensitivity and selectivity of these methods are rather high at the primary levels (molecular and cellular), they decrease at the higher levels (populations, ecosystems, and communities) of the biological hierarchy. Figure 87 shows this in a simplified form. At the same time, detecting effects at the higher levels is essential to reveal an integral biological response to the combined impacts of all factors.

Due to compensatory and adaptive processes that neutralize the stress responses at each level shown in Figure 87, detecting the first signs of adverse effects is possible mostly at biochemical and cellular levels. The combination of chemical and biological methods in such cases seems to be the most useful and effective way to reveal actual mechanisms and causes of different symptoms of ecological stress.

Published materials on the methodology of ecological monitoring in areas of offshore development are very limited [Grogan, Blanchard, 1992; Emerson, 1994]. They indicate that no commonly accepted systems are available to organize the monitoring observations in such areas. In different countries, conditions, and situations, these issues are approached differently. The decisions regarding monitoring programs depend on established traditions of environmental protection, regulatory requirements, legislative base, and other specific circumstances. In most cases, two types of independent systems of

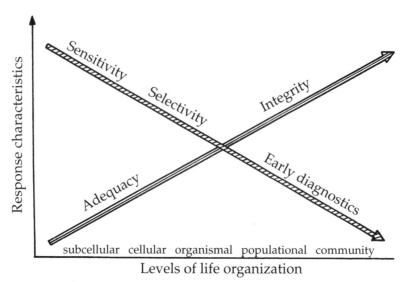

Figure 87 Changes in major characteristics of biological monitoring methods depending on the level of life organization in the sea [Addison, 1991].

ecological control operate simultaneously. Sometimes they supplement each other. The first type is conducted by the oil companies and operators themselves. It mainly controls the volumes and composition of wastes discharged into the sea. The other type is conducted by independent research organizations and experts. It usually includes more complex systems of environmental observations, uses a wider array of chemical and biological methods, and generally meets modern requirements for marine ecological monitoring.

One of the possible variants of the system approach to monitoring is realized on the British shelf of the North Sea. Table 66 shows it in matrix form. The system is based on a numerical rating of the severity of impact on environmental components at different stages of oil and gas production. Such assessments are based on regular observations of both stress factors (e.g., discharges) and affected environmental parameters. Their results serve as the foundation for conclusions about the nature, degree, and scale of adverse effects in areas of oil production and help to make decisions regulating the environmental impact. These materials may also be used as the basis for revising and improving the methods and structure of monitoring itself.

Table 67 summarizes the chemical and biological parameters of biotic and abiotic components of the marine ecosystems that are used

Table 66
Summary Impact Matrix for Brae "A" Development in the North
Sea (from [Grogan, Blanchard, 1992] with changes)

Kind of Activity and Impact Factor	Objects of Monitoring* and Degree of Impact**									
	A	W	S	P	BFC	BC	F	FS	B	M
Exploratory Phase:										
Physical presence		0			+	0	0	0	0	0
Seawater uptake				0						
Atmospheric emissions (well fluid burning; exhaust and venting)	0							0		
Drilling discharges* (mud on cuttings; other drilling materials)		0	1	0	0	1	0	0		
Other discharges (sewage/canteen wastes; laundry/ showers, etc.)		0	0	0	+	0	0	0		
Accidental spills	1	1	1	0	0	0	0	0	0	0
Hook Up and Commissioning:										
Physical disturbance (anchoring)		0	0		1					
Construction/ commissioning discharges (test fluids/paints; dusts/wash downs)	0	0	0	0		0				
Other discharges (sewage/canteen wastes; laundry/ showers, etc.)		0	0	0	+	0				
Operational Phase:										
Physical presence		0			+	0	0	1	0	0
Seawater uptake				1						
Atmospheric emissions (flaring; atmospheric renting; exhausts fugitive emissions)	0							0		

Table 66 (*continued*)

Kind of Activity and Impact Factor	Objects of Monitoring* and Degree of Impact**									
	A	W	S	P	BFC	BC	F	FS	B	M
Development drilling discharges*** (cuttings; muds on cuttings, particularly oil base)	1	2	0	0		2	0	0		
Produced water discharge*	2	0	2	0		0	0	0		
Other discharges (sewage/ canteen wastes; laundry/showers; cooling water)	0	0	0	+		0	0	0		
Accidental spills	0	0	1	0		0	0	0	0	0
Abandonment:										
Physical removal					1			+		
Discharges	(Other impacts to be evaluated once exact abandonment strategy is known)									

Notes: *—Objects of monitoring: A—air; W—water; S—sediments; P—plankton; BFC—bio-fouling communities; BC—benthic communities; F—fish; FS—fisheries; B—birds; M—mammals. **—Scale of impact: +—positive impact; 0—nil to negligible; 1—minor; 2—moderate; 3—major; 4—very severe; 5—catastrophic. ***—Chemicals associated with discharges included, i.e. all components.

the most often during ecological monitoring in different regions, including areas of the offshore developments. Table 68 shows the possibilities of using marine organisms for bioassays in major marine biotopes (water, surface microlayer, bottom sediments, and their fractions).

Certainly, simultaneous use of all these parameters and indicators would be unrealistic. However, the attempts to integrate chemical, biological, and experimental ecotoxicological methods and include all major biotopes and groups of marine organisms in the sphere of observations are clearly seen in many modern monitoring programs [Chapman et al., 1991; Gray, 1992; Morrisey, 1993; Martin, Richardson, 1995; ICES, 1995; GESAMP, 1995].

During environmental monitoring and control in areas of offshore oil production, the preference is given to the analysis of bottom

Table 67
Recommended Objects and Parameters for Multiple Ecological
Monitoring in the Areas of the Offshore Oil and Gas Fields
Developments

Discharges, Biotopes, and Biota Groups	Parameters (Levels, Concentrations, Effects)	Nature and Purpose of Collected Information								
		1	2	3	4	5	6	7	8	9
Liquid and Solid Discharges (drilling muds, cuttings, produced waters)	Volumes and regime of discharges; chemical composition (oil and oil products, PAHs, heavy metals, etc.)		+	+	+				+	+
	Toxicity (general, fractional, componential)		+			+	+	+		+
Water Column	Temperature, salinity, structure of water column, currents, turbulence	+		+					+	
	Suspended substances	+	+	+		+			+	
	Slicks and floating substances		+							
	Dissolved oxygen	+	+							
	Biochemical oxygen demand	+	+							
	Oil, oil products, PAHs, components of drilling and other wastes, heavy metals	+	+	+		+		+	+	+
	Toxicity	+	+	+		+		+		+
Surface Microlayer (about 1 mm)	Oil, oil products, PAHs heavy metals					+	+	+		
	Toxicity (biotesting)					+	+	+		
Bottom Sediments	C_{org}, size particles	+								
	Oil, oil products, barium, heavy metals	+			+		+	+	+	+
	Toxicity				+		+	+	+	+
Bottom Sediments (interstitial waters, water eluate)	Toxicity				+		+	+	+	+
Plankton: Bacterial communities	Abundance, biomass, domination of oil-degrading bacteria	+							+	

Table 67 (*continued*)

Discharges, Biotopes, and Biota Groups	Parameters (Levels, Concentrations, Effects)	Nature and Purpose of Collected Information								
		1	2	3	4	5	6	7	8	9
Phytoplankton and zooplankton	Biomass, domination, and other structural characteristics	+				+	+			
Ichthyoplankton	Survival, morpho-logical and other developmental anomalies					+		+	+	+
Necton:										
Juvenile fish	Abundance, distribution, physiological condition							+		+
Ichthyofauna	Abundance, species composition, stock, and other parameters							+		+
Benthos:										
Macrobenthos, meiobenthos, epybenthos	Biomass, abundance, species composition, diversity, and other communities characteristics	+					+	+	+	+
Bivalves and crustaceans	Accumulation of PAHs, heavy metals, and other toxicants	+			+		+	+		
	Symptoms of stress and pathology						+	+		+
Bottom fish	Histological, bio-chemical, and other subcellular changes						+	+		+
	Signs of diseases and pathologies						+	+		
	Accumulation of PAHs, heavy metals, and other toxicants	+	+				+	+	+	
	Oil odors and flavors		+							

Notes: 1—background characteristics; 2—accordance to requirements; 3—shape and size of the direct impact zone in the water column (zone of mixing); 4—shape and size of the direct impact zone in bottom sediments; 5—actual and potentially possible stress effects in plankton; 6—the same in benthos; 7—actual and potentially possible consequences for bioresources; 8—spatial and temporal features of ecological disturbances; 9—information needed to make regulatory decisions.

Table 68
Groups and Species of Marine Organisms and Their Test-responses Recommended for Biotesting and Systems of Complex Ecological Monitoring

Groups and Species of Test Organisms	Tested Environment	Test-response and Parameters
Heterotrophic microplankton		
Bacteria	Water, SML*	Changes in BOD dynamics, species domination, rates of substrate degradation; mutagenic effects
Protozoa (*Stylonichia mytilus, Tintinnopsis biroidea, Noctiluca seintillans, Cristigera*)	Bottom sediments**, cuttings, liquid discharges	Reduced survival, reproduction, and growth; developmental anomalies
Unicellular algae regional dominants (*Coscinodiscus, Ditylum, Gyrodinium, Exuviella,* etc.)	Water, liquid discharges	Changes in the rate of cell division and cell abundance; disturbances of photosynthesis and fluorescence intensity; anomalies of pigment compositions
Macrophytes (*Laminaria, Macrocystis pyrifera,* etc.)	–"–	Changes in the growth rate; disturbances of zoospore settling; morphological and electrophysiological anomalies
Zooplankton filtrators (*Acartia, Euretymora, Tigriopus, Calanipeda, Artemia salina,* etc.)	Water, SML, liquid discharges	Reduced survival and fertility; disturbances of reproduction, behavior, and trophic activity; morphological and other anomalies
Fish (eggs, larvae, juveniles) (*Salmo gairdner, Trachurus trachurus, Limanda limanda, Gadus morhua, Scophthalmus maximus, Sprattus sprattus, Spicara smaris,* etc.)	–"–	Increased mortality and frequency of morphological anomalies; disturbances of feeding, growth, respiration, behavior, and physiological parameters
Macrobenthos (adults, embryos, larvae) (*Mytilus edulis, Crossostrea gigas, Macoma, Echinocardium, Arenicola,* etc.)	Water, SML, bottom sediments, liquid discharges, cuttings	Reduced survival; disturbances of reproduction; delays in growth; behavioral, physiological, and other deviations from the norm

Notes: *—SML—surface microlayer (about 1 mm). **—Both samples of bottom sediments and their liquid fractions (interstitial waters and eluates) should be tested.

sediments and benthic communities. Biological, chemical, and toxicological techniques are used to reveal changes in the affected areas [Chapman et al., 1991; Daan et al., 1992]. These methods include observations of structure modifications in the benthic communities [Gray, 1992], detecting histological pathologies in benthic invertebrates [Ellis et al., 1994], bioassay of bottom sediments [MacDonald et al., 1993], and other methods and their combination [ICES, 1994].

Certainly, no standardized and universal methodological system is available to conduct ecological monitoring suitable in all regions. Everything depends on the specific situation and the ultimate goals of monitoring. Long-term experience [ICES, 1994; 1995] indicates that the effectiveness of ecological monitoring is often determined by the choice of adequate methods. Such methodological requirements as reliability and comparability of collected data, right selection of criteria for their interpretation, and applying the system of statistical assessments are critical to any program of ecological monitoring.

8.5. International and National Aspects of Regulation

At present, over 70 international conventions and agreements are directly concerned with protecting the marine environment. However, none of them is specially devoted to regulating offshore oil and gas development exclusively. In practice, these problems are solved either at the national level or within the framework of international conventions. The latter include:

- United Nations Convention on the Law of the Sea (1982). This declared the general principles and requirements for any activities that exploit the ocean's resources, including oil, gas, and fish. It served as a foundation for the concrete guidelines and standards developed by the International Maritime Organization [IMO, 1989] about removing offshore installations and structures after termination of the oil production activities (see Chapter 3, Section 3.6.1).

- International Convention for the Prevention of Pollution from Ships (1973) modified in 1978 (MARPOL, 73/78). It defines fixed or floating platforms as vessels and includes them into the sphere of regulations for discharges of oil and other hazardous substances into the sea.

- Convention for the Prevention of Marine Pollution by Dumping of Wastes and other Matter (1972) is similar to the previous convention. It regulates certain types of dumping in areas of offshore oil and gas production.

- International Convention for Prevention of Pollution of the Sea by Oil (1954, amended 1962, 1969 and 1971). This mainly regulates activities connected with tanker oil transportation.

A number of other global and regional agreements to a different extent include (and sometimes duplicate) the rules and restrictions regarding development of offshore oil and gas resources. For instance, the International Convention on Oil Pollution Preparedness, Response and Cooperation adopted by IMO in 1990 initiated considerable efforts in many countries to allocate funds, establish special agencies, and develop technical means and methods for the rapid and effective response to oil spills in the sea.

Regional agreements define the requirements for the offshore developments in the basins of Baltic, Black, Mediterranean, and other seas. Among them, the Convention for the Protection of the Marine Environment of the North East Atlantic (OSPAR Convention, 1992) deserves a special mention. Regular ecological monitoring and numerous activities to prevent coastal pollution are conducted within the framework of this agreement. This convention pays increased attention to controlling and regulating the oil and gas production on the shelf. It introduces especially strict requirements for this kind of activity [Read, Reiersen, 1992; ICES, 1995]. Unfortunately, Russia does not participate in this agreement.

A wide spectrum of regulatory (e.g., legal, legislative, administrative) measures and requirements are used at the national level in different countries to reduce ecological hazards and risks during offshore oil and gas production. Countries such as the United States [Warren, 1994], Canada [Wells, Rolston, 1991], Great Britain [Grogan, Blanchard, 1992], Norway [Gray, 1992], and Australia [May, 1992] have accumulated especially large experience in this area. These countries have developed a solid legislative base and strict regulatory programs. They also have corresponding structures to ensure the compliance of offshore activities with environmental requirements. Significantly, revising the regulatory standards in these countries is practically always oriented toward making them more rigid and strict (see Section 8.3).

Many countries are cautious about admission of the oil and gas industry to the developing resources on the shelf. They try using the preventive approach to regulate and control the offshore industry's activity. For example, in the United States, leasing and operating activities on the continental shelf are subjected to about 30 federal laws [MMS, 1995]. The rigid system of licensing oil developments requires the conducting of a detailed analysis of the ecological situation in each area of the continental shelf and assessing its vulnerability to the possible impact [Emerson, 1994; Warren, 1994]. If the proposed activities would cause serious harm to the marine environment and its living resources, the oil and gas exploration, development, and production plan in the area is denied. This procedure provided the basis for the U.S. Congress to declare a moratorium on drilling operations in the shallow and fish-rich Georges Bank [NRC, 1991]. Norway has also introduced legislative restrictions on oil exploration and production in the highly productive shelf areas [Gray, 1992].

In some countries, cartographic, geographic-informational, ecological-economic, and other systems for reflecting the ecological situation, vulnerability, and fisheries value of different areas of the shelf are developed. They are used for reaching balanced decisions about exploiting the marine resources, including oil and gas, in the coastal zone. These methods are applied, for example, in Canada [Ricketts, 1992], New Zealand [Tortell, 1992], and Argentina [Rodriguez, Vila, 1992]. Such approaches are quite similar to ecofisheries cartography of the continental shelf described in Section 8.3.3. They deserve serious attention from specialists in ecological and fishery science as well as from environmental protection agencies and organizations. Their implementation will help to ensure effective and well-founded protection of the coastal zone.

Conclusions

1. The main principles of modern environmental protection strategy that should be taken into account during ecological regulation of offshore oil and gas development include:

- ecocentric orientation of developing natural resources, i.e., the priority is given to ensuring the stability of natural ecosystems and self-reproduction of biological resources; and
- preventive orientation of environmental protection, i.e., the priority is given to measures that would prevent developing the degradation symptoms in the ecosystems. This is especially important for protection of the World Ocean, which has high accumulation ability and high inertia of response.

2. Systems of environmental control and management of the offshore oil and gas industry should include the following stages:

- setting goals and priorities;
- assessing the background state of the environment;
- establishing the regulatory base for its protection;
- identifying environmental hazards at the different stages of the offshore development,
- evaluating the possible consequences and risks during different impacts;
- conducting ecological monitoring; and
- implementing optimal managerial decisions based on a combination of all available information.

3. National and international environmental requirements for the offshore oil and gas industry include two groups of standards. The first group regulates the volumes and composition of discharges into the sea. The other one defines the limits of changes in the quality of the marine environment. The common practice of establishing ecological standards and regulating anthropogenic impact on water bodies uses methodology based on the concept of assimilation capacity of the water ecosystems.

4. Ecofisheries water quality-based standards and discharge guidelines are the most compatible with the environmental requirements to protect the marine ecosystems and biological resources. One promising direction to balance the interests of oil and gas and fisheries activities on the continental shelf is based on the ecofisheries mapping of coastal areas. This distinguishes the most productive and ecologically vulnerable areas where oil and gas activities should be restricted or totally prohibited.

5. Lately, many national and international programs of marine ecological monitoring have introduced a number of new features that should be taken into account during environmental control of offshore oil and gas developments. These features include:

- monitoring observations are becoming the main tool in regulatory systems controlling the sources of environmental hazard;
- the role of biological methods (especially methods of bioassay and early stress diagnostic) for assessing the state of the marine environment is on the increase; and

- the general scheme of monitoring has started including control of pollution sources (effluents, solid waste) with the help of both chemical and biological (toxicological) methods.

6. International and national requirements for the offshore oil and gas industry include a variety of environmental standards and criteria. A typical feature of the changes, recently introduced in many countries, involves stricter regulations. These regulations prohibit offshore activities in certain areas, include more rigid standards for discharges (up to zero discharges), and require broad activities to assess the risks and hazards of offshore oil and gas development.

References

Addison, R. F. 1992. Detecting the effects of marine pollution. In *Science Review 1990–1991.*—Dartmouth, Nova Scotia, pp.9–12.

Adler, U. P., Markov, E. V., Granovski, U. V. 1976. *Planning of experiment and search for the optimal conditions.* Moscow: Nauka. (Russian)

Bagotski, S. V. 1995. Is it possible to establish a standard for pollution? *Nauka* 6:121–123. (Russian)

Bewers, J. M., Wells, P. J. 1992. Challenges for improved marine environmental protection. *Mar.Pollut.Bull.* 25(1–4):112–118.

Buchan, G., Allan, R. 1992. The impact on the fishing industry. In *North Sea oil and the environment: developing oil and gas resources, environmental impacts and responses.*—London and New York: Elsevier Applied Science, pp.459–480.

Burke, C. J., Veil, J. A. 1995. Synthetic-based drilling fluids have many environmental pluses. *Oil and Gas Journal* (Nov.):59–64.

Chapman, P. M., Power, R. N., Dexter, R. N., Andersen, H. B. 1991. Evaluation of effects associated with an oil platform, using the sediment quality triad. *Environ.Toxicol. and Chem.* 10:407–424.

Daan, R., Van het Groenewoud, H., De Long, S. A., Mulder, M. 1992. Physico-chemical and biological features of a drilling site in the North Sea, I year after discharges of oil-contaminated drill cuttings. *Mar.Ecol.Prog.Ser.* 91(1–3):37–45.

Dorn, P. B., van Compernol, R. 1995. Effluents. In *Fundamentals of aquatic toxicology (second edition).*—Taylor and Francis, pp.903–938.

Ellis, M. S., Wilson-Ormond, E. A., Powell, E. N. 1994. Visual and histological semi-quantitative reproductive scales developed for shrimps and crabs as part of the Gulf of Mexico Offshore Operations Monitoring Experiment (GOOMEX). *J.Shellfish.Res.* 13(1):302–303.

Emerson, R. 1994. Assessing environmental effects: a detailed description of the analytic process. In *Environmental risk assessment for oil and gas development on the continental shelf of the Russian Far East.—Seminar Presentations (February 1994, Magadan, Russia).*—MMS Alaska OCS Region, pp.41–54.

EPA. 1991. *Environmental Protection Agency. Proposed development document for effluent limitation guidelines and standards for the offshore subcategory of the oil and gas extraction point source category.* EPA 440/1-91-055. Washington, DC: EPA, Industrial Technology Division.

Gavouneli, M. 1995. *Pollution from offshore installations.* Liumer Academic Publishers, Hingham. MA 02018-0358 (USA), 320 pp.

GESAMP. 1990. *The state of the marine environment. GESAMP Reports and Studies No.39.* UNEP, 112 pp.

GESAMP. 1991. *Global strategy for marine environmental protection. GESAMP Reports and Studies No.45.* London: IMO, 36 pp.

GESAMP. 1993. *Impact of oil and related chemicals and wastes on the marine environment. GESAMP Reports and Studies No.50.* London: IMO, 180 pp.

GESAMP. 1995. *Biological indicators and their use in the measurements of the condition of the marine environment GESAMP Reports and Studies No.55.* UNEP, 56 pp.

GOOS. 1995. *Global Ocean Observing System. Strategic development of the Oceans Module.* 10C, WWO, UNEP, 55 pp.

Gorshkov, V. G., Kotlyakov, V. M., Losev, K. S. 1994. Economical growth, state of the environment, wealth and poverty. *Izvestia RAN. Ser.geogr.* 1:7–13. (Russian)

Gramme, P. 1991. New technology offshore reduction of oil in produced and displacement water. In *ENS'091: Environment Northern Seas. Abstracts of Conference papers (Stavanger, Norway, Aug.1991).*—Stavanger: Industritrykk.

Gray, J. S. 1992. Biological and ecological effects of marine pollutants and their detection. *Mar.Pollut.Bull.* 25 (1–4):48–50.

Grogan, W. C., Blanchard, J. R. 1992. Environmental assessment. In *North Sea oil and the development: developing oil and gas resources, environmental impacts and responses.*—London and New York: Elsevier Applied Science, pp.363–402.

Hinssen, J., Van der Schans, J. W. 1994. Co-governance: a new approach of North Sea policy-making? *Mar.Pollut.Bull.* 28(2):69–72.

ICES. 1994a. *International Council for the Exploration of the Sea. Report of the ICES Advisory Committee on Marine Environment.* Copenhagen: ICES, 84 pp.

ICES. 1994b. *International Council for the Exploration of the Sea. Report of the Joint meeting of the Working Group on Marine Sediments in relation to pollution and the Working Group on Biological Effects on Contaminants (Nantes, France, 21–22 March 1994).* ICES C.M.1994/ENV:2, 19 pp.

ICES. 1995. *International Council for the Exploration of the Sea. Report of the ICES Advisory Committee on Marine Environment.* Copenhagen: ICES, 80 pp.

IMO. 1989. *International Maritime Organization. Guidelines and standards for the removal of offshore installations and structures on the continental shelf and the Exclusive Economic Zone.* London: IMO.

Izrael, U. A. 1984. *Ecology and control of the state of the environment.* Moscow: Gidrometeoizdat, 560 pp. (Russian)

Jackson, T., Taylor, P. J. 1992. The precautionary principle and the prevention of marine pollution. Papers from the First International Ocean Pollution Symposium (1991, Puerto Rico, USA). *Chem.Ecol.* 7(1–4):123–134.

Johnston, C. 1993. Problems mount at both ends of UK field life timescale. *Offshore Eng.* (Aug.):31–34.

Komedchikov, N. N., Luti, A. A., Asoyan, D. S., Berdnikov, K. V., Loginova, L. V., Narskikh, R. S. 1994. Ecology of Russia in maps. *Izvestia RAN. Ser.geogr.* 1:107–118. (Russian)

Kreneva, S. V. 1993. System of ecological control of the state of the natural waters. *Gidrobiologicheski zhurnal* 29(3):88–95. (Russian)

Laskorin, B., Lukyanenko, V. 1993. Strategy and tactics of the water bodies protection from pollution. *Mir nauki* 37(2):13–17. (Russian)

Levich, A. P. 1994. Biological concept of the environmental control. *Doklady RAN* 337(2):280–282. (Russian)

Long, E. R., Morgan, L. G. 1990. *The potential for biological effects of sediment-sorbed contaminants tested in the National Status and Trends Program.* Seattle (Washington): NOAA Ocean Assessment Division, 175 pp.

Lukyanenko, V. I. 1994. Ecological foundation of the general concept of the water bodies protection from pollution and the ways of its realization. *Zhurnal obshchei biologii. Ser.biol.* 3:463–467. (Russian)

MacDonald, D. D. 1993. *Development of an approach to the assessment sediment quality in Florida coastal waters. Report prepared by MacDonald Environmental Sciences, Ltd. Ladysmith, British Columbia for the Florida Department of Environmental Regulation.* Tallahassee, Florida, 133 pp.

Maksimov, V. N. 1991. Problems of the complex assessment of the quality of natural waters (ecological aspects). *Gidrobiologicheski zhurnal* 27(3):8–12. (Russian)

Marrah, K., Kamal, S. 1992. The use of oil base muds in offshore drilling from environmental and cost point of view. In *Proceedings of the International Seminar on Emerging Trends in Offshore Technology and Safety (New Delhy, India, February 1992).*—New Dehly: Engineers, pp.310–318.

Martin, M., Richardson, B. J. 1995. A paradigm for integrated marine toxicity research? Further views from the Pacific rim. *Mar.Pollut.Bull.* 30(1):8–13.

May, R. F. 1992. Marine coastal reserves, petroleum exploration and development, and oil spill in coastal waters of western Australia. *Mar.Pollut.Bull.* 25(5–8):147–154.

McIntyre, A. D. 1995. Balance in conservation. *Mar.Pollut.Bull.* 30(3):174–175.

Melberg, B. 1991. Reduction of pollution from drilling operations. In *ENS'091: Environment Northern seas. Abstracts of Conference papers (Stavanger, Norway, Aug.1991).*—Stavanger: Industritrykk.

Minton, R. C., Last, N. 1994. Downhole injection of OBM cuttings economical in the North Sea. *Oil and Gas Journ.* 92(22):75–79.

MMS. 1995. *Mineral Management Service. Outer continental shelf natural gas and oil resource management program: cumulative effects 1987–1991. OCS report MMS 95-0007.* Herdon, Va.: U.S. Department of the Interior (USDOI).

Morrisey, D. J. 1993. Environmental impact assessment—a review of its aims and recent developments. *Mar.Pollut.Bull.* 26(10):540–545.

Neff, J. M. 1993. *Petroleum in the marine environment: regulatory strategies and fisheries impacts.* Battele Ocean Science Laboratory. Duxbury, 13 pp.

Neff, J. M. 1998. Fate and effects of drilling mud and produced water discharged in the marine environment. In *U.S.-Russian Government Workshop on Management of Waste from Offshore Oil and Gas Operation (April 1998, Moscow).*

Neff, J. M., Rabalis, N. N., Boesch, B. F. 1987. Offshore oil and gas development activities potentially causing long-term environmental effects. In *Long-term environmental effects of offshore oil and gas development.*—London: Elsevier Applied Science, pp.149–173.

NRC. 1991. *National Research Council. Adequacy of environmental information for outer continental shelf oil and gas decisions: Georges bank.* Washington, DC: NRC, Board on Environmental Studies and Toxicology, 98 pp.

Patin, S. A. 1988. Establishing fisheries standards for the water environment quality. In *Water toxicology and optimization of bioproductive processes in aquaculture.*—Moscow: VNIRO, pp.5–18. (Russian)

Patin, S. A. 1991. Eco-toxicological aspects of studying and control of the water environment quality. *Gidrobiologicheski zhurnal* 27(3):75–77. (Russian)

Patin, S. A. 1995. Global pollution and biological resources of the World Ocean. In *World Fisheries Congress Proceedings, Athens, 1994.*—New Delhy: Oxford IBH Publishing Co, pp.69–75.

Patin, S. A., Lesnikov, L. A. 1988. Main principles of establishing the fisheries standards for the water environment quality. In *Methodical guidelines on establishing the maximum permissible concentrations of pollutants in the water environment.*—Moscow: VNIRO, pp.3–10. (Russian)

Ramsay, C. G., Grant, S. 1992. Hazard and risk. In *North Sea oil and the development: developing oil and gas resources, environmental impacts and responses.*—London and New York: Elsvier Applied Science, pp.559–584.

Read, A. D., Reiersen, L. O. 1992. International initiatives and statutory controls on the prevention of pollution from offshore operations. In *North Sea oil and the development: developing oil and gas resources, environmental impacts and responses.*—London and New York: Elsevier Applied Science, pp.403–416.

Reynalds, P. 1993. Norway prepares for tougher legislation for environmental protection in the North Sea oil and gas industry. *Offshore Eng.* (Aug.):26–28.

Ricketts, P. J. 1992. Current approaches to geographic information systems for coastal management. *Mar.Pollut.Bull.* 69(2):47–53.

Rodriguez, E., Vila, L. 1992. Ecological sensitivity atlas of the Argentine continental shelf. *Int.Hydrogr.Rev.* 69(2):47–53.

Rules for water protection. 1991. *Rules for the surface water protection (general principles).* Moscow: Goskompriroda SSSR, 34 pp. (Russian)

Sdasyuk, G. V., Shestakov, A. S. 1994. Eco-geographical situations and the need to transition to the sustainable development. *Izvestia RAN. Ser.geogr.* 1:42–51. (Russian)

Shvebs, G. I. 1993. Concept of the complex monitoring of the environment. *Izvestia Russkogo geograficheskogo obshchestva* 125(6):14–20. (Russian)

Smirnov, L. V., Shumova, O. V. 1994. Principles of eco-geographic cartography. *Izvestia Russkogo geographicheskogo obshchestva* 126(2):58–63. (Russian)

Somerville, H. J., Shirley, D. 1992. Managing chronic environmental risks. In *North Sea oil and the environment: developing oil and gas resources, environmental impacts and responses.*—London and New York: Elsevier Applied Science, pp.643–664.

Thain, J., Matthiessen, P., McMinn, W. 1994. Assessing sediment quality by bioassay in UK coastal water and estuaries. In *Proceedings of the Scientific Symposium on the 1993 North Sea Quality Status Report.* Pp.2–10.

Thorson, J. A. 1991. Stricter marine pollution standards accelerate move to zero discharge rigs. *Oil and Gas Journ.* 89(52):94–97.

Tortell, Ph. 1992. Coastal zone sensitivity mapping and its role in marine environmental management. *Mar.Pollut.Bull.* 25(1–4):88–93.

Veil, J. A. 1992. Review of the cost-effectiveness of EPA's offshore oil and gas effluent guidelines. In *Produced water. Technological and environmental issues and solutions.* New York: Plenum press, pp.23–24.

Volkov, I. V., Zalicheva, I. N., Ganina, V. S., Ilmast, T. B., Kaimina, N. V., Movchan, G. V. 1993. On principles of regulation of anthropogenic impact on the water ecosystems. *Vodnye resursy* 20(6):707–713. (Russian)

Warren, T. 1994. Introduction of Minerals Management Service: its basis in law and its responsibilities. In *Environmental risk assessment for oil and gas development on the continental shelf of the Russian Far East. Seminar Presentations (February 1994, Magadan, Russia).*—MMS Alaska OCS Region, pp.4–18.

Wells, P. G., Rolston, S. J., eds. 1991. *Health of our oceans. A Status report on Canadian marine environmental quality.* Dartmouth and Ottawa: Conservation and Protection, Environment Canada, 166 pp.

Wynne, B., Mayer, S. 1993. How science fails the environment. *New Scientist (UK),* 5 June 1993:33–35.

Zaitsev, U. P. 1970. *Marine neustonology.* Kiev, 250 pp. (Russian)

General Conclusion

This book has discussed the major environmental problems provoked by the exploration, development, and production of offshore oil and gas resources. It used numerous materials and data obtained in many countries where, in contrast with Russia, hundreds of oil platforms in the sea became a usual attribute of the industrial activity on the shelf long ago. These countries have accumulated extensive and expensive experience in ecological research, control, and regulation of this activity. For example, in the United States, the federal Environmental Studies Program alone spent over 114 million dollars from 1987 through 1991. This program provided decision makers with information to predict, assess, and manage environmental impacts of offshore activities on coastal areas.

Generally, the previously discussed materials suggest that the offshore oil and gas industry can comply with rather strict ecological standards defined by modern environmental protection requirements at the national and international levels. The experience of some countries indicates that industry is also able to ensure an acceptable balance of interests for all other parties, including fisheries and marine aquaculture. Certainly, this refers to a fundamental possibility that is not realized to its full extent. This possibility has already revealed itself in many countries as the tendency to achieve zero discharges, develop more effective technologies, and introduce stricter legislative regulatory measures. It should also be remembered that offshore oil and gas production is not the leading cause of ecological hazards in the coastal zone. The tremendous amounts of oil and other toxicants have been released for decades into the coastal areas from numerous land-based pollution sources. The average contribution of offshore oil and gas production does not exceed a few percent of the total oil input in the sea (see Chapter 4).

At the same time, the often-expressed opinion that the offshore oil and gas industry's activity cause no adverse changes in marine ecosystems seems to be too optimistic. We gave enough evidence of various (physical, chemical, and biological) impacts on the marine environment and biota at all stages of offshore oil and gas activity.

Their consequences range from short-term hazardous effects during seismic surveys to persistent modifications of the benthic communities in the areas of oil production. The contribution of the drilling wastes in oil pollution at the regional level can reach up to 30% of the background of all other oil sources in the sea.

Especially alarming are the so-called nonobvious (subtle) consequences of chronic impacts of low levels of chemical contamination by polycyclic aromatic hydrocarbons (PAHs) and other persistent toxicants that have mutagenic and carcinogenic properties. Modern sensitive methods reveal subcellular responses of marine organisms to the presence of trace amounts of these substances. However, the question about biological significance of these subtle effects at the subcellular level for the higher levels of the life hierarchy remains open.

In general, in spite of the long-term and extensive research devoted to environmental aspects of the offshore oil and gas activity, modern science is still mostly posing questions rather than giving much-needed practical answers. The issues of critical importance include assessing the environmental hazard of different impact factors, estimating the fisheries losses due to oil production operations, and establishing permissible limits of ecological changes. The answers to these questions would directly define the effectiveness of national and international efforts to protect the shelf zone under increasing anthropogenic press.

Solving these and many other problems discussed in this book is especially important for the oil and gas activities on the Russian shelf. Russia, having the richest reserve of offshore oil and gas hydrocarbons, is taking only the first steps to industrially exploit them. At first, such a delay seems beneficial since Russia could use other countries' experiences in solving ecological problems and ensuring environmental safety of the offshore activities. Unfortunately, data on current Russian practices of oil and gas developments on land raise many doubts about the capability of the Russian oil industry to use this advantage effectively. Remember just two official estimates: 10^{-4}–10^{-3}% and 1–2%. These reflect the relative amounts of oil hydrocarbons (at the background of the total extracted volumes) lost during oil production on the shelf of the North Sea and in Western Siberia, respectively.

Would the Russian oil and gas industry be able to reduce its current volumes of oil losses by several orders of magnitude when it starts the developments on the northern and far-eastern shelves? Would the prize paid for the offshore oil be too high even if Russian industry succeeds and cuts oil losses hundreds and thousands of times? Should development of the offshore oil and gas resources in the Arctic and Far East be postponed until the energy resources already available are managed more wisely? Would such restrictions better suit the sustainable economic development of Russia and its environmental safety? The answers to these and many other similar questions are beyond the framework of environmental science. They should be addressed to Russian strategists in the areas of economics and politics.

Developing offshore oil and gas resources provokes emerging long-term, complex, and diverse environmental problems in the sea. These issues can not be ignored. They should become the concern of not only scientific circles but of all levels of society. A rapid and adequate response is essential if we want to preserve the living self-reproducing resources of the continental shelf. These resources should be saved for future generations, even after depleting offshore oil and gas resources.

Appendix

Maximum Permissible Concentrations (MPC) and Approximate Safe Impact Limits (ASIL) of Chemicals and Wastes from the Oil and Gas Production Adopted in Russia for Natural Waters [List of MPC, 1995]

Agents and Wastes	MPC ASIL (mg/l)	LHP	Hazard Rating
Acetaldehyde (ethanale), C_2H_4O; component of cleaners	0.25	org	4
Alkamon OS-2, Grindril SP; drilling composition	0.12	tox	4
Alkylsulfate (up to 16% of sodium sulfate), R_2SO_4 (R = C_nH_{2n+2}, n = 12–14)	0.2	–"–	4
Alplex; component of drilling fluids	0.3**	–"–	–
Ammonium lignosulfonate; thickener	1.0	san–tox	–
Ammonium-ion, NH_4^+; component of discharges	0.5; 2.9*under 13–34 $^0/_{00}$	tox; –"–	4; 4
Amphicor, $RHPO_2^- NH_4^+$ R = C_nH_{2n+1}, n = 8–10; corrosion inhibitor	0.2	san–tox	4
Arsenic, As (all water-soluble forms); component of formation waters	0.05; 0.01*	tox; –"–	3; 3
Baracat; component of drilling fluids	0.002**	–"–	–
Baracol-140; component of drilling fluids	0.1**	–"–	–
Baracol-155; component of drilling fluids	0.5**	–"–	–
Baracol-156; component of drilling fluids	0.2**	–"–	–
Baracol-351; component of drilling fluids	0.12**	–"–	–

Appendix (*continued*)

Agents and Wastes	MPC ASIL (mg/l)	LHP	Hazard Rating
Baracol-728; component of drilling fluids	0.01**	tox	–
Baracor-100; component of drilling fluids	0.05**	–"–	–
Bara-Defoam; component of drilling fluids	0.003**	–"–	–
Barafible; component of drilling fluids	0.13**	–"–	–
Barafilm; component of drilling fluid	0.1**	–"–	–
Baranex; component of drilling fluids	0.64**	–"–	–
Barazan; component of drilling fluids	0.2**	–"–	–
Barium, Ba (all water-soluble forms); component of formation waters and drilling fluids	0.74; 2.0* under 12–18 $^0/_{00}$	–"–	4
Barium sulfate, $BaSO_4$; weighting agent in drilling fluids	2.0	–"–	4
Baroid drilling polymer-potassium fluid No 1	2.5**	tox	–
Baro-Trol; component of drilling fluids	0.25**	–"–	–
Benex; component of drilling fluids	0.15**	–"–	–
Benzene, C_6H_6	0.5	–"–	4
Beryllium, Be (all water-soluble forms), component of formation waters	0.0003	–"–	
Bioagent "Destroil" (*Acinetobacter sp.* IN-2); oil destructor	0.5 5×10^3 cells/ml	san	3
Bioagent "Leader" (*Rhodococcus maris*); oil destructor	0.001 1.7×10^4 cells/ml	tox org	4
Bioagent "Valentis" (*Acinetobacter valentis*); oil destructor	1.0 2.5×10^7 cells/ml	tox org	4
Biodril; component of drilling fluids	0.2**	–"–	–
Biomul, $CH_2 = CH\text{-}COOC_nH_{2n+1}$ C_6H_5, n = 10–12; component of drilling fluids	0.5**	–"–	–
Block copolymer GDPE, $RO(C_3H_6O)_m (C_2H_4O)_kH$ $(R = C_nH_{2n+1}, n = 7–12)$	0.1*	san	–

Appendix (*continued*)

Agents and Wastes	MPC ASIL (mg/l)	LHP	Hazard Rating
Boron, B (ionic forms except boron hydrides) component of formation waters	0.5; 10.0* under 12–18 $^0/_{00}$	tox	4
Bromide-anion, Br^-; component of formation waters	1.35; 12.0* in addition to natural bromide content	–"–	4
Butanol-1, $C_4H_{10}O$, $CH_3CH_2CH_2CH_2OH$; solvent	0.03	san–tox	3
C-10; emulsifier	0.1	san–tox	–
Cadmium, Cd (all water-soluble forms); component of formation waters	0.005; 0.01*	tox –"–	2 2
Caesium, Cs (all water-soluble forms); component of formation waters	1.0	–"–	4
Calcium, Ca (all water-soluble forms); component of formation waters	180; 610* under 13–18 $^0/_{00}$	san–tox	4
Carbomix; component of drilling fluids	0.3**	tox	–
Carboxyl-methyl-cellulose-500; component of drilling fluids	12.0*	tox	4
Cat-300; component of drilling fluids	0.12**	–"–	–
Cat-thin; component of drilling fluids	0.04**	–"–	–
Chloride-anion; component of formation waters	300.00; 11,900 under 12–18$^0/_{00}$	san–tox tox	
Chromium, Cr^{3+}; component of formation waters	0.07	–"–	3
Chromium, Cr^{6+}; component of formation waters	0.02	tox	–
Cobalt, Co (all water-soluble forms), component of formation waters	0.01	–"–	3
Condet, cleaning additive in drilling fluids	0.01**	–"–	–
Copper, Cu (all water-soluble forms); component of formation waters	0.001; 0.005*	tox –"–	3 3
Correxit 7664 in oil (Correxit 7664—10%, oil—90%), dispersant for removing oil spills	0.002	–"–	3

Appendix (*continued*)

Agents and Wastes	MPC ASIL (mg/l)	LHP	Hazard Rating
Correxit 7664; dispersant for removing oil spills	0.2	tox	4
Correxit 9527; dispersant for removing oil spills	0.05*	–"–	4
Corrosion inhibitor EKB-2-2	0.005	–"–	–
Corrosion inhibitor EKB-6-2	0.0001	–"–	–
Cyclohexane, C_6H_{12}; component of discharges 0.01	–"–	3	
Descaler EOMS-1	0.1	–"–	4
Descaler EST-1	0.1	–"–	3
Desco chrome free; component of drilling fluids	0.3**	–"–	–
Diethanolamine, $C_4H_{11}NO_2$ $(HOCH_2CH_2)_2NH$; adsorbent of H_2S and CO_2	0.01	–"–	3
Dispersant 124 d; for removing oil spills	absent (0.00001)	–"–	1
Dispersant 124 v; for removing oil spills	absent (0.00001)	–"–	1
Dispersant DN-75; emulsifier for removing oil spills	0.15; 0.005*	–"–	3
Dispersant NF; for removing oil spills	0.25	–"–	4
Dissolvan HOEF 1877-4; technological agent	0.05	–"–	4
DK-drill; component of drilling fluids	0.0001	–"–	–
DKS-extender (modified polyacrylamide), component of drilling fluids	0.001	–"–	3
DKS-extender (modified polyacrylamide); viscosifier in drilling fluids	0.001	–"–	3
DMP-410; component of drilling fluids	0.2	–"–	3
Drilling biopolymer fluid No 4	4.5**	–"–	–
Drilling fluid "Quadrill" (IDVIS—0.23%, IDFFLR—0.3%, IDFFLRXL—0.25%, IDFLO—0.12%, IDBONDP—0.29%, NaOH—0.15%, KCl—5.0%, Na_2CO_3—0.07%, glycerin—5-7%, water—up to 100%)	2.2**	–"–	–

Appendix (*continued*)

Agents and Wastes	MPC ASIL (mg/l)	LHP	Hazard Rating
Drilling fluid No 1, polymer-based, used at the Uchtaneftegasologia fields	50	tox	4
Drilling fluid No 2, polymer-based ***; used at the Uchtaneftegasologia fields	100	–"–	4
Drilling fluid No 3 ***, used at the Uchtaneftegasologia fields	50	–"–	4
Drilling fluid TU-6-01-166-81 (hydrolyzed polyacrylamide, hydrolyzed polyacrylnitrile, emultane, etc.)	5.0	san–tox	3
Drilling fluid VISPLEX (bentonite—3%, VISPLEX agent—0.03%, FLORPLEX agent—1.15%, NaOH—0.15%, water—up to 100%)	25.8**	tox	–
Drilling fluid, treated ***, used at the Bavanenkovskoe field	1.25	–"–	4
Drilltreat; component of drilling fluids	0.1**	–"–	–
Duratone HT; component of drilling fluids	0.03**	–"–	–
E z Mud L; component of drilling fluids	0.004**	–"–	–
E z Spot; emulsifier and lubricator in drilling fluids	0.004**	–"–	–
Ecolan (product of special wood treatment) adsorbent for removing oil spills	1,000	san	4
Ecos-B-3, lubricant in drilling fluids	0.005	tox	3
Emultal TU-6-14-1035-79, $(HOCH_2CH_2)_2NCH_2CH_2OCOR$ $R = C_{17}H_{33}, C_{17}H_{31}, C_{17}H_{29}$; emulsifier in drilling fluids	0.03	–"–	–
Enviro-thin; component of drilling fluids	0.4**	tox	–
Ethylbenzene, $C_6H_5C_2H_5$; component of discharges	0.001	–"–	–
Ethyleneglycol, $C_2H_6O_2$, $HOCH_2CH_2OH$	0.25	san–tox	4
FCLS-H, ferrochromiumligno-sulfonate; component of drilling fluids	50.0	tox	2

Appendix (*continued*)

Agents and Wastes	MPC ASIL (mg/l)	LHP	Hazard Rating
FLORPLEX; component of drilling fluids	1.4**	san–tox	4
Foaming agent PO-A	0.01*	tox	3
Formaldehyde and sodium bisulfite adduct, Rongalit, $NaHSO_3xCH_2Ox2H_2O$; biocide	0.01	san–tox	3
Free dissolved chlorine, Cl_2	Absent (0.00001)	tox	1
GDPE-0.64; components of surfactants	0.1*	tox	4
GDPE-106; components of surfactants	0.2	san	4
Gelton II; component of drilling fluids	0.3**	tox	–
GKZH-11, CH_5O_3NaSi, regulator of viscosity and reological properties of drilling fluids	1.0	san–tox	–
Hexane, $C_6H_{14}CH_3(CH_2)_4CH_3$; solvent	0.5	tox	3
IDF drilling fluid	5	–"–	4
IDFLOC P (high-molecular-weight polyacrylamide); flocculent in drilling fluid VISPLEX	0.87	san–tox	4
Invermul-NT (oil-based emulgator), component of drilling fluids	0.02**	tox	–
Iron, Fe (all water-soluble forms); component of formation waters	0.1 0.05*	–"– –"–	4 2
Isopropylbenzene, kumol, C_9H_{12}, component of discharges	0.3	org	3
Katapin TU-6-01-1026-75, biocide	0.0007	tox	1
Lead, Pb (all water-soluble forms); component of formation waters	0.1;	tox	0.01*
Magnium, Mg (all water-soluble forms); component of formation waters	40.0; 940* under 13–18 $^0/_{00}$	san–tox tox	4 4
Manganese, Mn^{2+}; component of formation waters	0.01; 0.05*	tox –"–	4 4
Mercury, Hg (all water-soluble forms); component of formation waters	absent (0.00001)	tox	1
Methanol, CH_3OH; inhibitor of hydrate-formation	0.1	–"–	4

Appendix (*continued*)

Agents and Wastes	MPC ASIL (mg/l)	LHP	Hazard Rating
Methylcellulose, component of drilling fluids	3.00**	tox	4
Micatex, K_2/Mg, $Fe_2/$ $6Al_2Si_6O_{20}$ $(OH,F)_4$; component of drilling fluids	0.25**	–"–	–
Mil-free (surfactant, diesel fuel—45% by weight), component of drilling fluids	0.7**	–"–	–
Milgard; sulfide-containing additive in drilling fluids	0.25; 0.75**	san–tox	–
Milgard-L (zinc-ammonium suspense containing 40% of $C_8H_{10}N_2O_6Zn$); sulfide-containing additive in drilling fluids	0.3**	tox	–
Milpark FSL; component of drilling fluids	0.5**	–"–	–
ML-6, cleaner	0.5	–"–	4
Morpen, 4% water solution, foaming agent	0.001* under 10–13 $^0/_{00}$	–"–	3
Mudrac; sulfide-containing additive in drilling fluids	0.06**	–"–	–
Naphthalene, $C_{10}H_8$, component of discharges	0.004	–"–	3
Neonol 1020-3; emulgator	0.0001*	–"–	3
Neonol 2B 1317-12; $C_nH_{2n+1}O$ $(C_2H_4O)_{12}H_n = 13–17$; component of surfactants	0.32; 0.1* under 34 $^0/_{00}$	–"– –"–	4
Neonol A-1214-5; component of surfactants	0.005*	–"–	3
Neonol A-1620-4, defoamer P; component of surfactants	0.01*	–"–	3
Neonol AF 14; C_8H_{17}-C_6H_4-O $(C_2H_4O)_{14}H$; component of surfactants	0.25; 0.1* under 34 $^0/_{00}$	–"– –"–	4
Neonol AF 9-10; C_9H_{19}-C_6H_4O $(C_2H_4O)_{10}H$; component of surfactants	0.1*	–"–	4
Neonol AF 9-4; C_9H_{19} C_6H_4O $(C_2H_4O)_4H$; component of surfactants	0.01*	–"–	–
Neonol AF 9-6; $C_9H_{19}C_6H_4O$ $(C_2H_4O)_{10}H$; component of surfactants	0.05*	–"–	3

Appendix (*continued*)

Agents and Wastes	MPC ASIL (mg/l)	LHP	Hazard Rating
Newdrill NP; component of drilling fluids	0.5**	tox	–
Newdrill; component of drilling fluids	0.07**	tox	–
Newthin; component of drilling fluids	0.7**	–"–	–
New-trol; component of drilling fluids	0.7**	–"–	–
Nickel, Ni (all water-soluble forms); component of formation waters	0.01	–"–	3
No-Sule (mixture of inorganic compounds of zinc); component of drilling fluids	0.12**	–"–	–
Oil and oil products in dissolved and emulsified state	0.05	fish	3
Oil products	0.05*	tox	3
OM-6; dispersant for removing oil spills	0.005*	–"–	3
OM-84; dispersant for removing	0.25*	–"–	–
OMTI-2K, turbine oil TU VTE 43.004-89; lubricator oil spills	0.0001	–"–	–
OP-10; surfactant	0.5;	–"–	4
	0.1*	–"–	4
OXZIL; component of drilling fluids	6.0*	–"–	4
PAF-13A A-3; polyelectrolyte	0.2	san–tox	4
PAF-13A; nitrogen-containing polyelectrolyte	0.1	tox	4
PAF-32; polyelectrolyte	0.03	san	3
PAF-41; polyelectrolyte	0.2	san–tox	4
Peat crumb; for removing of oil spills	57.0 for the dry weight	san–tox	–
Phenol, hydroxybenzene, carbolic acid, C_6H_5OH	0.001	fish	3
Phosphoxit-7; corrosion inhibitor and descaler	0.005*	tox	3
Phthalic acids, $C_6H_4(COOH)_2$; components of discharges	2.0*	–"–	–
Polyacrylamide AK-617 cationic; viscosifier	0.08	–"–	–
Polyacrylamide anionic; agent improving the oil recovery	0.2	–"–	3

Appendix (*continued*)

Agents and Wastes	MPC ASIL (mg/l)	LHP	Hazard Rating
Polyacrylamide non-ionic DR1-4973; viscosifier	0.01	tox	3
Polyacrylamide partially hydrolyzed (24%); Valshale; component of drilling fluids	0.05	–"–	3
Polyacrylamide partially hydrolyzed (up to 50%); GPAA TU-6-01-1049-81, Valswell, Greendrill FP; regulator of viscosity in drilling fluids	0.8	–"–	–
Polyacrylamide partially hydrolyzed AK-618; viscosifier	0.04	–"–	4
Poly-Kem-D (partially hydrolized polyacrylamide); drilling fluid additive	0.025	–"–	4
Polyphos 108H, surfactant	0.05*	–"–	3
Polyphos 126-T, surfactant	3.0	san	4
Polyvinylacetate emulsion PVA-E, viscosifier	0.3	tox	4
Polyvinylmethoxymethylacrylamide PVS-MOL, thickener	0.5	san–tox	
Potassium bichromate, $K_2Cr_2O_7$	0.05 by substance; 0.02 by Cr^{+6}	tox	–
Potassium, K (all water-soluble forms); component of formation waters	50; 10 for water bodies with mineralization up to 100 mg/l; 390* under 13–18 $^0/_{00}$	san–tox, tox	–
Q-broxin; component of drilling fluids	0.5**	–"–	–
Saidril; component of drilling fluids	0.001	–"–	–
Saipan; component of drilling fluids	0.001	–"–	–
Sintanol ALM-7; surfactant component	0.002	–"–	3
Sintanol DS-10; surfactant component	0.0005	–"–	3
Sintanol DS-6; surfactant component	0.1*	–"–	–
SN-5; lubricants	0.25* under 34 $^0/_{00}$	–"–	3
SNPK-1003 (1-alkyl-2-methyl-5-etrhyl-perid inebromide)	0.1	san–tox	3

Appendix (*continued*)

Agents and Wastes	MPC ASIL (mg/l)	LHP	Hazard Rating
SNPK-1004 anticorrosion, multi-purpose agent	0.05	tox	–
SNPK-44; demulsifier	0.1	–"–	4
SNPK-7214 P; descaler	0.01	–"–	3
SNPK-7215 (oxyethylized alkylphenols, nitrogen-containing additive, butylbenzene-based solvent); descaler	0.01	–"–	3
SNPK-7215 M; descaler	0.01	–"–	3
Sodium carbonate, Na_2CO_3; component of drilling fluids	Discharges into water bodies before the hydrolysis is completed is forbidden	–	
Sodium dimethyl-ditiocarbomate, carbomate MN TU 6-M-540-83, $C_3H_6NS_2Na$; biocide	0.00005	–"–	–
Sodium hydroxide, NaOH; corrosion inhibitor and descaler, technological agent	pH standard	–"–	
Sodium lignosulfonate	3.0	–"–	4
Sodium polyacrylate, Kem-Pa-S, Valspers; additive in drilling fluids	0.0125	–"–	4
Sodium salt of carboxyl-methyl-cellulose, Finnfix; component of drilling fluids	1.0	–"–	4
Sodium sulfide, $Na_2S \times 9H_2O$; biocide	0.01; 0.05 by S^{2-} for oligotrophic bodies of water; 0.001 and 0.0005 by S^{2-}	–"–	3
Sodium, Na^+ (all water-soluble forms); component of formation waters	120 7,100* under 13–18 $^o/_{oo}$	san–tox tox	
Sprint-33, lubricant	0.25	org, tox	4
Straw oil (mixture of hydrocarbons); lubricating component	0.01	tox	3
Sulfanol NP-1	0.2	–"–	4
Sulfate-anion, SO_4^{2-}, component of formation waters	100; 3,500* under 12-18%	–"–	–
Therma-chek; component of drilling fluids	0.06**	–"–	–

Appendix (*continued*)

Agents and Wastes	MPC ASIL (mg/l)	LHP	Hazard Rating
Therma-thin DP; component of drilling fluids	0.5**	tox	–
Thin absorbent (mixture of aromatic hydrocarbons—up to 50%; benzene—5%; toluene—20–25%; xylene—15–20%); technological agent	0.01	san–tox	4
Toluene, methylbenzene, $C_6H_5CH_3$	0.5	org	3
Torg-trim II; component of drilling fluids	0.14**	tox	3
VISPLEX; stabilizer in the drilling fluid VISPLEX	10.0**	–"–	4
VMT-L TU 38 101960-83; agent for soil stabilization during pipelaying operations	0.3	–"–	–
Water-soluble thickener based on monoethanolamine; technological agent	0.1	–"–	4
XC-207; component of drilling fluids	0.00015**	–"–	–
Xylene, ortho-xylene, 1,2-dimethylbenzene, $C_6H_4(CH_3)_2$; component of discharges	0.04	org	3
Zinc, Zn (all water-soluble forms); component of formation waters and inhibition solutions	0.01; 0.05*	–"– –"–	3 3

Notes: 1. MPC—maximum permissible concentration. 2. ASIL—approximate safe impact limit; in contrast with MPC, is established for temporary use (2 years). 3. LHP—limiting hazard parameter—the most sensitive test-parameter used to establish MPC: tox—toxicological, san—sanitary, san–tox—sanitary-toxicological, org—organoleptic, fish—fishery. 4. Hazard rating of chemicals and wastes is based on their toxicity: 1—extremely high, 2—high, 3—medium, 4—moderate. *—MPC established for seawater. **—ASIL. ***—In case of using these drilling fluids for drilling wells in other fields, additional studies are required.

Index

Order Form

Postal Orders:

> EcoMonitor Publishing
> PO Box 866
> East Northport, NY 11731 USA

Online Orders: sales@offshore-environment.com

URL: http://www.offshore-environment.com   A.

Patin/ Environmental impact of the offshore oil and gas industry

_____ $199 ISBN 0-9671836-0-X

Company name: _____

Name: _____

Address: _____

City: _____

State: _____ Zip: _____

Phone: _____

Sales tax:

Please add $16.00 sales tax (8%) for books shipped to New York addresses.

Postage and handling:

Please add $4.00 for the first copy and $2.00 for each additional copy.

Enclosed: check or money order for $ _____

(must include applicable postage and handling)

Order Form

Postal Orders:

       EcoMonitor Publishing
       PO Box 866
       East Northport, NY 11731 USA

Online Orders: sales@offshore-environment.com

URL: http://www.offshore-environment.com

Patin/ Environmental impact of the offshore oil and gas industry

_____ $199 ISBN 0-9671836-0-X

Company name: _____

Name: _____

Address: _____

City: _____

State: _____ Zip: _____

Phone: _____

Sales tax:

Please add $16.00 sales tax (8%) for books shipped to New York addresses.

Postage and handling:

Please add $4.00 for the first copy and $2.00 for each additional copy.

Enclosed: check or money order for $ _____

(must include applicable postage and handling)

Order Form

Postal Orders:

EcoMonitor Publishing
PO Box 866
East Northport, NY 11731 USA

Online Orders: sales@offshore-environment.com

URL: http://www.offshore-environment.com   ʌ.

Patin/ Environmental impact of the offshore oil and gas industry

_____ $199 ISBN 0-9671836-0-X

Company name: _____

Name: _____

Address: _____

City: _____

State: _____ Zip: _____

Phone: _____

Sales tax:

Please add $16.00 sales tax (8%) for books shipped to New York addresses.

Postage and handling:

Please add $4.00 for the first copy and $2.00 for each additional copy.

Enclosed: check or money order for $ _____

(must include applicable postage and handling)